The World According to Monsanto

Also by Marie-Monique Robin

Voleurs d'organes: Enquête sur un trafic

Les Cent Photos du siècle

Le Sixième Sens: Science et paranormal (with Mario Varvoglis)

Escadrons de la mort, l'école française

L'École du soupçon: Les dérives de la lutte contre la pédophilie

The World According to Monsanto

Pollution, Corruption, and the Control of the World's Food Supply

MARIE-MONIQUE ROBIN

Translated from the French by George Holoch

NEW YORK

The New Press gratefully acknowledges the Florence Gould Foundation and Furthermore: a program of the J. M. Kaplan Fund for supporting the publication of this book.

This work, published as part of a program providing publication assistance, received financial support from the French Ministry of Foreign Affairs, the Cultural Services of the French Embassy in the United States and FACE (French American Cultural Exchange).

www.frenchbooknews.com

First published as *Le monde selon Monsanto* by Éditions La Découverte, Paris, 2008
Published in the United States by The New Press, New York, 2010
Distributed by Perseus Distribution

LIBRARY OF CONGRESS CATALOGING-IN-PUBLICATION DATA
Robin, Marie-Monique.
[Monde selon Monsanto. English]
The world according to Monsanto : pollution, corruption, and the control of the world's food supply / Marie-Monique Robin ; translated by George Holoch.
p. cm.
Includes bibliographical references and index.
ISBN 978-1-59558-426-7 (hardcover : alk. paper)
1. Monsanto Company. 2. Agricultural innovations—United States—Social aspects. 3. Agricultural chemicals industry—Social aspects. 4. Agricultural chemicals—Environmental aspects. 5. Agricultural chemicals industry—Corrupt practices. 6. Social responsibility of business. 7. Food supply. 8. Human rights and globalization. I. Title.
HD9482.U64M6613 2009
338.7'6600973—dc22 2009008621

The New Press was established in 1990 as a not-for-profit alternative to the large, commercial publishing houses currently dominating the book publishing industry. The New Press operates in the public interest rather than for private gain, and is committed to publishing, in innovative ways, works of educational, cultural, and community value that are often deemed insufficiently profitable.

www.thenewpress.com

Composition by NK Graphics
This book was set in Fairfield

Printed in the United States of America

2 4 6 8 10 9 7 5 3

To my parents, Joël and Jeannette,
farmers who taught me to love the good things of the earth,
and so to love life.

Contents

PART II: GMOs: THE GREAT CONSPIRACY

PART III: MONSANTO'S GMOs STORM THE SOUTH

Preface

A Book for Public Health

Nicolas Hulot

As I progressed in my reading of Marie-Monique Robin's book, a flood of weighty questions overwhelmed me, filling me with anxiety that I might sum up in a single question: how is this possible? How can Monsanto, that emblematic firm of global agrochemistry, have made so many fatal mistakes, and how can it have marketed products so harmful to human health and to the environment? How has the company succeeded in conducting its business as though nothing had happened, constantly extending its influence (and its wealth), despite the tragedies its products have caused? How has it so quietly managed to continue its activities unconcerned and fool everyone? Why has it been able to carry on despite the heavy legal penalties it suffered and despite the bans imposed on some of its products (unfortunately, after they had already caused a good deal of irreparable harm)?

This book discloses a reality that hurts the eyes and grips the heart: that of a calmly arrogant company heedlessly profiting from the suffering of victims and the destruction of ecosystems. As the pages go by, the mystery is revealed. They show the prospering of a company whose history "constitutes a paradigm of the aberrations in which industrial society has become mired." You may often shake your head in disbelief, but the demonstration is limpid, and we understand where Monsanto gets its power, how its lies have won out over the truth, and why many of its allegedly miraculous products in the end turned out to be nightmares. In other words, at a time when the North American company has taken on an even more totalizing ambition than before—imposing genetically modified organisms (GMOs) on farmers and food consumers around the world—this indispensable book raises the question, while there is still time, of whether it is necessary to allow a company such as Monsanto to hold the future of humanity in its test tubes and to impose a new world agricultural order.

Consider the facts. How did Monsanto become one of the major industrial empires on the planet? By accomplishing nothing less than the large-scale production of some of the most dangerous products of modern times: PCBs, which were used as coolants and lubricants, are devastatingly harmful to human health and the food chain and were banned after massive contamination was observed; dioxin, a few grams of which is enough to poison a large city and the manufacture of which was also banned, was developed from one of the company's herbicides, which was the basis for the grimly famous Agent Orange, the defoliant dropped on Vietnamese jungles and villages (which enabled Monsanto to secure the largest contract in its history from the Pentagon); bovine growth hormones—the first test products for GMOs—are intended to cause the animal to produce beyond its natural capacities regardless of the known consequences for human health; the weed killer Roundup used to be presented in endless advertising as biodegradable and favorable to the environment, a claim flatly contradicted by legal decisions in the United States and in Europe.

I have had serious doubts regarding certain practices of this company, particularly its use of police tactics against farmers. Marie-Monique Robin's book not only confirms them but reveals both a company driven by the engine of business alone, which is hardly surprising, and, more troubling, a company whose actions are based on an extraordinary sense that it can do as it likes. She sketches a portrait of a firm that is expert at slipping through the cracks and persisting in its practices against all comers, no doubt convinced that it knows better than anyone else what is good for humanity, persuaded that it is accountable to no one, appropriating the planet as its playing field and profit center. In Monsanto's position outside democratic control, it is hard to tell whether it is commercial blindness, scientific arrogance, or pure and simple cynicism that dominates.

Robin's investigation is both dense and laser sharp; testimony is abundant and concordant, documents are revealed, and archives are deciphered. Her book is not a pamphlet filled with fantasy and gossip. It brings to light a terrifying reality. For, in the course of many years of marketing its products—PCBs, herbicides, dioxin, bovine growth hormones, Roundup—Monsanto was fully aware of their harmfulness. The documents that the book reveals leave no doubt on that subject. Monsanto developed the habit of publicly asserting the opposite of what was known inside the company. Thanks to Robin, we now know what Monsanto knew. The company was aware of the toxic effects of its products. It persevered nonetheless, and it was allowed to go on.

Monsanto is now coming back in force and claims that the GMOs for which it is the principal seed producer have been developed out of its concern to help the farmers of the world to produce healthier food while at the same time reducing the impact of agriculture on the environment. The company claims that it has changed and that it has broken with its past as an irresponsible chemist. I don't have the scientific competence to assess the toxicity of certain molecules or the risks incurred by genetic manipulation. I only know that the scientific community is sharply divided about the effects of transgenesis and that the results of experiments with cultivated GMOs have not provided proof that they cause no harm to health or the environment or that they are able to intensify food production to conquer hunger. The balance sheet Robin draws up for Mexico, Argentina, Paraguay, the United States, Canada, and India is in any case distressing. I also know that the use of Monsanto 810 corn seeds, the only variety grown in France for commercial purposes, was wisely suspended by the French government in January 2008, after an administrative authority set up in the wake of the major environmental conference held in October 2007 pointed to new scientific findings and raised troubling questions. More generally, I know, as does any citizen on Earth with a grain of common sense, that one has to call a halt when it is obvious that industrial and commercial considerations have gone beyond the limits of the most basic precautions.

Today, while a real scientific, economic, and social debate is stirring France and Europe about the health and environmental effects of GMOs, as well as their consequences for the condition of farmers and the question of patents of living things, Marie-Monique Robin's book is timely. It should be considered a work promoting public health and read with that in mind.

The global ecological crisis calls for a major transformation of the economic and social organization of human communities. It calls into question the capacity of world agriculture to provide sufficient food resources for the future nine billion inhabitants of the planet. There is no doubt that scientific and technological innovation can play a dynamic role—but not in just any way and not in everyone's hands.

Indeed, what exactly would the world according to Monsanto be like?

Introduction

The Monsanto Question

"You ought to do an investigation of Monsanto. We all need to know what this American multinational really is, which is in the process of seizing control of seeds, and so of world food." The scene was the New Delhi airport in December 2004. I was being addressed by Yudhvir Singh, the spokesman of the Bharatiya Kisan Union, a peasant organization in northern India with twenty million members. I had just spent two weeks with him traveling around Punjab and Haryana, the two emblematic states of the "green revolution," where almost all Indian wheat is produced.

A Necessary Investigation

At the time I was making two documentaries for the Franco-German television station Arte in a series dealing with biodiversity, entitled "Seizing Control of Nature."[1] In the first, *The Pirates of the Living*, I describe how the advent of the techniques of genetic manipulation provoked a veritable race for genes, in which the giants of biotechnology have not hesitated to seize the natural resources of developing countries through abuse of the patent system.[2] For example, a Colorado farmer who claimed to be an independent spirit secured a patent for a variety of yellow bean that had long been grown in Mexico; claiming to be its American "inventor," he demanded royalties from all Mexican peasants wishing to export their crops to the United

States.[3] And the American company Monsanto got a European patent for an Indian variety of wheat used in making chapati (unleavened bread).

In the second documentary, titled *Wheat: Chronicle of a Death Foretold?* I recounted the history of biodiversity and the threats it is under through the saga of the golden grain from its domestication ten thousand years ago up to the advent of genetically modified organisms (GMOs), in which Monsanto is the world leader. At the same time, I produced a third film for Arte Reportage titled *Argentina: The Soybeans of Hunger*, which presented the disastrous results of transgenic agriculture in the land of beef and milk. It turned out that the GMOs in question, which covered half the arable land in the country, consisted of "Roundup-ready" soybeans, which had been manipulated by Monsanto to resist Roundup, the best-selling herbicide in the world since the 1970s—manufactured by Monsanto.[4]

For these three films—which present several complementary facets of a single problem, namely, the consequences of biotechnology for world agriculture and, beyond that, for the production of food for humankind—I traveled around the world for a year: Europe, the United States, Canada, Mexico, Argentina, Brazil, Israel, India. Everywhere lurked the ghost of Monsanto, almost like the Big Brother of the new world agricultural order, the source of much anxiety.

This is why the recommendation from Singh as I was about to leave India solidified a vague feeling that I did in fact have to look more closely at the history of this North American multinational, founded in St. Louis, Missouri, in 1901, which now owns 90 percent of the patents for all GMOs grown in the world and became the world's largest seed company in 2005.

As soon as I got back from New Delhi, I turned on my computer and typed "Monsanto" into my favorite search engine. I found more than 7 million references, painting the portrait of a company that, far from enjoying universal favor, was considered one of the most controversial of the industrial age. In fact, adding to "Monsanto" the word "pollution" produced 343,000 hits. With "criminal" the number was 165,000. For "corruption," it was 129,000; Monsanto falsified scientific data produced 115,000 answers.

From there, I navigated from one Web site to another, consulting a great many declassified documents, reports, and newspaper articles, which enabled me to assemble a picture that the firm itself would prefer to conceal. Indeed, on the home page of Monsanto's Web site, it has presented itself as "an agricultural company" whose purpose is "to help farmers around the world produce healthier food, while also reducing agriculture's impact on our environment." But what it did not say is that before getting involved in agriculture, it was one of the largest chemical companies of the twentieth century, specializing particularly in plastics, polystyrenes, and other synthetic fibers.

Under the heading "Who We Are"/"Company History," one finds not a word about all the extremely toxic products that made its fortune for decades: the PCBs, chemicals used as insulators in electrical transformers for more than fifty years, sold under the brands Aroclor in the United States, Pyralène in France, and Clophen in Germany, the harmfulness of which was concealed by Monsanto until they were banned in the early 1980s; 2,4,5-T, a powerful herbicide containing dioxin, which was the basis for Agent Orange, the defoliant used by the American army during the Vietnam War, the toxicity of which Monsanto knowingly denied by presenting falsified scientific studies; 2,4-D (the other component of Agent Orange); DDT, which is now banned; aspartame, the safety of which is far from established; and bovine growth hormone (banned in Europe because of the risks it poses to animal and human health).

These were all highly controversial products, and they have simply disappeared from the firm's official history (except for bovine growth hormone, which I will deal with at length in this book). A careful examination of internal company documents reveals, however, that this nefarious past continues to burden its activities, forcing it to set aside considerable sums to cover the judgments that regularly cripple its profits.

A Quarter Billion Acres of GMOs

These discoveries led me to propose a new documentary to Arte entitled *The World According to Monsanto*, the research for which forms the basis for this book. The idea was to narrate the history of the multinational company and to try to understand to what extent its past might shed light on its current practices and what it now claims to be. In fact, with 17,500 employees, revenue of $7.5 billion in 2007 (and profit of $1 billion), and facilities in forty-six countries, the St. Louis company claims to have been converted to the virtues of sustainable development, which it intends to promote through the marketing of transgenic seeds, which in turn are supposed to extend the limits of ecosystems for the good of humanity.

Since 1997, with extensive publicity and an effective slogan—"Food, Health, and Hope"—it has managed to impose its GMOs, principally soybeans, corn, cotton, and rapeseed, over vast territories. In 2007, transgenic crops (90 percent of which, it should be recalled, have genetic traits patented by Monsanto) covered about 250 million acres: more than half were located in the United States (136.5 million acres), followed by Argentina (45 million), Brazil (28.8 million), Canada (15.3 million), India (9.5 million), China (8.8 million), Paraguay (5 million), and South Africa (3.5 million). This "surge in GMO land" has spared Europe, with the exceptions of Spain and Romania.[5] It should be noted that 70 percent of the GMOs cultivated in the world were at the time resistant to Roundup, Monsanto's prize herbicide, which the firm used to claim was "biodegradable and good for the environment" (an assertion that earned it, as we shall see, two official legal findings of false advertising), and 30 percent were manipulated to produce a toxic insecticide known as Bt toxin.

As soon as I began this long-term investigation, I contacted the company's management to ask for a series of interviews. The St. Louis headquarters sent me to Yann Fichet, an agronomist who is the director of institutional and industrial affairs of the French subsidiary located in Lyon. He set up an interview with me on June 20, 2006, in a Paris hotel near the Palais du Luxembourg (seat of the French Senate), where he told me he spent "a good deal of time." He listened to me at length and promised to transmit my request to Missouri headquarters. I waited three months and then got in touch with him again, when he ended up telling me that my request had been rejected. During filming in St. Louis, I called Christopher Horner, head of public relations for the firm, who confirmed the rejection in a telephone conversation on October 9, 2006: "We appreciate your persistence in, in asking, but, uh, you know we've had several conversations internally about this and, uh, have not changed our position. So there's no reason for us to participate."

"Is the company afraid about the questions I could ask?"

"No . . . you know, that it's certainly not a question of, you know, 'Do we have the answers or not?' It's a question of what the end product is going to be and do we give legitimacy to the end product by participating. Our suspicion is that it would not be positive."

Confronted with this refusal, I did not give up on presenting the firm's views. I got hold of all the written and audiovisual archival materials available in which its representatives spoke and also made extensive use of the documents it has placed online, in which it justifies the benefits that GMOs are supposed to bring to the world. "Farmers who planted biotech crops used significantly less pesticides and realized significant economic gains compared to conventional systems," said the company in its 2005 *Pledge Report*, a kind of ethical charter the company has been publishing regularly since 2000, in which it presents its commitments and its results.[6]

As a daughter of farmers, I am very aware of the difficulties that the agricultural world has experienced since I was born on a farm in Poitou-Charentes in 1960, and I have no difficulty imagining the impact that this kind of language can have on farmers who are struggling every day, in Europe and elsewhere, merely to survive. At a time when globalization is impoverishing the rural North and South, those who work on the land no longer know where to turn. Would the genius of St. Louis save their lives? I wanted to learn the truth because what is at stake concerns us all; it's a question of who in the future will produce food for humankind.

"Monsanto Company is helping smallholder farmers around the world become more productive and self-sufficient," the *Pledge Report* also says.[7] "The good news is that practical experience clearly demonstrates that the coexistence of biotech, conventional, and organic systems is not only possible, but is peacefully occurring around the world."[8] This sentence drew my attention because it touches on one of the major questions raised by GMOs, namely, that of possible risks to human health: "Consumers around the world are living proof of the safety of biotech crops. In the 2003–2004 crop year, they purchased more than $28 billion of biotech crops from U.S. farmers."[9] I thought of all the consumers who are nourished by the labor of farmers and who can, through their choices, affect the evolution of agricultural practices and of the world beyond—on the condition that they are informed.

All these quotations from Monsanto's *Pledge Report* are at the center of the polemic that pits defenders of biotechnology against its opponents. For the former, Monsanto has turned the page of its past as an irresponsible chemical company and is now offering products able to resolve the problems of hunger in the world and of environmental contamination by following the "values" that are supposed to direct its activity: "integrity, transparency, dialogue, sharing, respect," as its 2005 *Pledge Report* proclaims.[10] For the latter, all these promises are nothing but smoke and mirrors masking a vast plan for domination that threatens not just the food security of the world but also the ecological balance of the planet, and which follows in a straight line from Monsanto's nefarious past.

I wanted to decide for myself. To do that, I followed two paths. First, I worked on the Internet day and night. In fact, the great majority of the documents I cite in this book are available on the Web. All one has to do is look for them and connect them, which I invite the reader to do, because it is truly fascinating: everything is there, and no one—least of all those charged with writing the laws that govern us—can reasonably say that we didn't know. But, of course, that is not enough. And that is why I took up my pilgrim's staff again. I traveled to the United States, Canada, Mexico, Paraguay, India, Vietnam, France, Norway, Italy, and Great Britain. Everywhere, I compared Monsanto's words to the reality on the ground, meeting dozens of witnesses whom I had previously identified via the Web.

Many in the four corners of the world have raised the alarm, denouncing here a manipulation, there a lie, and in many places human tragedies—often at the cost of serious personal and professional difficulties. As the reader will find out in the course of this book, it is not a simple matter to oppose the truth of the facts to the truth of Monsanto, which is trying to "seize control of the seeds and hence of the food of the world," as Yudhvir Singh said to me in 2004—a goal that the firm seems on the way to achieving in 2009, unless European farmers and consumers decide otherwise, bringing the rest of the world in their wake.

PART I

One of the Great Polluters of Industrial History

1

PCBs: White-Collar Crime

We can't afford to lose one dollar of business.

—"Pollution Letter," declassified Monsanto document, February 16, 1970

Anniston, Alabama, October 12, 2006: With trembling hands, David Baker put the cassette into his VCR. "It's an unforgettable memory," the six-foot-tall man murmured, furtively wiping away a tear. "The greatest day in my life, the day when the people of my community decided to take back their dignity by making one of the largest multinationals in the world, which had always despised them, give in." On the screen were images filmed on August 14, 2001, of thousands of African Americans who walked silently and firmly in the golden late-afternoon light toward Anniston's cultural center on 22nd Street. The *Anniston Star* reported the next day that at least five thousand residents attended the meeting, the largest group many had ever seen in Anniston.

Asked why she had come, a fifty-year-old woman explained, "Because my husband and my son died of cancer."

A man pointed to a little girl perched on his shoulders. "She has a brain tumor. We had lost hope of getting Monsanto to pay for all the harm its factory has done us, but if Johnnie Cochran is working for us, then it's different."

The name was on everyone's lips. In 1995, the United States had held its breath as the celebrated Los Angeles lawyer defended O. J. Simpson against the charge of murdering his ex-wife and her friend in 1994. After a long and highly publicized trial, Simpson had been acquitted, thanks to the skill of his lawyer, the great-grandson of a slave, who had argued that his client was the victim of a racist police frame-up. From then until his death in 2005, Cochran was a hero to the American black community. "A god," David Baker said to me. "That's why I knew that by persuading him to come to Anniston, which he didn't even know existed, I had practically almost won the fight."

"Johnnie!" the crowd roared as the elegantly dressed lawyer climbed onto the stage. And Cochran spoke to a reverently silent audience. He was able to find the words that would resonate in this little southern town that had long been torn by the civil rights struggle. He spoke of the historic role of Rosa Parks, an Alabama native, in the struggle against racial segregation in the United States. He quoted the Gospel of Matthew: "Inasmuch as ye have done it unto one of the least of these my brethren, ye have done it unto me." Then he spoke of the story of David and Goliath, paying tribute to David Baker, the man who had made this unlikely meeting possible. "I look at this audience and I see a lot of Davids," he said with passion. "I don't know if you know what power you have. Every citizen has the right to live free from pollution, free from PCBs, from mercury and lead—that's a constitutional principle! You will rise up against the injustice Monsanto has done you, because the injustice done here is a threat to justice everywhere else! You are doing a service to the country that must no longer be ruled by the private interests of the giants of industry!"

"Amen!" cried the crowd, giving him a standing ovation. In the course of the next few days, 18,233 inhabitants of Anniston, including 450 children with neurological defects, filed through the small office of the Community against Pollution organization, set up by Baker in 1997 to bring legal action against the chemical company. They joined the 3,516 other plaintiffs, including Baker himself, who were already engaged in a class action suit that had been filed four years earlier. After a half century of silent suffering, almost the entire black population of the town was challenging a company with a decades-long history as a major world polluter, and would soon receive the largest known settlement paid by an industrial company in U.S. history: $700 million.

"It was a tough battle," commented Baker, still stirred by emotion. "But how could we imagine that a company could act so criminally? You understand? My little brother Terry died at seventeen from a brain tumor and lung cancer.[1] He died because he ate the vegetables from our garden and the fish he caught in a highly contaminated stream. Monsanto turned Anniston into a ghost town."

The Origins of Monsanto

Yet Anniston had had its glory days. Long known as the "model city," or the city with the "world's best sewer system" because of the quality of its municipal infrastructure, the little southern town, rich in iron ore, was long considered a pioneer of the industrial revolution. Officially chartered in 1879 and named after the wife of a railroad president, "Annie's Town" was celebrated as "Alabama's magnificent city" in the *Atlanta Constitution* in 1882. Run by a minority of white industrialists who were smart enough to reinvest their money locally to foster social peace, it competed with the nearby state capital, Birmingham, to attract entrepreneurs. In 1917, for example, Southern Manganese Corporation decided to establish a factory there for the manufacture of artillery shells. In 1925, the company changed its name to the Swann Chemical Company, and four years later it launched production of PCBs, universally hailed as "chemical miracles," which would soon make Monsanto a fortune and bring disaster to Anniston.

PCBs, or polychlorinated biphenyls, are chlorinated chemical compounds that embody the great industrial adventure of the late nineteenth century. While working to improve the techniques for refining crude oil to extract the gasoline needed for the infant automobile industry, chemists identified the characteristics of benzene, an aromatic hydrocarbon that would later be widely used as a chemical solvent in the manufacture of medicines, plastics, and coloring agents. In the laboratory, the sorcerer's apprentices mixed it with chlorine and obtained a new product that turned out to be thermally stable and to possess remarkable heat resistance. Thus PCBs were born, and for half a century they colonized the planet: they were used as coolants in electric transformers and industrial hydraulic machines, but also as lubricants in applications as varied as plastics, paint, ink, and paper.

In 1935, the Swann Chemical Company was bought by a rising enterprise from St. Louis, the Monsanto Chemical Works. Established in 1901 by John Francis Queeny, a self-taught chemist who also wanted to honor his wife, Olga Mendez Monsanto, the small company, set up with a $5,000 personal loan, began by manufacturing saccharin, the first artificial sweetener, which it then sold exclusively to another rising company in Georgia, Coca-Cola. It soon began supplying the soft drink company with vanillin and caffeine, and then started manufacturing aspirin, of which it was the largest American producer until the 1980s. In 1918, Monsanto made its first acquisition, buying an Illinois company that made sulfuric acid.

This shift to basic industrial products led to the purchase of several chemical companies in the United States and Australia after its shares went on sale at the New York Stock Exchange in 1929, one month before the crash, which the company survived, renamed the Monsanto Chemical Company. In the 1940s, it became one of the world's major producers of rubber, followed by plastics and synthetic fibers such as polystyrene, as well as phosphates. At the same time, it reinforced its monopoly in the international PCB market, guaranteed by a patent that enabled it to sell licenses almost everywhere in the world. In the United States and the United Kingdom (where the company had a factory in Wales), PCBs were marketed under the name Aroclor, while they were known by the name Pyralène in France, Clophen in Germany, and Kanechlor in Japan.

"That's how Anniston became the most polluted city in the United States," Baker explained to me as we got into his car for a tour of the area. First came Noble Street downtown, which was the pride of the city in the 1960s, with two movie theaters and many stores, most now closed. We then drove through the east side, dotted with pleasant houses where the white minority traditionally lived. Finally, on the other side of the tracks, came the west side, the home of the city's poor, mostly black, in the middle of an industrial area. That was where David Baker was born fifty-five years ago.

We were going through what he had rightly called a ghost town. "All these houses have been abandoned," he told me, pointing to dilapidated and tumbledown houses on both sides of the street. "People ended up leaving because their vegetable gardens and water were highly contaminated." We turned the corner from a lane full of potholes onto a wide thoroughfare with the sign "Monsanto Road." It ran alongside the factory where the company had produced PCBs until 1971. A fence surrounded the site, which now belongs to Solutia (motto: "Applied Chemistry, Creative Solutions"), an "independent" company also based in St. Louis, to which Monsanto turned over its chemical division in 1997, in one of the company's typical sleights of hand likely intended to protect it from the storm that its irresponsible conduct in Anniston was about to unleash.

"We weren't fooled," Baker said. "Solutia or Monsanto, it's all the same to us. Look, here's the channel of Snow Creek, where the company dumped its waste for more than forty years. It ran from the factory through the town, and flowed into the surrounding creeks. It was poisoned water. Monsanto knew it but never said anything."

According to a declassified report, secretly prepared in March 2005 by the Environmental Protection Agency (EPA), 680 million pounds of PCBs were produced in Anniston from 1929 to 1971.[2] Sixty thousand pounds of PCBs were emitted into the atmosphere, 1.8 million pounds were dumped in streams such as Snow Creek (following facility-cleaning operations), and 68 million pounds of contaminated wastes were deposited in an open pit located on the site, in other words, in the heart of the city's black community.

Half a Million Pages of Secret Documents

As we started to go around the site on foot, we met a hearse that honked its horn and stopped alongside us. "This is Reverend Jeffrey Williams," Baker explained. "He runs an Anniston funeral home. He succeeded his uncle, who recently died from a rare cancer, typical of PCB contamination."

"Unfortunately, he's not the only one," said Reverend Williams. "This year I've buried at least a hundred people who died of cancer, many young people between twenty and forty."

"I learned about the tragedy that's affecting all of us from his uncle," Baker went on. "For decades we accepted the deaths of our family members as a mysterious fate."

When his seventeen-year-old brother Terry collapsed and died in front of the family home, Baker was living in New York, where he was working as an officer of the American Federation of State, County, and Municipal Employees. After twenty-five years of good and faithful service, he decided in 1995 to go back home, where his experience as a union leader would soon be of great help to him. By chance, he was hired by Monsanto, which was then recruiting "environmental technicians," responsible for decontaminating the factory site. "It was in the mid-1990s," he said, "and we weren't yet informed of the pollution dangers, but the company was quietly starting to clean up. That was where I heard about PCBs for the first time, and I began to suspect that they were hiding something."

At the same time, Donald Stewart, an Anniston lawyer who had briefly been a United States senator, was contacted by a black resident of the west side of town, who asked him to come to the Mars Hill Baptist Church, located directly opposite the PCB factory. Accompanied by his congregants, the pastor informed him that Monsanto had offered to purchase the church from the community as well as a number of houses in the neighborhood. The lawyer understood that something was going on and agreed to represent the interests of the small church. "In fact," said Baker, "the company was in the process of clearing the ground around itself to avoid having to compensate property owners." Baker thought he knew why Monsanto was doing this, explaining that "it sensed that sooner or later pollution would come out into the open."

In any event, people started to talk in Anniston. The former union organizer from New York set up a first meeting in the funeral parlor of Russell "Tombstone" Williams, Jeffrey's uncle, which fifty people attended. They spoke late into the night of the deaths and illnesses that were devastating families (including those affecting young children), repeated miscarriages, and learning-related problems for the younger children. From this meeting came the idea of setting up an organization called Community against Pollution, presided over by Baker.

In the meantime, the Mars Hill Church affair had progressed: Monsanto offered a settlement, putting a million dollars on the table. During a meeting with the small Baptist community, Stewart found out that Monsanto's offer to buy several of its members' houses was contingent upon them promising never to take the company to court. The lawyer understood that Monsanto was hiding something big, and he suggested that they file a class action suit. Baker's committee was asked to recruit the plaintiffs, with the maximum number set by Stewart at 3,500.

Stewart had caught a whiff of the case of his life, but he also knew that it was likely to be long and costly. To deal with legal costs, he decided to contact the New York firm Kasowitz, Benson, Torres, and Friedman, famous for its litigation against the tobacco industry. The joint adventure would last more than seven years and would involve an investment of $15 million, with lawyers' fees sometimes running as high as $500,000 per month. The first stage consisted of organizing blood tests and fatty tissue analyses of the 3,500 plaintiffs, to measure their PCB levels. These tests, which could only be conducted by specialized laboratories, cost about $1,000 each.

While the complaint was being prepared under the title *Abernathy v. Monsanto*, Stewart moved heaven and earth to get his hands on company documents proving that it had known of the toxicity of PCBs. He knew that without this incriminating evidence, the fight would be hard to win, because the company could always offer the defense of ignorance. Intuitively, he was convinced that a multinational full of scientists would operate in a very bureaucratic fashion, with a hierarchy that controlled everything through a very sophisticated document system; the slightest event or decision, he thought, had to have left written traces. He minutely scrutinized the depositions of Monsanto representatives, and he came across a pearl: according to a company lawyer, a "mountain of documents"—500,000 pages that had disappeared from the St. Louis offices—had been deposited in the library of a New York law firm that represented Monsanto.

Stewart asked to consult them, but he was told that the documents were inaccessible because they were protected by the work product doctrine, which allows attorneys to keep documents secret before a trial in order to avoid providing ammunition for the opposing party. Stewart turned to Judge Joel Laird of the Calhoun County court, who was handling *Abernathy v. Monsanto*: in a crucial decision, the judge ordered Monsanto to open up its internal archives.

Monsanto Knew, and Said Nothing

The "mountain of documents" is now accessible on the Web site of the Environmental Working Group, an NGO dedicated to environmental protection and headed by Ken Cook, who met with me in his Washington office in July 2006.[3] Before meeting with him, I spent many nights combing through this mass of memoranda, letters, and reports drafted over decades by Monsanto employees with truly Kafkaesque precision and coldness.

Indeed, there is something I still have trouble understanding: how could people knowingly run the risk of poisoning their customers and the environment and not stop to think that they themselves or their children might be the victims of, to put it mildly, their negligence? I am not speaking of ethics or morality, abstract concepts foreign to the logic of capitalism, but merely of the survival instinct: was it lacking in the managers of Monsanto?

"A company like Monsanto is a world of its own," Cook told me, admitting that he had been plagued by the same questions. "The pursuit of profit at any price anesthetizes people devoted to a single purpose: making money." He showed me a document that summed up this way of operating. Entitled "Pollution Letter," it was dated February 16, 1970. Drafted by N.Y. Johnson, who worked in the St. Louis office, this internal document was addressed to the company's marketing staff to explain to them how to answer their customers who had learned of the first public disclosures of the potential dangers of PCBs: "Attached is a list of questions and answers which may be asked of you by customers receiving our Aroclor-PCB letter. You can give verbal answers; no answers should be given in writing. . . . We can't afford to lose one dollar of business."

What is absolutely breathtaking is that Monsanto knew that PCBs presented a serious health risk as early as 1937. But the company carried on regardless until the products were finally banned in 1977, the date when its W.G. Krummrich plant in Sauget, Illinois (an eastern suburb of St. Louis, the site of Monsanto's second PCB production facility), was closed down.

In 1937, Dr. Emmett Kelly, Monsanto's medical director, was invited to a meeting at the Harvard School of Public Health, also attended by PCB users such as Halowax and General Electric, along with representatives of the U.S. Public Health Service. At this meeting, Cecil K. Drinker, a Harvard researcher, presented the results of a study he had conducted at the request of Halowax: a year earlier, three employees of that company had died after being exposed to PCB fumes, and several had developed a terribly disfiguring skin disease, which was then unknown but later named chloracne. I will come back in the next chapter to this serious pathology, which is characteristic of dioxin poisoning, sometimes resulting in an eruption of pustules all over the body, and which may last for several years or indeed never go away.

In a panic, Halowax management asked Cecil Drinker to test PCBs on rats. The results, published in the *Journal of Industrial Hygiene and Toxicology*, were conclusive: the test animals had developed severe liver lesions. On October 11, 1937, an internal Monsanto report tersely noted that "experimental works in animals shows that prolonged exposures to Aroclor vapors . . . will lead to systemic toxic effects. Repeated bodily contacts with the liquid Aroclor may lead to an acne-form skin eruption."

Seventeen years later, the problem of chloracne was the subject of an internal report written in chillingly technical language: "Seven workers developed chloracne in a plant using Arochlor," a Monsanto manager reported, and then calmly explained: "The fact that air tests, even in the presence of vapors, showed only negligible amounts of chlorinated hydrocarbons indicates that this type of intermittent but fiercely long continued mild exposure is not innocuous."

On February 14, 1961, the head of production of Hexagon Laboratories, another Monsanto customer, sent a letter to Kelly in St. Louis: "In reference to our recent telephone conversation, I would like to further discuss the incident wherein two of our plant personnel were exposed to hot Arochlor (1248) vapors generated by a broken pipe connection. For your information and records the two men developed symptoms of hepatitis as you predicted and were confined to a hospital for approximately two weeks. . . . Since we are dealing with a highly toxic material . . . it is felt that a more thorough and clearly written description of the hazards should be described under Safety of Handling."

Monsanto did not follow its customer's recommendation; it had only begrudgingly complied with labeling laws passed in 1958 intended to strengthen safety precautions in the handling of toxic products. "It is our desire to comply with the necessary regulations, but to comply with the minimum and not to give any unnecessary information which could very well damage our sales position in the synthetic hydraulic fluid field."

Sometimes, confronted with urgent questions from their customers, Monsanto managers lost themselves in circumlocutions that might provoke a smile if the stakes were not so serious. For example, in August 1960, a manufacturer of compressors in Chicago was concerned about the possible environmental consequences of the discharge of wastes containing PCBs into rivers. "I would like to say that if small quantities of these materials are accidentally spilled into a receiving stream there would probably be no harmful effect," a representative of the Monsanto medical department answered. "If, on the other hand, a great deal of the material was spilled some readily identifiable damage might ensue."

As the years went by, however, the tone changed, probably because the threat of legal action brought by its own customers was weighing ever more heavily on the company. In 1965, an internal memo reported a telephone conversation with the head of an electrical company that used Aroclor 1242 as an engine coolant. The manufacturer had apparently said that in his own plant Aroclor spills on the floor were common. The memo noted: "I was brutally frank and told him that this had to stop before he killed somebody with liver or kidney damage."

"Criminal" Conduct

In the face of the alarming reports coming from the field, there were very few voices who spoke up against the general inertia, including Dr. J.W. Barrett, a Monsanto scientist based in London, who suggested in 1955 that studies be conducted to rigorously evaluate the toxic effects of Aroclor. Kelly responded curtly: "I don't know how you would get any particular advantage in doing more work." Two years later, the head of the medical department, with the same self-assurance, commented on the results of an experiment conducted by the U.S. Navy with Pydraul 150, a PCB used as a hydraulic fluid in submarines. "Skin applications of Pydraul 150 caused the death of all the rabbits tested. . . . No matter how we discussed the situation, it was impossible to change their thinking that Pydraul 150 is just too toxic for use in a submarine."

It is surprising when reading these documents to see the extent to which the company's position was apparently immune to challenge. It conscientiously collected alarming data, which it hastened to lock in a drawer, keeping its eyes riveted on sales instead: "2.5 million pounds per year," crowed the author of a 1952 document. But there were moments when I began to dream of a possible change in the company's behavior.

For example, on November 2, 1966, the report of an experiment conducted at Monsanto's request by Professor Denzel Ferguson, a zoologist from Mississippi State University, arrived in St. Louis. His research team had immersed twenty-five caged fish in Snow Creek, where waste from the plant was dumped and which, as we have seen, flowed through the city of Anniston. "All 25 fish lost equilibrium . . . and all were dead in 3½ minutes and . . . blood issues from the gills after 3 minutes exposure," the scientist reported. He went on to say that at certain points the water was so polluted that it "kills fish in less than 24 hours when diluted 300 times." In their final report, the Mississippi State scientists made several recommendations: "Do not release untreated waste in the future! Clean up Snow Creek." And the conclusions pointed out: "Snow Creek is a potential source of future legal problems. . . . Monsanto needs to monitor the biological effects of its effluents as a protection against future accusations."

Late in November 1966, the Brussels office of Monsanto Europe received a letter from a correspondent in Stockholm reporting on a scientific meeting concerning research conducted by a Swedish scientist, Soren Jensen. Published in *New Scientist*, this work had caused a great stir in Sweden.[4] While analyzing DDT in samples of human blood, Jensen had accidentally discovered a new toxic substance, which turned out to be PCB. The irony of the story is that DDT, a powerful insecticide discovered in Switzerland in 1939, was also a chlorinated chemical product that Monsanto sold widely until it was finally banned in the early 1970s, in particular because of its human health effects. Jensen discovered that PCBs had already extensively contaminated the environment even though they were not manufactured in Sweden: he found significant quantities in salmon caught near the coast and even in the hair of his own family (his two children, ages three and six, his wife, and his five-month-old infant, who must have been contaminated by breast milk). He concluded that PCBs "accumulated in certain organs of animals and the food chain. They are said to be related to DDT and equally poisonous."

And yet Monsanto management did not change its attitude: one year later it allocated an additional $2.9 million to further development of Aroclor products in Anniston and Sauget. "The company's irresponsibility was staggering," said Ken Cook. "It had all the data at its fingertips, but it did nothing. That's why I say it was guilty of criminal conduct." In fact, no specific measures were taken to protect the workers in the Anniston plant. "At Anniston no special protecting clothing is provided for the Arochlors operators," a 1955 document notes. "A daily change of clothing was provided in the past but this practice ceased before the war." The only clearly announced recommendation was not to eat in the Aroclor department.

But the company was discreetly collecting data that would be used against it twenty years later: "The effects of exposure of PCBs on our employees have been reviewed by our medical Department and a consultant from the Eppley Institute," explained William Papageorge, known as the "PCB czar" because he supervised their production for several decades. "In summary there is no evidence that our employees have been adversely affected by PCBs. . . . We have no program underway to study these "effects." Similarly, technicians in St. Louis confirmed by firsthand observation that toxic products persisted in the environment for at least thirty years. In 1939, in fact, PCBs had been buried in patches of ground to test their effectiveness as termite poison: "There is still visual evidence of the presence of Aroclor," noted an "officer" in 1969.

"The worst thing about all of this," said Cook, "is that Monsanto never warned the residents of Anniston that the water, the soil, and the air of the west side of town was highly contaminated. As for state and local authorities, not only did they close their eyes, but they covered up the company's actions. It's really scandalous. I think one of the explanations of this tragedy is the racism of the leaders at the time: after all, they were only blacks."

Complicity and Manipulation

In the spring of 1970, just after the Nixon administration, with great fanfare, had announced that the Environmental Protection Agency would be established later that year to meet the "public's growing demand for clean air, water, and soil," Monsanto made a preemptive strike: a note from May 7 marked "confidential" describes a meeting between company representatives and Joe Crockett, the technical director of the Alabama Water Improvement Commission (AWIC), the public body responsible for the state's water quality. The purpose of the meeting was to "inform the AWIC representative of the situation" and "to *build confidence* that Monsanto intends to cooperate with governmental agencies to define the effects of Aroclor on the environment" (emphasis added). This was simply a public relations exercise, which in fact succeeded, since Crockett recommended that no statements be given "which would bring the situation to the public's attention." The note concludes: "The full cooperation of the AWIC on a confidential basis can be anticipated."

At the same time the Food and Drug Administration (FDA) was conducting tests on fish caught at the confluence of Snow Creek and Choccolocco Creek. They determined that PCB levels in the fish were at 277 parts per million (ppm), whereas the safe level for consumption had been set at 5 ppm.

Curiously, the FDA took no steps to issue an advisory against fishing in the incriminated waterways nor against Monsanto, which thus had an opportunity to put the "cooperation" of the AWIC to the test: "We are now discharging 16 pounds of PCBs per day (compared to 250 in 1969) into Snow Creek," according to an August 1970 document marked "Confidential. FYI and Destroy." "Joe Crockett will try to handle the problem quietly without release of the information to the public at this time." The residents of Anniston therefore continued to consume fish caught in contaminated streams until 1993, when the FDA issued its first order warning against the practice.

But Monsanto's negligence, which some would call cynicism, did not stop there. I have already noted that the company was discharging some of its wastes in a dump near the factory that, when it rained, produced runoff into neighboring gardens. In December 1970, a neighborhood resident was allowing one of his pigs to forage in a vacant lot next to the dump. He was approached by a representative of Monsanto who offered to buy his animal. As an internal memo reveals, the animal was slaughtered and analyzed: its fat contained 19,000 ppm of PCBs.[5] But in this case as well, no information was ever provided to the residents, who continued to allow their pigs to forage in the vacant lot for many years.

In fact, everything indicates that the company's single obsession was to carry on its business come what may. In August 1970, when PCBs were increasingly gaining attention in the media, company management decided to set up an ad hoc committee to consider the situation. The committee issued a report marked "confidential," which began by listing its objectives: "permit continued sales and profits of aroclors" and "protect image of . . . the Corporation." There followed a long list of all cases of contamination recorded in the country. It turns out, for example, that a University of California researcher had detected elevated levels of PCBs in fish, birds, and eggs in the coastal region.[6] A study conducted by the FDA had revealed that PCBs had been found in milk from herds in Maryland and Georgia; another study conducted by a laboratory of the Commercial Fisheries Bureau of the U.S. Interior Department in Florida had showed that juvenile shrimp did not survive in water containing 5 ppm of PCBs, and so on. Reading the report leads to the conclusion that PCBs were everywhere: they were used as lubricants in turbines, pumps, and food distribution equipment for cows, they were a component of the paint used for the walls of reservoirs, grain silos, swimming pools (particularly in Europe), and road markings and were used in the manufacture of oils used in metal fabrication, solder, adhesives, carbonless copy paper, and more.

"As the alarm concerning the contamination of the environment grows it is almost certain that a number of our customers or their products will be incriminated. The company could be considered derelict, morally, if not legally, if it fails to notify *all* customers of the potential implication," the committee stated. It concluded that the company was faced with an "extraordinary situation. There can not be too much emphasis given to the threat of curtailment or outright discontinuance of the manufacture and sales of this very profitable series of compounds. If the products, the Division, and the Corporation are to be adequately protected, adequate funding is necessary."

To put it plainly, Monsanto was proposing not to confess its mistake and simply withdraw its Aroclor product line from the market, but on the contrary to do everything possible to keep it on sale. The first stage of the battle plan was to finance a toxicological study to test PCBs on rats. To that end, the company signed a contract with Industrial Bio-Tech Labs (IBT) in Northbrook, Illinois, one of whose new directors was Dr. Paul Wright, a toxicologist from Monsanto recruited for the occasion. A few months later, the preliminary results of the study reached company headquarters: "PCBs are exhibiting a greater degree of toxicity in this chronic study as we had anticipated. . . . We have additional interim data which will perhaps be more discouraging." A letter to Joseph Calandra, the head of IBT, followed: "I think we are surprised (and disappointed?) at the apparent toxicity at the levels studied. We would hope that we might find a higher 'no effect' level with this sample as compared to the previous work." In July 1975, Monsanto's manager of environmental assessment and toxicology attempted to correct the results by strongly suggesting that the phrase "slightly tumorigenic" be replaced by the phrase "does not appear to be carcinogenic."

A Poison as Toxic as Dioxin

Professor David Carpenter, director of the Institute for Health and the Environment at the University of Albany, told me: "We all have PCBs in our bodies. They belong to a category of twelve very dangerous chemical pollutants known as persistent organic pollutants (POPs), because, unfortunately, they are resistant to natural biological decay and they accumulate in the living tissue through the entire food chain.

"PCBs have contaminated the whole planet, from the Arctic to the Antarctic, and regular exposure can lead to cancer, namely, liver, pancreatic, intestinal, breast, lung, and brain cancer, cardiovascular disease, hypertension, diabetes, immune deficiency, thyroid disorders, sexual hormone imbalances, reproductive problems, and serious neurological disturbances, because some PCBs belong to the dioxin family."

He went on to explain that PCBs are biphenyl molecules in which one or more of the ten hydrogen atoms are replaced by chlorine atoms. There are 209 possible combinations, and hence 209 different PCBs, known as congeneric PCBs, the toxicity of which varies depending on the location and number of chlorine atoms in the molecule.

Writing these lines reminded me of an article in *Le Nouvel Observateur* of August 23, 2007, which, following *Le Monde*, *Libération*, and *Le Figaro*, reported on what *Le Dauphiné libéré* had called a "French Chernobyl."[7] According to the weekly, "The Rhône is polluted to its mouth. It contains levels of PCBs that are five to twelve times above European health norms!* Analysis after analysis, orders from the prefects came down like guillotines: the ban on the consumption of fish, decreed first north of Lyon and then applied as far as Drôme and Ardèche, was extended on August 7 to the departments of Vaucluse, Gard, and Bouches-du-Rhône. It may soon reach the Camargue marshes, which are fed by water from the river, and even coastal fishing in the Mediterranean and the harvesting of shellfish and crustaceans near the coast."

The alarm was raised fortuitously by a professional fisherman who was the victim of his own good faith. "In late 2004, dead birds were found upstream from Lyon," he explained to a journalist. "While they were being analyzed, as a precautionary measure, the veterinary services prohibited all consumption of fish. It was only a case of strictly avian botulism, but no one wanted my fish afterward. I asked for complete analyses to prove that they were good. And bingo! They were stuffed with PCBs!"

Since then, government services have been struggling to determine the origin of the pollution that is said to have affected hundreds of thousands of tons of sediment in the Rhône. I have already noted that the purchase and

*According to *Le Monde* of June 26, 2007, "the most contaminated fish had a level forty times above the daily acceptable level."

sale of PCBs or equipment containing them have been prohibited in France since 1987. A decree issued January 18, 2001, incorporated into French law a European directive enacted nearly five years earlier, on September 16, 1996, concerning the elimination of existing PCBs, a process that is supposed to be definitively completed by December 31, 2010, at the latest.[8] A national plan for the decontamination and elimination of equipment containing PCBs was established only in 2003. According to the French Environment and Energy Management Agency (ADEME), 545,610 pieces of equipment containing more than five liters of PCBs had been inventoried in France by the end of June 2002 (450,000 of which belonged to Électricité de France (EDF)), amounting to 33,462 tons of PCBs to be eliminated. But according to the association France Nature Environnement, the goal is far from being reached in light of the fact that the declaration of equipment to be treated was voluntary. "Our fear was of seeing diffuse PCB pollution in the environment due to incomplete elimination of these wastes, with the risk that they would be dumped on industrial wastelands or in improvised dumpsites, or simply used as scrap metal," the association wrote in its February 2007 newsletter.[9]

"The problem," Carpenter explained, "is that PCBs are very difficult to destroy. The only way is to burn them at very high temperatures in special incinerators also able to treat the dioxin produced by their combustion." Two factories in France are certified to carry out this delicate task: one is located in Saint-Auban in Alpes-de-Haute-Provence, the other in Saint-Vulbas in Ain, on the banks of the Rhône. According to a report in *Le Nouvel Observateur*, until 1988 the Saint-Vulbas installation was authorized to discharge three kilos daily of PCB residues into the river (the maximum quantity is now three grams a day). To this possible source of contamination should probably be added discharges by the numerous companies in the "chemical corridor" that use Pyralène: oils containing PCBs were allowed to leak into the ground and from there into the water table and nearby streams. "For decades, in the United States and around the world, public authorities preserved the silence organized by Monsanto about the toxicity of PCBs," said Carpenter. "Everyone closed his eyes to this poison, which is as dangerous as dioxin."

One merely has to read a report presented to Congress by the U.S. Public Health Service and the EPA in 1996 to understand that the "health implications of exposure to PCBs" are extremely serious.[10] The thirty-page report enumerates no fewer than 159 scientific studies conducted in the

United States, Europe, and Japan that all reached the same conclusion: the three principal sources of human contamination by PCBs are direct exposure in the workplace, living near a polluted site, and, most important, the food chain, with the consumption of fish being by far the riskiest. In addition, all the studies found that contaminated mothers transmitted PCBs to their infants in breast milk and that the substances could cause irreparable neurological damage in the babies, who would be affected by what doctors have labeled "attention deficit disorder" and would have significantly lower than average IQs.

The devastating toxicity of PCBs could be studied in detail because of an accident in Japan in 1968, when thirteen hundred people on the island of Kyushu used rice bran cooking oil contaminated by PCBs because of a leak in a refrigeration system. They were affected by a disease at first called *yusho*, meaning "skin disease caused by oil," characterized by severe skin eruptions, discoloration of the lips and nails, and swelling of the joints. When it turned out that the source of the mysterious disease was PCBs, researchers undertook long-term medical follow-up of the victims. The results showed that children born to mothers contaminated during pregnancy had an elevated early mortality rate and/or significant mental and behavioral impairment; in addition, the rate of liver cancer was fifteen times higher among the victims than in the normal population, and average life expectancy was considerably reduced. Finally, PCBs were still detectable in the blood and sebum of contaminated people twenty-six years after the accident.

These results were confirmed by a study of two thousand people in Taiwan contaminated in 1979 in circumstances similar to those of their Japanese neighbors (the "Yu-Cheng accident").[11] These two dramatic events explain the panic that seized Belgian authorities in January 1999 when the "dioxin chicken" crisis erupted. The cause was again the accidental contamination by PCBs of cooking oil that was then added to animal feed supplied to chickens, pigs, and cattle.

From the litany of studies listed in the EPA report, I will take note of two others that were particularly dramatic. One concerned 242 children whose mothers (of Amerindian origin or the wives of recreational fishermen) had regularly consumed fish from Lake Michigan over a period of six years before and during their pregnancies; all the children had low birth weight and persistent cognitive deficits. The other concerned Inuits of Hudson Bay, who were particularly exposed because of their heavy reliance on the meat of sea mammals at the top of the food chain, such as seals, polar bears, and whales, where the highest levels of contamination were found. (In fact, some species of sea mammals, including killer whales, are threatened with extinction caused by PCBs.)[12]

Denial Now and Forever

"There is no consistent, convincing evidence that PCBs are associated with serious long-term health effects," declared John Hunter, CEO of Solutia, on January 14, 2002, in a conference he called with investors and representatives of the press.[13] He was attempting to reduce the impact of an article in the *Washington Post* titled "Monsanto Hid Decades of Pollution," published on January 1, 2002, just before the trial of *Abernathy v. Monsanto* opened.[14] "Despite the extent of the scientific evidence, internal documents, and witness testimony, the manufacturers in St. Louis have continued to deny the responsibility of the firm in the ecological and health disaster of Anniston," stated David Carpenter, called as an expert witness at trial. "They have never showed the slightest compassion for the victims," Ken Cook confirmed to me, "not a word of excuse or a sign of regret, denial now and forever! Their line of defense can be summed up like this: 'We didn't know that PCBs were dangerous before the late 1960s, but as soon as we found out, we acted quickly to rectify the problem with government agencies.'"

The arrogance revealed by some company representatives in the trial transcript is truly chilling, and they do anything but make amends. An example is this excerpt from the testimony of William Papageorge, the "PCB czar," given on March 31, 1998, in the Calhoun County court. "To your knowledge, sir, did Monsanto ever disclose to the residents of Anniston in 1968 or 1969 that twenty-seven pounds of organics and acid waste from the Aroclor and HCl departments were being lost from the plant?" asked the attorney.

"There was no reason to talk those numbers. They were meaningless," answered Papageorge.

"But the answer is no?"

"That is correct."

"Thank you. Did anyone ever tell the residents of Anniston at that time that Monsanto was visually checking Snow Creek and Choccolocco Creek to determine the effects of the PCBs in the plant effluent water?"

"Sir, this is no different than a service station man telling his neighbors he has got motor oil on the curb by his service station. Those things are just nonproductive comments that one can make to others."

"I'm going to move to strike. But the answer, though, is no? Is that right?"

"Yeah."

"Did Monsanto ever provide the residents of Anniston with any data concerning the health hazards of PCBs in humans?"

"Why would they?"

On February 23, 2002, after deliberating for five hours, the jury delivered its verdict: it unanimously found Monsanto and Solutia liable for having polluted "the Anniston area and people's blood with PCBs."[15] The legal grounds for the verdict were "negligence, wantonness, fraud, trespass, nuisance, and outrage," and it included a harsh judgment of Monsanto's conduct, which was "so outrageous in character and extreme in degree as to go beyond all possible bounds of decency, so as to be regarded as atrocious and utterly intolerable in civilized society." The firm soon filed an appeal with the Alabama Supreme Court, asking that Judge Joel Laird be removed from the case, but the appeal was rejected. The jury then undertook the difficult task of evaluating the damages that each victim would recover on the basis of the PCB blood level measured and the cost of a program for decontaminating the site. Fifteen percent of the 3,516 plaintiffs had a PCB blood level higher than 20 ppm (the acceptable level was 2 ppm), with spikes as high as 60 or even 100 ppm. David Baker had a level of 341 ppm and was awarded damages of $33,000. The highest award was $500,000.

A month after the verdict, the EPA, which had been conspicuously inactive on the issue for more than twenty years, announced that it had signed an agreement with Solutia to decontaminate the site. This decision, very favorable to the polluter and nullifying the jury's work, provoked the anger of Alabama senator Richard Shelby, who brought the matter before a Senate subcommittee, which pointed out that Linda Fisher, the number two staffer at the EPA, was a former Monsanto executive.

At the same time, the federal district court in Birmingham announced that the case of *Tolbert v. Monsanto*, a class action filed by Johnnie Cochran, would open in October 2002. Solutia's share price on the New York Stock

Exchange collapsed. Judge U.W. Clemon, who wanted to avoid a costly trial, then undertook the tedious task of persuading the parties to negotiate an overall settlement covering the two cases. The company had until then rejected that solution, probably hoping to financially exhaust the plaintiffs by multiplying technical legal motions and delaying tactics. "In fact," Baker explained to me, "the prospect of a highly publicized trial, with Johnnie Cochran in court, made Monsanto give up and negotiate to reduce publicity." Finally, the polluter offered $700 million: $600 million divided into two equal funds to indemnify victims, and $100 million to decontaminate the site and finance a specialized clinic.[16]

"Who will pay?" wondered the *St. Louis Post-Dispatch* on February 7, 2004. The problem was indeed intricate: Monsanto had gotten rid of its chemical division in 1997 by selling it to Solutia. And in December 1999, the company, which then had a pharmaceutical branch and an agricultural branch (transgenic seeds and Roundup), announced its merger with Pharmacia and Upjohn under the name Pharmacia. In the summer of 2002, Monsanto recovered its independence, retaining only its agricultural division, and Pharmacia was purchased by the pharmaceutical giant Pfizer. As a result, the $700 million would finally be paid by Solutia ($50 million), Monsanto ($390 million), and Pfizer ($75 million), with the remainder covered by insurance.

The lawyers pocketed 40 percent of the damage award, which provoked some complaints. "That's how the American system works," Baker explained to me. "In this kind of case, the lawyers are paid only if they win, and Johnnie Cochran, for example, had spent $7 million preparing the trial. That means if you don't find a Johnnie Cochran, you can't do anything against a company like Monsanto. The thing I regret is that none of the company executives was sentenced to prison."

The status of corporations as "persons" in United States law generally shelters company officials from individual liability. "In the American legal system," said Cook, "it is very rare for executives or managers of companies to be found criminally liable. On the other hand, companies can be sued in civil court, and they are made to pay. But in fact, the damages they pay decades later are only a fraction of their profits. So it pays to keep secrets. I wonder what secrets Monsanto is keeping now. You can never trust a big company like Monsanto to tell us the truth about a product or a pollution problem. Never."

PCBs Are Everywhere

According to accepted estimates, 1.5 million tons of PCBs were produced from 1929 to 1989, a significant portion of which ended up in the environment. How much exactly? It is hard to know. The fact remains that PCBs are everywhere and they are a nightmare for us as citizens, but they are also a nightmare for Monsanto (and the subsidiary it used, Solutia, which declared bankruptcy in 2003 largely because of the litigation it had inherited).

Here is a brief, not exhaustive, summary: In January 2003, the Environment Department of Oslo fined Bayer, Kaneka, and Solutia €7 million for contaminating the fjord on which the harbor is located with PCBs used in ship paints. (It should be noted in passing that many experts, including David Carpenter, strongly advise against consuming salmon raised in Norway and Scotland.) In January 2006, 590 workers in a General Electric factory in New York sued Monsanto for PCB contamination.[17]

In 2007, as France was discovering that the Rhône was polluted by PCBs, Wales was shaken by a scandal that had been suppressed for more than forty years.[18] Monsanto had a subsidiary in Newport that until 1978 produced 12 percent of the PCBs manufactured in the world. From 1965 to 1971, the factory dumped into the Brofiscin quarry, an extremely porous former limestone quarry, some 800,000 tons of waste contaminated with PCBs. The activity had been denounced at the time by farmers who had noticed that their cattle were dying mysteriously. The decontamination of the site could cost more than €200 million. For now, Monsanto and Solutia are blaming the company that the Newport factory contracted with to transport and dump the wastes.

At a time when concern for the environment is in the headlines, it is likely that the ghost of PCBs will haunt Monsanto for a long time to come, just like dioxin, of which it was an experienced producer.

2

Dioxin: A Polluter Working with the Pentagon

> We believe that a human rights policy to guide our actions as a global corporate citizen is a logical next step to our Monsanto pledge—to address the importance of showing and advancing respect for our people and those affected by our actions.
>
> —Monsanto, *Pledge Report*, 2005

"Dioxin? That poison has given me nightmares for twenty-five years," said Marilyn Leistner as she parked her car in front of the Route 66 State Park visitor center about twenty miles from St. Louis. "Take a look, there's nothing left of Times Beach. Who could imagine that this was once a town of fourteen hundred where more than eight hundred families lived?"

It is indeed hard to imagine. It was October 2006, and we were in front of a newly refurbished building that housed a kitschy tribute to the mythical Route 66. One of the original U.S. highways, the "Mother Road" ran for 2,400 miles from Chicago through eight states, including Missouri, to Los Angeles. Next to the building stood a Western-style wooden sign with the words "Route 66 State Park." "They wiped Times Beach off the map and put a state park on the decontaminated site to bury one of the worst dioxin scandals in the United States," said Leistner, the last mayor of the vanished town.

A Town Wiped Off the Map

"Times Beach" and "dioxin" were long joined together in the headlines of American newspapers, to the dismay of the residents of the little town established in 1925 as a summer resort for people working in St. Louis. At first, no one lived in the town. People set up trailers and came on weekends to swim in the Meramec River, to fish, or to picnic. Known as "the Beach," the town attracted permanent residents who built wooden houses on stilts because the idyllic spot was frequently flooded. Times Beach gradually became a real town, with stores, a gas station (owned by Marilyn Leistner's husband), a church, thirteen saloons, and a town council.

In the early 1970s, the not very wealthy town decided to try to deal with the problem of the dust that covered its unpaved roads and made residents' lives miserable. The town decided to call on the services of the Bliss Waste Oil Company, a company that collected waste oil and industrial wastes from gas stations and chemical plants in Missouri. The company's head, Russell Bliss, suggested spraying mud mixed with residual oil on the streets of Times Beach.

"By the summer of 1971, we had noticed the deaths of many cats, dogs, birds, and even a raccoon," said Leistner. "One resident contacted the EPA, and they told him to freeze some dead animals and an agent would come to get them. But no one ever came." But the EPA had already been alerted. In March 1971, the owner of a horse arena located northwest of St. Louis had been troubled by the unexplained death of fifty horses after Bliss Waste employees had sprayed the stable floors with brownish mud. A few weeks later, his two children, who often played in the stables, fell seriously ill and had to be hospitalized. When the Centers for Disease Control (CDC) were contacted, they conducted tests of the suspect material and found alarming levels of toxic products: 1,590 ppm of PCBs, 5,000 ppm of 2,4,5-T (a powerful herbicide), and 30 ppm of dioxin.[1]

"There literally were bushel baskets full of those dead wild birds," Dr. Patrick Phillips, a Missouri State Health Department veterinarian, told the *New York Times* after being called to various places around the state where Bliss's company had sprayed.[2] *Science* published an article in 1975 on this mysterious deadly pollution, but for years the authorities did nothing.[3] However, the EPA was quietly conducting an investigation. It was focused primarily on Missouri factories that produced toxic waste, as revealed by an exchange of letters in September 1972 between agency officials and William Papageorge, Monsanto's "PCB czar"; apparently, samples had been taken from Russell Bliss's oil tanks and analyses sent to Monsanto.

At length, the "black autumn" of 1982 arrived. "It was a nightmare," said Leistner, who was then a town alderman. "On November 10, I was informed by a local reporter that Times Beach was on a list of one hundred sites contaminated by dioxin that had been drawn up by the EPA. On December 3, agency technicians came to collect soil samples. Two days later, the town experienced the worst flood in its history and many families had to be evacuated. On December 23, when residents were just coming back to their houses, the EPA informed us that the level of dioxin detected in the samples was three hundred times the level considered acceptable."[4]

There was panic in Times Beach. While teams of EPA technicians wearing hazmat suits and gas masks went through the town, reporters flooded in from around the country. "At the time, we knew very little about dioxin," Leistner recalled, "and it was by watching television news that we found out it was the most dangerous molecule ever invented by man. But that's all. No one was able to tell us what that might mean for our health." And there was good reason for this: the highly toxic effects of dioxin had been knowingly suppressed by its producers, particularly by a certain company in St. Louis.

In the meantime, the CDC set up an emergency unit in Times Beach. Residents were asked to come in for a health check. The television news broadcasts of the time that I was able to see show the anxiety in the people's faces, outbursts of tears, and their impotent anger at the silence of the doctors avoiding questions. "My whole family was examined," said Leistner. "My husband was suffering from delayed-onset cutaneous porphyria, a persistent skin disease.* Two of my daughters, my son, and I suffered from hyperthyroidism. I had had surgery for several benign tumors. One of my daughters suffered from severe allergies that produced hives all over her body; my second daughter was extremely thin, she had dizzy spells, and was losing her hair. When I asked the CDC officials if this was connected to dioxin, they said they didn't know."

In any event, panic had reached a peak in Times Beach. Severely de-

*This skin disease produces blisters, scabs, and scars, primarily on the backs of hands, the forearms, and the face. It can be triggered by exposure to dioxin.

pressed, the mayor resigned. At the same time, one of his assistants simply disappeared. According to Leistner, "he was a Monsanto executive who was working at St. Louis headquarters. When he found out the EPA had detected PCBs, he moved out." And so Marilyn Leistner found herself in charge of the little town, having to "face the storm." On February 22, 1983, Anne Burford, administrator of the EPA, announced that the government had decided to "buy Times Beach for the sum of $30 million." The extraordinary plan proposed to indemnify and relocate all the residents, tear down the town, then decontaminate the site and burn the contaminated soil in an incinerator.

Monsanto Avoids Responsibility

"See, this is where our houses are buried," said Marilyn Leistner, stopping for a few minutes in front of a large mound of grass-covered earth. "Everything we owned was bulldozed, the furniture, the appliances, even the children's toys, because the flood had spread the dioxin and the PCBs everywhere. We left like plague-stricken refugees, because no one wanted to have anything to do with us: people were convinced we were contagious."

"You didn't sue?"

"Of course we did, but the case was dismissed because the court determined that we couldn't prove the diseases we were suffering from were connected to dioxin contamination."

"And the PCBs?"

"Well, officially the EPA was never able to trace the source of the PCBs that Russell Bliss had mixed with his oil."

It is staggering, to put it mildly, that the EPA was unable to trace the source of the PCBs when the only manufacturer of those products owned a factory producing them in Sauget, Illinois, about twenty miles from Times Beach. "In fact," Leistner explained, "we later found out that Rita Lavelle, who was Anne Burford's assistant at the EPA, had destroyed documents that could have incriminated Monsanto."

The affair was a big story in the United States in 1983. In investigating a misuse of the Superfund Program budget, a fund allocated to the EPA to decontaminate sites polluted by industrial wastes, some of which had been fraudulently used to finance the electoral campaigns of Republican candidates, Congress discovered that documents that would compromise the companies had disappeared. The investigation proved that the Reagan administration, known for its unstinting support of big business, had ordered Anne Burford to "freeze" the Times Beach case. Appointed as EPA administrator shortly after Reagan's inauguration, she was forced to resign following the scandal, in March 1983. Her deputy, Rita Lavelle, had worse luck: she was sentenced to six months in prison on charges of perjury and obstructing a congressional investigation.[5] The investigation revealed that she had shredded a number of incriminating documents and that she frequently lunched with representatives of Monsanto. But the company didn't lose everything with the change: the new head of the EPA was William Ruckelshaus, who had been its first administrator in 1970 before briefly becoming acting director of the FBI in 1973 and then joining the boards of directors of Monsanto and Solutia.

"The problem," Gerson Smoger, an environmental lawyer from Oakland who represented some former Times Beach residents, explained to me in October 2006, "is that we were never able to get our hands on the contracts that Russell Bliss had signed with Monsanto, which had two factories in Missouri, one in Sauget and the other in Queeny, both in the St. Louis suburbs. He always claimed he didn't have them."

"Some people say he was paid to make them disappear," I commented.

"Anything is possible. What's certain is that he testified several times that Monsanto was one of his customers, but we have no written proof."

He proceeded to list the various converging pieces of evidence: on April 21, 1977, Russell Bliss confirmed in a sworn deposition that Monsanto was his principal supplier of industrial wastes; on October 30, 1980, Scott Rollins, a driver working for Bliss, testified to a representative of the state's attorney general that he frequently loaded barrels from the company's factories, and so on. "Monsanto has always denied that it worked with Russell Bliss," the lawyer said. "Furthermore, the company defended itself by saying that the PCBs came from other factories that used its hydraulic fluids. So the question of responsibility arises: at the time, we didn't yet know that Monsanto had concealed the toxicity of PCBs from its customers, so they were considered responsible for their wastes. That's how the PCBs were simply disregarded in the Times Beach case, and the authorities were concerned only with dioxin, forgetting in passing that most Monsanto products were also contaminated with dioxin."

In fact, only one company accepted responsibility for the pollution:

Syntex Agribusiness of Verona, Missouri. This subsidiary of Northeastern Pharmaceutical and Chemical Company (NEPACCO) made the herbicide 2,4,5-T, a powerful defoliant contaminated with dioxin, of which Monsanto was also a significant producer. But fortunately for Monsanto, it did not manufacture the defoliant in Missouri. In a settlement with the EPA, Syntex agreed to pay $10 million toward the decontamination of twenty-seven toxic dumps in eastern Missouri, including Times Beach. "The irony of the story," said Smoger, "is that at the very moment Syntex was designated as the guilty party, Monsanto was publishing falsified studies to conceal the toxic effects of its 2,4,5-T herbicide."

The Herbicide 2,4,5-T and Dioxin

To understand the tragic irony of this drama of modern times, we have to go back to the origin of dioxin, a toxic substance produced in the process of manufacturing certain chlorinated chemical compounds or during their combustion at high temperatures. The term "dioxin" covers a family of 210 related substances (as with PCBs, the term "congeneric" is used), the most toxic of which has the scientific name tetrachloro-p-dibenzodioxin or 2,3,7,8-TCDD, TCDD for short. Long unknown to the public, the existence of dioxin emerged from the secrecy of industrial and military laboratories on July 10, 1976, in an episode known to history as the "Seveso catastrophe."

On that day, an accident in the Italian chemical factory in Icmesa, owned by the Swiss multinational Hoffmann–La Roche, provoked the formation of an extremely harmful cloud that spread over the plain of Lombardy, particularly the town of Seveso. A few days later, more than three thousand domestic animals died of poisoning, while dozens of residents developed chloracne, the chronic disfiguring skin disease. In the face of the magnitude of the catastrophe and the emotion it provoked worldwide, the managers of Hoffmann–La Roche were obliged to reveal the agent responsible: dioxin, a product derived from the manufacture of the herbicide 2,4,5-T, the leading product of the Icmesa factory.

The identification of this molecule, a pure product of industrial activity, was closely related to the history of the defoliant, invented at about the same time in British and American laboratories during World War II. In the early

1940s, several researchers succeeded in isolating the hormone that controls plant growth, and they reproduced this molecule synthetically.* They found that in small doses the artificial hormone greatly stimulated plant growth, and in large doses it killed the plants. Thus were born two very effective herbicides, which brought about a veritable "agricultural revolution and the beginning of weed science," to use the words of the American botanist James Troyer.[6] They were 2,4-dichlorophenoxyacetic acid (2,4-D) and 2,4,5-trichlorophenoxyacetic acid (2,4,5-T), which, along with insecticides such as DDT, accompanied the green revolution following World War II. Their simultaneous discovery by four different laboratories brought about a patent war that has never been resolved, which explains how many chemical companies have taken advantage of this legal void to launch production on both sides of the Atlantic. The demand was soon huge, because these "selective" herbicides have a considerable advantage for work in the fields: used in the right amounts, they destroy weeds (dicots) and leave intact grains such as corn and wheat (monocots).

In 1948, Monsanto built a factory manufacturing 2,4,5-T in Nitro, West Virginia. On March 8, 1949, a leak on the production line caused an explosion, leading to the release of an unidentified material that covered the building interior and escaped in the form of a cloud. In the weeks that followed, the workers present at the time of the accident and those called in to clean up the site developed a skin disease that was then totally unknown; they also experienced nausea, vomiting, and persistent headaches. Monsanto management asked Raymond Suskind, a doctor at the Kettering Laboratory in Cincinnati, to conduct discreet medical follow-up of the affected personnel. He presented a report on December 5, 1949, that was not made public until the mid-1980s during the trial of *Kemner v. Monsanto*, which will be discussed below. "A total of seventy-seven persons employed at the plant have developed cutaneous and other symptoms probably as a result of this accident," the doctor conscientiously reported, and he attached a set of astonishing photographs showing shirtless men with faces disfigured by cracks and pustules and bodies covered with purulent cysts.

In April 1950, Suskind prepared a second report on six particularly af-

*In the United Kingdom, the researchers were William G. Templeton (Imperial Chemical Industries) and Philip S. Nutman (Rothamsted Agricultural Experiment Station); in the United States, Franklin D. Jones (American Chemical Paint Company) and Ezra Kraus and John Mitchell (University of Chicago).

fected workers who were still suffering from the mysterious skin disease a year after the accident and were also experiencing respiratory, central nervous system, and liver troubles, and impotence. The doctor went so far as to recommend "special treatment" for a worker who had developed an acute psychological pathology because his skin had darkened so much that he was "mistaken for a Negro and forced to conform with the racial segregation customs of the area on buses or in the theatres."[7]

In 1953, Suskind widened his study to thirty-six workers, ten of whom were exposed in the 1949 accident and twenty-six others working in the production unit. He reported that thirty-one of them had very severe skin lesions, accompanied by irritability, insomnia, and depression. Twenty-three years later, in a confidential report disclosed at the *Kemner* trial, he again reported, with equal detachment, that of the thirty-six workers, thirteen had already died at an average age of fifty-four.

Through all these years, Monsanto adopted the same attitude as for PCBs: it hid the data in a drawer and said nothing to the health authorities and certainly not to its workers. But it seems highly improbable that its managers were unaware of a study published in 1957 by Karl-Heinz Schulz, a researcher in Hamburg, who had done follow-up on the workers in a BASF factory that manufactured 2,4,5-T after an accident on November 17, 1953, similar to the one in Nitro.[8] This work had made it possible to identify the TCDD (dioxin) molecule and to provide a definitive name for the disease that characterized it: chloracne.

Long Live War

Not only did Monsanto fail to call into question the manufacture of 2,4,5-T, but the company did not hesitate to work closely with Pentagon strategists to develop its use as a chemical weapon. Following a Freedom of Information Act request to the Pentagon, the *St. Louis Journalism Review* revealed in 1998 that Monsanto had conducted a regular correspondence beginning in 1950 with the Chemical Warfare Service dealing with the military use of the herbicide.[9] According to Cary Conn of the National Archives and Records Administration, the file contained 597 pages divided into four sections, including "laboratory development" and "pilot plant demonstration." However, these documents, which posed no immediate danger to the security of the United States, could not be consulted because they had been classified "top secret" by the army on May 4, 1983. It will be clear later that the date was not an accident.

The co-founder of the Institute for Social Ecology and editor of a special issue of *The Ecologist* devoted to the questionable history of Monsanto, Brian Tokar, pointed out that it was not surprising that Monsanto management had been in contact with Pentagon officers.[10] Indeed, this was true of all major chemical companies in the twentieth century, who had greatly profited from the two world wars. In a 2002 article titled "Agribusiness, Biotechnology, and War," Tokar wrote: "Virtually all of the leading companies that brought us chemical fertilizers and pesticides made their greatest fortunes during wartime." These companies now control biotechnology, seed, and food production.[11] For example, during World War I, DuPont, which later became one of the world's largest seed companies, supplied the Allies with gunpowder and explosives. During the same period, Hoechst (which merged with the French company Rhône-Poulenc in 1999 to form Aventis, a biotechnology giant) was supplying the German army with explosives and mustard gas. In 1925, Hoechst joined with BASF and Bayer to form IG Farben, the largest chemical conglomerate in the world, which produced Zyklon B (the gas used in the death camps to exterminate Jews). As for Monsanto, established at the beginning of the century to produce saccharin, it multiplied its profits during World War I by selling chemical products used to make explosives and poison gas.

It was sometimes war itself that made it possible to launch new products that subsequently made profits for chemical multinationals for decades. DDT, for example, which had been synthesized in 1874, emerged from oblivion during World War II thanks to the American army, which decided to use this now-banned insecticide to combat a typhus epidemic that was decimating its troops in Western Europe and to eradicate the mosquitoes carrying the malaria parasite in the South Pacific.

Monsanto embarked upon large-scale production of DDT in 1944, at a time when its ties to Pentagon strategists had become extremely close. In 1942, in fact, its research director, Charles Thomas, had been invited by General Leslie R. Groves to participate in the top-secret project that issued in one of the greatest human and ecological catastrophes of the modern era. The Manhattan Project, as it was dubbed, was designed to make as quickly as possible the first atomic bombs in history, those that were to be dropped on Hiroshima and Nagasaki in August 1945. Endowed with a $2 billion budget, the Manhattan Project assembled the best American physicists in the Pentagon's nuclear weapons laboratory in Oak Ridge, Tennessee, while Monsanto chemists under Thomas's direction were given a delicate mission: to isolate and then purify plutonium and polonium, which would be used to trigger the nuclear bombs. Enjoying the Pentagon's absolute confidence, the company received consent for this important work to be carried out in its research laboratory in Dayton, Ohio.

Promoted to vice president of Monsanto after the war, Charles Thomas took charge of the Clinton Laboratories in Oak Ridge, where he was given the project of developing civil applications for nuclear power for the federal government, while maintaining his office in St. Louis. He ended his career as CEO of Monsanto in 1960, at a time when his company, which had become one of the most powerful chemical enterprises in the world, was about to secure the largest contract in its history: the production of Agent Orange for the Vietnam War.

Operation Ranch Hand and Agent Orange

"The Ranch Hand operation was unique in the history of American arms, and may remain so. In April 1975, President Ford formally renounced the first use of herbicides by the Unites States in future wars. 'As long as this policy stands,' Major [William] Buckingham writes, 'no operation like Ranch Hand could happen again.'"[12] These are the words of Richard H. Kohn in foreword to a book by Buckingham published by the Office of Air Force History in 1982 covering the use of herbicides in Southeast Asia from 1961 to 1971.

The advantage of this book, which carefully avoids considering the health and ecological consequences of the massive spraying of defoliants in South Vietnam, is that it presents in clinical technical detail the genesis of the chemical warfare waged by the United States under the euphemistic title Ranch Hand to the great benefit of multinationals such as Dow Chemical and, of course, Monsanto. It also tells us that "herbicides, or weed-killing chemicals, had long been used in American agriculture" and that the first experimental airborne spraying of pesticides took place near Troy, Ohio, on

August 3, 1921, to combat a sphinx caterpillar infestation; the pilot was Lieutenant John Macready, and he was accompanied by an entomologist named J.S. Houser. The experiment was repeated the following year on a cotton plantation in Louisiana, to exterminate boll weevils in similar circumstances. This is proof, if any were needed, that industrial agriculture never would have seen the light of day without close cooperation between the military and scientific establishments, whose respective goals are not exactly to produce healthy food that respects the environment.

In the 1940s, the aircraft industry perfected spraying tanks that were fixed onto military aircraft to spray DDT in Western Europe and the Pacific, "to save lives," according to Major Buckingham, who goes on to say that the Allies and Axis in World War II abstained from using chemical sprays either because of legal restrictions or to avoid retaliation in kind.[13]

The removal of the taboo seems in fact to have had two causes: the emergence of the Cold War, which justified the use of any means to confront the Communist menace, and the discovery of the revolutionary herbicides 2,4-D and 2,4,5-T. I have already noted that they were invented simultaneously by British and American laboratories. Researchers soon recognized the potential they represented in wartime because they made it possible to destroy crops and thereby starve enemy armies and populations. In 1943, the U.K. Agricultural Research Council launched a secret testing program that would be used in Malaysia in the 1950s when, for the first time in history, the British Army used herbicides to destroy the crops of Communist insurgents. In the United States at the same time, the Center for Biological Warfare at Fort Detrick, Maryland, was testing dinoxol and trinoxol, mixtures of 2,4-D and 2,4,5-T, precursors of Agent Orange. It is reasonable to think that these preliminary tests were conducted with the close cooperation of Monsanto.

The first tests in real-life conditions took place in South Vietnam beginning in 1959. They were apparently such a novelty that the American military thought it appropriate to film them for a period of two years. In this extraordinary document, which I have been able to consult, one sees a military aircraft flying at low altitude above virgin forest release a milky cloud in a straight line as the aircraft moves forward. "After two weeks, it is obvious that the treatment has been effective," the voice-over notes with satisfaction; "90 percent of the trees and bushes have been destroyed for two years." Aerial shots then show a hole slicing through the luxuriant vegetation for several miles. The image of the spraying of herbicides—like images of napalm victims such as the young Kim Phuc running naked on a road in Vietnam—became one of the symbols of one of the most controversial wars of the twentieth century.

Operation Ranch Hand began officially on January 13, 1962, one year after John F. Kennedy entered the White House. According to Buckingham's book, it was the president himself who made the decision, after bitter discussions between advisers from the Department of Defense under Robert McNamara, who urged the use of these "techniques and gadgets," and those from the State Department, who feared international reactions and that the defoliation program would be used by "Communist propaganda" to turn the population against the United States. American involvement in Vietnam was still limited at the time: officially it consisted of assisting the efforts of the army of South Vietnam, then ruled by the dictator Ngo Dinh Diem, to contain the pressure of the Viet Cong, who were supported by the Communist North Vietnam of Ho Chi Minh. The goals of Operation Ranch Hand were to clear the main roads, the waterways, and the borders of South Vietnam in order to control the movements of the Viet Cong and to destroy the harvests that were supposed to supply the "rebels."[14]

In July 1961, the first shipments of defoliants arrived at the military base in Saigon. They were delivered in fifty-five-gallon drums with a variety of colored stripes intended to facilitate the recognition of different products: "Agent Pink" contained pure 2,4,5-T, "Agent White" 2,4-D, and "Agent Blue" arsenic, while the most toxic, "Agent Orange," introduced in 1965, was made of half 2,4,5-T and half 2,4-D.

On January 10, 1962, a communiqué from the South Vietnamese government was printed in all the country's newspapers: "The Republic of Vietnam today announced plans to conduct an experiment to rid certain key communications routes of thick, tropical vegetation. U.S. assistance has been sought to aid Vietnamese personnel in this undertaking. . . . Commercial weed-killing chemicals will be used in experiments. These chemicals are used widely in North America, Europe, Africa, and the USSR. . . . The chemical will be supplied by the United States at the request of the Vietnamese Government. The Government emphasized that neither of the two chemicals is toxic, and that neither will harm wild life, domestic animals, human beings, or the soil."[15] What was left unsaid by the propaganda of

President Diem, whom the White House had asked to accept responsibility for Operation Ranch Hand, was that the doses of herbicide used by the American forces would be as much as thirty times higher than those used in the United States, where 2,4,5-T and 2,4-D were carefully diluted before agricultural use.

On January 13, 1962, a Fairchild C-123 of the U.S. Air Force took off from Tan Son Nhut Air Base carrying more than 200 gallons of Agent Purple. Between then and 1971, an estimated 20 million gallons of defoliants were sprayed on 8 million acres of forests and crops. More than three thousand villages were contaminated, and 60 percent of the defoliants used were Agent Orange, which is the equivalent of more than eight hundred pounds of pure dioxin.[16] According to a Columbia University study published in 2003, dissolving eighty grams of dioxin in a municipal water supply system could eliminate a city of 8 million inhabitants.[17]

The Conspiracy

The man I visited one day in October 2006 had the emaciated look of the gravely ill close to death. At sixty-seven, he looked fifteen years older. Sitting in his wheelchair, he pointed to the space where his legs used to be. Alan Gibson was vice president of the Vietnam Veterans of America, which has 55,000 members. "When I got back from Vietnam, I started to have problems with my eyes," he explained. "And then, three years later, I had the first symptoms of what the doctors call peripheral neuropathy. My bones started to dissolve and come out of my toes. One day, I was washing my feet, and a piece of bone fell into my hand."

"First they said it was gout," his wife, Marcia, interrupted. "Then they amputated his toes, then his feet, and finally both legs."

"Is this disease common among Vietnam veterans?" I asked.

"Yes," said Marcia. "I'm a nurse at the VA hospital, and the most common diseases are cancer, especially lung cancer, liver cancer, and leukemia, and neurological disorders. In our association, there are also many veterans who have children or even grandchildren with physical or mental handicaps."

Alan Gibson no longer recalls exactly when and where he first saw defoliants being sprayed. "It was so frequent," he said. "We were in the jungle and then suddenly it felt like rain. There was engine noise. They told us it was [the same] weed killers that our farmers used every day. I have buddies who washed in empty Agent Orange drums or used them to barbecue. They never told us that the herbicides contained dioxin. But the government knew."

Who knew what, and when? More than thirty years after the end of the Vietnam War, the question continues to divide experts. According to a report prepared by the General Accounting Office in November 1979, "At that time the Department of Defense did not consider herbicide orange toxic or dangerous to humans and took few precautions to prevent exposure to it."[18]

One frequently cited piece of evidence sheds light on the blindness of the military authorities. This is the testimony of Dr. James Clary, a scientist working in an Air Force chemical weapons laboratory in Florida, who played a key role in developing the spray tank designed to disperse Agent Orange. In a letter to Senator Tom Daschle, he wrote: "When we [military scientists] initiated the herbicide program in the 1960s, we were aware of the potential for damage due to dioxin contamination in the herbicide. We were even aware that the military formulation had a higher dioxin concentration than the civilian version due to the lower cost and speed of manufacture. However, because the material was to be used on the 'enemy,' none of us were overly concerned. We never considered a scenario in which our own personnel would become contaminated with the herbicide."[19]

Another statement seems to indicate that military leaders stationed in Vietnam were not informed of the extreme toxicity of the dioxin contained in Agent Orange. Admiral Elmo R. Zumwalt Jr., promoted to commander of naval forces in Vietnam in September 1968, headed the fleet patrolling the Mekong delta. To protect marines from being ambushed by the Viet Cong in this strategic zone, he ordered that the riverbanks be sprayed with Agent Orange. It turned out that the captain of one of the boats was his own son, Elmo R. Zumwalt III, who died of cancer at the age of forty-two, leaving an orphan with various handicaps. Thereafter, Admiral Zumwalt moved heaven and earth to dispel the secrecy surrounding dioxin. He was appointed special assistant to the secretary of Veterans Affairs, Edward J. Derwinski, and fought tirelessly to have victims of Agent Orange given adequate care.

"I think government authorities were not informed of the harmfulness of dioxin before the late 1960s," said Gerson Smoger, a lawyer for many Vietnam veterans, "for a very simple reason: the two principal manufacturers, Dow Chemical and Monsanto, deliberately concealed the data they had in order not to lose a very lucrative market. I'm not afraid to say that this was an out-and-out conspiracy."*

With offices in Oakland, California, Gerson Smoger has specialized in environmental pollution cases—as noted, he represented residents of Times Beach—and he has also been prominent in class actions against pharmaceutical giants and tobacco companies. But his most important work involves Agent Orange. For years he has been collecting thousands of documents in his office basement, carefully arranged in numbered boxes, a mind-boggling sight. "It takes months to consult them all," Smoger said with a smile at my look of distress. "But I have been able to find proof that the behavior of Dow Chemical and Monsanto was criminal. First, contrary to what their executives said, they regularly tested the dioxin content of their products, but they never transmitted their results to the public health or military authorities. The Monsanto case is particularly serious, because the Agent Orange the company produced in its Sauget factory contained the highest level of dioxin."

Smoger was referring to a memorandum from Dow Chemical dated February 22, 1965, describing a meeting of thirteen executives of the firm in which they discussed the toxicity of 2,4,5-T. They agreed to organize a meeting with other manufacturers of Agent Orange, including Monsanto and Hercules, "to discuss toxicological problems caused by the presence of certain highly toxic impurities" in samples of 2,4,5-T. "The meeting was kept strictly confidential," said Smoger. "Dow spoke of an internal study that showed that rabbits exposed to dioxin developed severe liver lesions. The question was whether the government should be informed. As a letter, of which I also have a copy, proves, Monsanto criticized Dow for wanting to reveal the secret. And the secret was kept for at least four years, the years when the spraying of Agent Orange reached a peak in Vietnam."

By late 1969, government authorities could no longer say that they were uninformed: a study conducted by Diane Courtney for the National Institutes of Health (NIH) found that mice subjected to significant doses of 2,4,5-T developed fetal malformations and produced stillbirths.[20] The news

*Seven companies produced Agent Orange: Dow Chemical, Monsanto, Diamond Shamrock, Hercules, T. H. Agriculture & Nutrition, Thompson Chemicals, and Uniroyal.

provoked strong feelings and a good deal of anxiety. On April 15, 1970, the secretary of agriculture announced on radio and television "the suspension of the use of 2,4,5-T around lakes, ponds, recreation areas, and houses, and on crops intended for human consumption, because of the danger the herbicide poses to health."[21]

This was the end of Agent Orange, but for American veterans it was the beginning of a long battle for recognition of the harm they had suffered.

Monsanto Organizes to Protect Itself

In 1978, Paul Reutershan, a veteran suffering from intestinal cancer, sued the manufacturers of Agent Orange. Thousands of veterans soon joined him in the first class action ever filed against Monsanto and its like. The following year, on January 10, 1979, a freight train carrying twenty thousand gallons of chlorophenol (a substance used in the manufacture of wood treatment products) derailed in Sturgeon, Missouri, spilling the entire cargo. It turned out that the shipment came from the Sauget factory where Monsanto had until recently been manufacturing its PCBs. Samples taken by the EPA found that the chemical product contained dioxin. Sixty-five residents of Sturgeon, among them Frances Kemner, who was the lead plaintiff in the class action *Kemner v. Monsanto*, sued Monsanto.

This was a serious case for the company, especially since, following the Seveso catastrophe in 1976, TCDD was subject to particular scrutiny from the public and the media. Monsanto understood that it had to react if it did not want to be involved in a multitude of trials in which the long-term effects of dioxin on human health, particularly with respect to cancer, could not fail to come up. But it also knew that it had two assets, on which it would constantly rely from the late 1970s on.

In the first place, as Greenpeace, one of its fiercest opponents, pointed out in a report published in 1990, whatever its origin, dioxin is "a ubiquitous contaminant in the U.S. population and environment."[22] It is therefore difficult to prove that the level of dioxin recorded in an individual is directly tied to exposure as the result of an accident like the one in Sturgeon or spraying in Vietnam. To guard against possible accusations, Monsanto executives stopped at nothing: with the complicity of employees at the St. Louis morgue, they secretly took tissue samples from the corpses of road accident victims and had them analyzed, revealing that the fatty tissue of the corpses contained dioxin. This action, which says a good deal about the company's practices, was brought to light at the trial of *Kemner v. Monsanto*, which will be discussed in Chapter Three.[23]

In the second place, as Greenpeace also pointed out, "epidemiological studies . . . are extremely difficult because dioxin is now ubiquitous in the human population," and it is therefore practically impossible to find a control group of people that one can be sure have never been exposed. "In other words, rather than comparing an exposed population to an unexposed population, epidemiologists studying dioxin can only compare a more exposed group with a less exposed group. Such study populations and control groups must have distinctly different degrees of exposure and be of sufficient size to make incrementally increased health effects statistically significant." Greenpeace therefore concluded that

> human populations available for epidemiological studies have been groups identifiably exposed to extraordinarily high concentrations of dioxin, such as the following:
>
> —communities accidentally or purposefully exposed to dioxin releases, such as those in Seveso, Italy, and Times Beach, Missouri, USA;
>
> —people exposed to 2,4,5-T or other dioxin-contaminated pesticides, such as pesticide applicators and Vietnam veterans exposed to Agent Orange; and
>
> —workers employed in industrial facilities with known dioxin releases, such as certain Monsanto and BASF chemical workers.

Monsanto understood by 1978 that it was the only entity that had health data going back to 1949, the date of the Nitro factory accident. To determine whether dioxin causes cancer, one would have to locate the Nitro workers who had been examined by Raymond Suskind and compare their state of health thirty years later with that of the normal population. Suskind was therefore asked to supervise three epidemiological studies, with the help of two Monsanto scientists. As the *Kemner v. Monsanto* trial revealed, Dr. George Roush, the company's medical director, reviewed the content of the studies before publishing them in peer-reviewed scientific journals in 1980, 1983, and 1984.[24] As might be expected, the studies concluded that there was no connection between exposure to 2,4,5-T and cancer.

"This is how the veterans in the first class action had their claim for reparations denied," Gerson Smoger explained. "When these studies were published, they were considered the absolute reference. And given their inability to prove that the cancers they were suffering from were connected to their exposure to dioxin, the veterans were forced to accept a settlement."

In fact, on May 7, 1984, when the opening of the trial initiated by Paul Reutershan in 1978 was imminent, the manufacturers of Agent Orange put $180 million on the table as a final settlement offer. Judge Jack Weinstein ordered that 45.5 percent of the amount be paid by Monsanto, because of the high dioxin content of its 2,4,5-T.[25] Placed in a trust fund, the money was supposed to indemnify veterans who could prove total disability not connected to war wounds, with a deadline of ten years. Forty thousand veterans received amounts ranging from $256 to $12,800. "A pittance," said Smoger. "Until it was discovered that the Monsanto studies had been manipulated."

3

Dioxin: Manipulation and Corruption

> Reliable scientific evidence indicates that Agent Orange is not the cause of serious long-term health effects.
>
> —Jill Montgomery, Monsanto spokesperson, 2004

In February 1984, a few months before the Vietnam veterans were forced to give up their chance to receive real reparations, the trial of *Kemner v. Monsanto* opened in Illinois. For more than three years, fourteen jurors would listen to 130 witnesses and try to evaluate the harm suffered by the residents of Sturgeon, as well as the responsibility of Monsanto. This was "the longest jury trial in the nation's history," wrote the *Wall Street Journal*, which went on to say that "Monsanto has 10 lawyers working on the case in four-hour shifts to stay fresh. . . . Courtroom observers say that by establishing a reputation as a fierce adversary with an apparently unlimited budget to battle, Monsanto may discourage similar cases in the future."[1]

Falsified Scientific Studies

The resources mobilized by the company matched the stakes: if all users of its products containing traces of dioxin were to turn against it, the company knew, it would be heading straight for bankruptcy. Hence, it had no hesitation about using every possible delaying tactic, at the risk of exasperating the court. "Justice delayed is justice denied. I believe that this court should refuse to be used as a pawn in such a waste of judicial resources."[2]

On October 22, 1987, after deliberating for eight weeks, the jury rendered a peculiar verdict: the plaintiffs were awarded only a symbolic $1 in compensatory damages, on the grounds that they had been unable to prove the link between their health problems and the accident, but $16 million in punitive damages were assessed against Monsanto because the jury had been outraged by its irresponsible conduct in the management of health risks associated with dioxin.*

Thanks to the rigorous work of the plaintiffs' attorney, Rex Carr, the jury heard that "Monsanto knew that simply redistilling its chlorophenols would eliminate or greatly reduce the Dioxin content" but that "this was not done until 1988." Moreover, it "could have rid its chlorophenol products of dioxins by testing every batch and not selling those it found contaminated."[3]

According to the testimony of Donald Edwards, a company engineer, Monsanto "was dumping daily 30 to 40 pounds of dioxin into the Mississippi River from its Krummrich Plant . . . getting in the St. Louis food chain through the river," without informing the authorities. Worse, the testimony of three executives, including a chemist and the marketing manager, indicated that the company knew that Santophen, a chemical used in the manufacture of Lysol, a cleaning product recommended for cleaning children's toys, was contaminated with dioxin. Out of fear of losing the market, it decided not to inform its customer, Lehn and Fink, going so far as to lie when that company raised questions. A letter from Clayton F. Callis, a Monsanto executive, to one of his colleagues confirms the offhand way in which the company treated the dioxin problem: "Dow has made an issue of the dioxin content of Penta," he wrote on March 3, 1978, regarding pentachlorophenol, a wood treatment product manufactured by both companies. "Our product has higher dioxin content. Therefore, the monkey is on Monsanto's back to prove that the dioxins are acceptable. This means studying the toxicology of not just one molecule, but many. This is close to an impossible task."

There were many other similar examples, but the highlight of the trial was the revelation that the three previously cited studies supervised by Suskind

*Monsanto appealed the verdict and won, the court deciding that it was not possible to award punitive damages if the plaintiffs had not been able to prove that their problems were connected to dioxin exposure (*U.S. News and World Report*, June 13, 1991).

and published between 1980 and 1984 were misleading. If they had been conducted properly, they would have produced a diametrically opposed conclusion, namely, that dioxin is a powerful carcinogen. Demonstrated by the lawyer Rex Carr, the fraud was later confirmed by several scientific bodies, including the National Institute for Occupational Safety and Health (NIOSH) and the National Research Council (NRC).[4] The NRC determined that the Monsanto studies were "plagued with errors in classification of exposed and unexposed groups . . . and hence [were] biased toward a finding of no effect."[5] It was also confirmed by Greenpeace, which presented a very detailed account in 1990 that was widely covered in the press, whereas the revelations produced at the *Kemner* trial had gone largely unnoticed.[6]

Greenpeace showed that the study published in 1980 by Raymond Suskind and his colleague from Monsanto, Judith Zack, suffered, to put it mildly, from a lack of rigor in the definition of individuals considered "exposed" or "not exposed." According to the explanations Suskind provided to the court, the two researchers had adopted as a preliminary hypothesis that "the group that was exposed to the runaway reaction and who could be identified as well by their development of chloracne was probably the most heavily exposed group in the Nitro population."[7] Hence, the group of those "exposed" included only the workers present on the day of the accident who had *also* contracted chloracne; those who had been present but had not gotten the disease were excluded from the group, whereas Suskind knew perfectly well that the absence of chloracne did not necessarily imply lack of exposure.

Conversely, anyone with skin problems (psoriasis, acne, and the like) was included in the cohort of the "exposed," whereas workers on the production line who were absent on the day of the accident were systematically placed in the control group of the "not exposed," even if they were suffering from chloracne. In a letter to *Nature* in 1986, the toxicologists Alastair Hay and Ellen Silberberg noted that "the total cohort of workers exposed to dioxin at Monsanto should be considered as a whole without making a distinction between workers exposed to dioxin in the process accident or when making 2,4,5-T." This was especially true because the data gathered by Suskind in his 1953 study showed that "the incidence of chloracne was approximately the same in the two groups" and that "notably serious diseases of long latency, such as cancer, may be expected to result from lower and more chronic exposure."[8]

The study published in 1983 by Judith Zack and William Gaffey, both Monsanto employees, was supposed to compare the state of health of 884 of the factory's employees, including those working on the 2,4,5-T production line (the "exposed" group) and "all the others" (the control group), including "employees holding a job having plant-wide responsibilities with the potential for exposure to 2,4,5-T were, for the purposes of this study, considered to be non-exposed," as the two authors acknowledged.[9] The result was that rates of cancer were lower in the exposed group than in the non-exposed group. The trick was having included in the study only employees working in the factory and/or having died between January 1, 1955, and December 31, 1977. In other words, those who had worked at Nitro between 1948 and 1955 were excluded, as were those who died after 1977. This arbitrary protocol made it possible to exclude from the study twenty workers who Monsanto knew had been exposed (notably in the 1949 accident), nine of whom had died of cancer and eleven of heart disease. Furthermore, four workers who had died of cancer and had been classified as "exposed" in the 1980 study were placed in the control group in the 1983 study.[10]

But it was the last study, the one published in 1984 by Raymond Suskind and Vicki Hertzberg, a colleague at the Kettering Laboratory, in the prestigious *Journal of the American Medical Association* that crossed all bounds. At a hearing in the *Kemner* case, Roush acknowledged that instead of the four cases of cancer recorded in the exposed group, there were twenty-eight (the other twenty-four had been omitted for some reason).[11] When Suskind was subsequently questioned, he was, according to the plaintiffs, "shown to be such a fraud that he refused to return to the State of Illinois for completion of his cross-examination."[12]

Hunting Down the Whistle-Blowers

Meanwhile, Greenpeace sent its file to Cate Jenkins, a chemist who had been working at the EPA since 1979. At the time, her assignment was to detect toxic industrial wastes and develop regulations to control them. Known for her intransigence toward polluters, this undisputed expert on dioxin had already clashed with her superiors, who thought that she had pushed a little too hard in her investigation of Penta, the wood treatment product. The production of Penta "releases 75 different dioxins," including "TCCD and hexadioxin, which is 5000 times more toxic than arsenic," she explained to the Canadian magazine *Harrowsmith* in 1990.[13]

As soon as she became aware of the materials put together by Rex Carr and picked up by Greenpeace, Jenkins grasped the implications that these revelations might have for U.S. government regulation of dioxin. In fact, relying on the only epidemiological studies then available, that is, the ones conducted by Monsanto, the EPA had concluded in 1988 that "the human evidence supporting an association between 2,3,7,8-TCDD [dioxin] and cancer is considered inadequate."[14] The agency had therefore decided to classify dioxin as a type B2 carcinogen in the EPA system, a "probable human carcinogen" for which there is "sufficient" evidence that it is a carcinogen from animal studies.* Consequently, dioxin was not considered a priority pollutant and thus was not subject to regulation on atmospheric emissions provided for under the Clean Air Act. It seemed apparent to Jenkins that if the Monsanto studies had not been manipulated, the EPA's conclusions (as well as those of the rest of the world, which had adopted the American position) would have been different.

As a conscientious public servant, she therefore decided to prepare a confidential memorandum entitled "Newly Revealed Fraud by Monsanto in an Epidemiological Study Used by the EPA to Assess Human Health Effects from Dioxins," which she sent on February 23, 1990, to the chairperson of the Executive Committee of the Science Advisory Board of the agency, as well as to the office of the EPA administrator.[15] She attached a portion of a brief from the *Kemner* case and asked that a scientific audit of the Monsanto studies be conducted, an initiative that would soon cause one of the stormiest episodes in her career.

Unfortunately, I was unable to meet Cate Jenkins, who refused to grant me an interview. When I contacted her in May 2006, she was in charge of coordinating EPA analyses of toxic waste in the ruins of "ground zero," the site of the World Trade Center towers in New York destroyed in the attack of September 11, 2001. "It's a very delicate issue," she explained in a rather enigmatic e-mail, "and I'd rather concentrate on that." She strongly sug-

*The EPA classification adopts the one recommended by the International Agency for Research on Cancer. It includes five groups: group A (human carcinogen); group B (probable human carcinogen), with two categories, B1 (limited human evidence) and B2 (no human evidence but sufficient animal evidence); and so on, down to group E (not a human carcinogen).

gested that I get in touch with William Sanjour, one of the most highly visible EPA managers before he was shunted aside until his retirement in 2001. In September 2007, as I was writing this, he was on the cover of *Fraud Magazine* as the recipient of the Sentinel Award from the Association of Certified Fraud Examiners for his work at the EPA.[16]

"Cate's afraid," he told me when I reached him by phone in the spring of 2006. "When you know what she's been through, you can understand that." Like Jenkins, Sanjour is a whistle-blower—someone employed in government or private enterprise who realizes that his or her employer is endangering the public interest by violating a law or regulation, an offense sometimes coupled with fraud or corruption. Provoking the fury of their superiors, whistle-blowers are harassed, marginalized, defamed, and often fired. For them, the fall is all the harsher because they were really convinced that their efforts to reveal the truth had meaning. Pragmatists call them idealists; for companies such as Monsanto, they interfere with the smooth running of the production process. From this perspective, the story of William Sanjour is exemplary.

After studying physics at Columbia, he joined the EPA at its founding in 1970. He was soon appointed to head the Hazardous Waste Management Division, in charge of supervising the treatment and storage of toxic industrial waste. His work led Congress to enact the Resource Conservation and Recovery Act (RCRA) in 1976, which he worked to enforce, even if it meant provoking the fury of the polluters and his own superiors. "Unfortunately," he says now, "the EPA is more concerned with protecting the interests of the companies it is supposed to regulate than with defending the public interest." Sanjour's superiors were unhappy with his testimony to Congress and his public speeches, in which he openly denounced the collusion of the agency with major industrial companies.

A lover of sailing, Sanjour arranged to meet me on July 14, 2006, at a small marina not far from the nation's capital. "Let me tell you how the EPA dreamed up a law especially for me," he said with mock pleasure. Indeed, to silence its black sheep, with the cooperation of the Office of Government Ethics the agency issued a regulation barring its employees from having their travel expenses paid when they were invited to speak for no fee and outside work time by activist or citizen organizations. "My expertise was frequently called on all over the country," he said. "From one day to the next, I had to decline all invitations, because it would have cost me too much." He took the matter to court and, with the help of the National Whistleblower Center, an organization based in Washington, D.C., that supports whistle-blowers and helps protect their careers, succeeded in having the offending regulation overturned and establishing a precedent affirming the right of whistle-blowers to denounce employers who are obviously violating the law.[17]

This EPA "sentinel," who had been serving as a policy analyst for several years, decided in July 1994 to draw up a report on the affair of Cate Jenkins and dioxin, entitled *The Monsanto Investigation*.[18] In it he provided an "analysis of the failure of EPA to investigate allegations that the Monsanto Company had falsified scientific studies on the carcinogenicity of dioxin."

"I sifted through the whole file," he explained. "Here it is." He took out of a trunk a stack of documents at least twenty inches thick, secured in the course of various proceedings against Monsanto (such as *Kemner v. Monsanto*) or the EPA (such as Jenkins's complaint to the Department of Labor). Spread before me were hundreds of documents, including, according to Sanjour, many internal memoranda from Monsanto showing how the company concealed the toxicity of one of the most dangerous products ever inflicted on the planet, not stopping at trying to crush the woman who dared to reveal the scandal.

The EPA Obeys Orders

"Here's the proof that the EPA was infiltrated by Monsanto," Sanjour told me, handing me a five-page letter from James H. Senger, vice president of Monsanto, to Raymond C. Loehr, who chaired the Executive Committee of the agency's Science Advisory Board. "It's dated March 9, 1990, just two weeks after Cate sent her confidential memo to the committee. How had the company found out about it?"

"Monsanto has learned of the EPA's receipt of highly inflammatory and inaccurate information pertaining to epidemiology studies involving Monsanto's Nitro, West Virginia plant," the executive wrote. "The allegations of fraud are not credible. . . . We are very disturbed by the false charges being made against Monsanto and Dr. Suskind." Less than three weeks later, the CEO of Monsanto, Richard J. Mahoney, penned a letter to William Reilly, the EPA administrator, attaching an article from the *Charleston Gazette*.[19] "Unfortunately, that internal EPA memorandum has now found its way to the news media and is being treated as the official EPA position," he complained. "This is creating a serious problem for Monsanto that we simply don't deserve. We request a prompt statement from your office to the effect that Ms. Jenkins does not speak for the US EPA on this issue and that her views are hers alone and not the official position of the agency." This was followed by an answer from Don R. Clay, deputy administrator, whose servile tone is perplexing: "The opinions expressed in the internal EPA memorandum were those of Dr. Jenkins not the EPA. . . . We regret any problem that Monsanto may have had as a result of the news media's use of this memorandum. If I can be of further assistance, please do not hesitate to contact me."

As a sensible whistle-blower, Jenkins had arranged to leak her memorandum to the media so that a trace of it would remain in the event the EPA decided to bury her request. The affair had created a stir at the top of the *Journal of the American Medical Association* (*JAMA*), the most widely read medical weekly in the world, which six years earlier had unquestioningly published Monsanto's third study. What follows is an excerpt from a letter from the vice president of the American Medical Association, published on April 13, 1990, in reply to the questions quite properly raised by a doctor who was worried about the reliability of studies published by the scientific journal, considered the bible of medical research: "*JAMA* is very concerned about the reliability of the scientific studies we publish. However, when allegations of possible fraud arise, the editors of scientific journals are in no position to carry out the necessary investigations. *We lack access to the necessary records and individuals involved* [emphasis added]. Thus, the conduct of such investigations becomes the responsibility of the institution employing the authors of the study (usually a university), or the private or governmental agency that sponsored the research, or both."*

In other words, *JAMA* published what was sent to it without verifying the validity of the data, even when the author of the article is paid by a major industrial company. Yet publication in such a prestigious medical research journal is a guarantee of seriousness, which the vice president of Monsanto did not hesitate to make use of in his March 9, 1990, letter defending Suskind, where he pointed out that Suskind's conclusions had been "peer reviewed." Lies are thereby propagated in the international scientific com-

*All internal documents quoted in this section come from the file given to me by William Sanjour.

munity thanks to a damaged system that affects all areas of research, including biotechnology.

Meanwhile, at the EPA Cate Jenkins's memo became a hot potato that no one knew how to get rid of. Curiously, the agency's advisory board, which had used the Monsanto studies to recommend that dioxin be classified in category B2 in 1988, acknowledged its incompetence to conduct the scientific audit Jenkins had asked for, and turned the issue over to another agency, NIOSH. At the same time, no doubt to put on a good face for the media, the head of the agency asked the Office of Criminal Enforcement (OCE), its criminal investigation division, to assess the validity of the accusations of fraud.

"That was the best way to bury the matter," complained Sanjour, "because who was going to risk making a judgment about fraud if no one had first carried out the scientific audit that Cate had asked for?" On August 20, 1990, the criminal investigation was officially opened, and two detectives, John West and Kevin Guarino, were specially appointed from Denver. They were supposed to verify the "alleged violations of the federal environmental laws by the Monsanto Chemical Company, its offices and employees," who may have infringed on the Toxic Substances Control Act, which required that companies inform the EPA of the toxicity of their products, and who might be guilty of "conspiracy to defraud EPA" and of "false statements."[20]

"The investigation never happened," Sanjour told me. "No one ever established whether Monsanto had committed fraud; the only investigation was of Cate Jenkins, the whistle-blower, who was mistreated, harassed, and had her life turned into a living hell." In the file Sanjour had given me, I was able to consult the monthly activity reports prepared by the agency's two detectives. Most of them were simply otherwise blank pages with the notation "There was no significant investigative activity to report this month." A short "interview report" dated November 14, 1990, states that the two Keystone Kops had met Jenkins in her office. The next day, obviously concerned by their lack of curiosity, she sent out a second detailed report supporting her arguments on "Monsanto's fraud," specifying on the last page that she had sent copies to sixteen organizations and individuals, including Greenpeace, Admiral Zumwalt, and the National Vietnam Veterans Coalition (NVVC), which encompassed sixty-two veterans organizations.

Three days later, the unrepentant whistle-blower was invited to an NVVC ceremony where she received the association's medal honoring her courage and the quality of her work. Jenkins publicly confirmed that the EPA was conducting a criminal investigation of Monsanto's fraudulent studies. "From that point on," said Sanjour, "Monsanto constantly intervened at the EPA to block the investigation and to force the agency to discipline or even fire Cate. All these internal documents prove it," he said, showing me a sheaf of letters on Monsanto letterhead. "And this is only the tip of the iceberg. Monsanto is one of the most powerful companies in the United States: it has privileged access to the White House, the Congress, the press. Get this: not only was the investigation buried, but a lawyer for Monsanto ended up drafting a letter for the EPA in which the agency apologized."

I read the letters in question very carefully, and I must admit that I was staggered by the self-assurance of the Monsanto executives: far from making amends, they evidenced unshakable arrogance, presenting themselves as victims, adopting the stance of offended virgins, or expressing thinly veiled threats, as though they were addressing humble subordinates. "Monsanto recognizes that the Agency did issue a corrective statement indicating that the employee was acting in her individual capacity . . . and that her memorandum does not reflect the Agency's official position," Monsanto's Senger wrote, for example, on October 1, 1990, to Clay at the EPA. "However, that statement has not remedied the problems caused by continual references to the accusations repeated in the memorandum which was prepared on Agency stationery. . . . Given the Company's science-based focus, preservation of Monsanto's reputation for performing accurate world-class research is of critical importance to our business and research operations."

In 1991, James Moore, a lawyer for Monsanto, appeared on the scene. He had not been chosen at random: he worked for Perkins Coie, the law firm of William Ruckelshaus, who, as noted, had twice been EPA administrator, with a long stint on the board of Monsanto and Solutia in between. "For all the reasons I have previously discussed with you, there is no basis for a conclusion that fraud was perpetrated," Moore asserted on March 12, 1992, to Howard Berman, associate director of the OCE. "The inquiry by EPA's criminal unit should be concluded expeditiously so that Monsanto can clear its good name and such references to the alleged criminal nature of the studies will cease."

The admonition bore fruit. On August 7, 1992, a final "investigative report" concluded: "The investigation is complete. The original allegations regarding the submission of fraudulent studies by Monsanto to the EPA were reviewed. The [word deleted] EPA Office of Health and Environmental Assessment, [word deleted] stated that even if the studies were falsified, they would have little implication in the end because the Monsanto studies were immaterial to EPA's decision making regarding dioxin."

But this was still not enough for Monsanto, as demonstrated by a note written on August 26, 1992, by an official at the EPA (whose name has been deleted) recounting a recent telephone conversation with Moore: "Now Jim Moore wants to talk about what EPA might or should say to set the record straight by untarnishing the company's reputation. He understands the sensitivity of the request. . . . At the very least, I would think that he would be entitled to a letter saying that the investigation was closed for lack of evidence sufficient to support a criminal prosecution. . . . I suggested to Jim that he consider writing a letter to [word deleted] proposing what Monsanto wants and why."

Collusion Between Government and Industry

While Monsanto was dictating its orders to the EPA, Cate Jenkins was suffering the trials of the whistle-blower. She was relieved of most of her responsibilities on August 30, 1990, and remained essentially unoccupied until she was involuntarily transferred to an administrative position—a job as a "pencil pusher," in her words—especially created for her. Her fate had in fact been sealed by late February 1990, as revealed in a memorandum from her superior, Edwin Abrams: "I don't think Cate should be involved with anything that puts her in direct contact with the regulated community or the general public, because she has extremist views on dioxin. If we insist on retaining her, place her in some administrative or staff position (like Bill Sanjour) and not worry about whether she is happy."[21] The reference made Sanjour chuckle, but he quickly got serious: "We criticized Soviet methods a lot in this country," he said angrily. "I think there's an atmosphere like the KGB inside the EPA."

On April 21, 1992, Jenkins filed a complaint with the Labor Department against the EPA. A month later a judge ordered that she be reinstated in her original position on the grounds that her involuntary transfer was discriminatory and illegal. The EPA appealed. The order of reinstatement was upheld two years later by the secretary of labor, who also criticized the conduct of the EPA, which "on more than one occasion, has punished its whistle-blowers by transferring them to undesirable positions."[22]

"Despite her ordeal, Cate can be proud of her work," Sanjour says today. "It's thanks to her that the Vietnam veterans were finally heard and that the collusion between Monsanto and the government was uncovered. Unfortunately, it was too late for my friend Cameron Appel, who died of cancer in 1976, at just thirty, leaving two orphans. He was a captain in the U.S. Air Force during the Vietnam War. I dedicated my report on dioxin to him, because I think the story needs to be given human faces. That's what Monsanto seems to forget, because it's interested only in dollars."

As the "sentinel of the EPA" says, Jenkins's courageous memorandum opened Pandora's box and led to an outpouring of revelations and decisions that benefited first of all the American victims of Agent Orange. "We were able to get some new legislation in late 1991 and she was responsible for it," testified John Thomas Burch, chair of the NVVC, before an administrative law judge of the Labor Department on September 29, 1992. "This study was blocking us, because it was specially mentioned by those who controlled the legislation. And once we were able to show there were defects in the study, we could get by that and we could move forward. . . . It meant thousands of men getting medical care who wouldn't have gotten it otherwise."[23]

In fact, the first person to react to Jenkins's report was Admiral Elmo Zumwalt Jr., who after the death of his son had been appointed special assistant to Veterans Affairs secretary Edward Derwinski. In an interview in the *Washington Post*, he said he was "shocked by some studies I felt were dishonestly done, funded by the chemical industry," and by the fact that "the Centers for Disease Control failed to focus on veterans, who had sustained the greatest exposure to the chemical. . . . I had no inkling how difficult it was to get at the truth."[24] On May 5, 1990, the admiral presented a confidential report in which he asserted that Monsanto's fraud was part of a vast government plot intended to prevent compensation of the victims of Agent Orange and of dioxin in general.[25]

For instance, the U.S. Congress in 1982 had appropriated $63 million for the Veterans Administration (the former name of the Department of Veterans Affairs) to conduct a study of the effects of dioxin on veterans. When the VA was inordinately slow in beginning work, Congress transferred responsibility for the study to the CDC, to which the Pentagon was supposed to provide details on the Air Force's spraying program and archives detailing troop movements during the Vietnam War. Four years later, Dr. Vernon Houk, director of the Center for Environmental Health and Injury Control at the CDC, announced that he had canceled the study "on strictly scientific grounds," because his researchers had been unable to find "a population of people who were exposed in sufficient numbers" to be able to conduct it.[26] In his report, Admiral Zumwalt denounced the "purposeful effort to sabotage any chance of a meaningful Agent Orange exposure analysis." He went on to say: "Unfortunately, political interference in government sponsored studies associated with Agent Orange has been the norm, not the exception. In fact, there appears to have been a systematic effort to suppress critical data or alter results to meet preconceived notions of what alleged scientific studies were meant to find."

In support of his criticism, Admiral Zumwalt cited another example of manifest fraud, which had been revealed by Senator Tom Daschle during a congressional hearing on Agent Orange.[27] In the course of the hearing, Senator Daschle demonstrated that the U.S. Air Force had deliberately concealed the results of a study of the effects of spraying on pilots in the Ranch Hand program: contrary to the "reassuring" results it had published in 1984, it appeared that the children of the pilots involved had twice as many birth defects as those in the control group.

The former commander of the U.S. Navy in Vietnam drove home the point: "Shamefully, the deception, fraud and political interference that has characterized government sponsored studies . . . has not escaped studies ostensibly conducted by independent reviewers, a factor that has only further compounded the erroneous conclusions reached by the government." He referred, of course, to the Monsanto studies, but could have also been referring to those conducted by its German counterpart BASF, one of whose factories had had an explosion in 1953 similar to the one in Nitro four years earlier. In a troubling parallel, scientists paid by the German company had published research asserting that the workers exposed during the accident had developed no particular pathology.[28] An article in *New Scientist* seven years later revealed that the study had been falsified in the same crude manner as the Monsanto studies: twenty supervisory employees who had not been exposed to 2,4,5-T had been placed in the exposed group, thereby masking the elevated level of cancers of the lung, trachea, and digestive system.[29]

Last but not least, at the very time the admiral was preparing his report, two studies provided further ammunition: the first, published in the journal

Cancer, demonstrated that Missouri farmers who had used chlorinated herbicides such as 2,4,5-T and 2,4-D had an abnormally high rate of cancer of the lips, bones, nasal cavities, sinuses, and prostate as well as of non-Hodgkin's lymphoma and myeloma.[30] These results were confirmed by similar research on Canadian farmers published the following year.[31]

Corruption: The Richard Doll Affair

Faced with this avalanche of revelations produced by one of the country's most prestigious military officers, the administration of George H.W. Bush could only yield. On February 2, 1991, Congress passed a law (P.L. 102-4) requiring the National Academy of Sciences to draw up a list of illnesses attributable to dioxin exposure. Sixteen years later, that list included thirteen serious diseases: several types of cancer (some of which are very rare), type 2 diabetes, peripheral neuropathy (from which Alan Gibson, the veteran I met, was suffering), and chloracne. This evolving list enabled the Department of Veterans Affairs to compensate and provide medical care for thousands of veterans, out of the 3.1 million soldiers who had served during the Vietnam War.

This radical change of course also affected the EPA, which had to go back to the drawing board at a time when the international community was eagerly awaiting a study of the initial health effects of the Seveso catastrophe. This study, conducted by Dr. Pier Alberto Bertazzi, confirmed unusual levels of soft tissue sarcoma, non-Hodgkin's lymphoma, and myeloma among the exposed population.[32] This new study was "one more nail in the coffin for dioxin," confidently declared Dr. Linda Birnbaum, head of the Experimental Toxicology Division of the EPA, as she announced that the agency was in the process of reevaluating its classification of the substance, which was likely to be shifted to category A (a human carcinogen) considering the "overwhelming" evidence. The crowning irony was that in support of her argument, Birnbaum cited four studies published by Swedish scientists between 1979 and 1988 that previously had been deliberately ignored by the EPA, to the great benefit of Monsanto, once again pulling strings in the background.

The story is so incredible that it's worth taking a second look at it, because it says so much about the company's practices. Completely by chance, in 1973 a young Swedish researcher named Lennart Hardell discovered the deadly effects of dioxin on human health. He had been consulted by a sixty-three-year-old man in the Norrlands University Hospital at Umea University. Suffering from liver and pancreatic cancer, he had been a forest worker in northern Sweden who for more than twenty years had been employed spraying a mixture of 2,4-D and 2,4,5-T on deciduous trees. Thus began a long research process in collaboration with three other Swedish scientists that led to the publication of studies pointing to the association between soft-tissue sarcoma and dioxin exposure.[33]

In 1984, Hardell was asked to testify before an investigative commission that had been set up by the Australian government, which was then facing demands for reparations from soldiers who had participated in the Vietnam War alongside the Americans. The Royal Commission on the Use and Effects of Chemical Agents on Australian Personnel in Vietnam presented its report in 1985, and it provoked sharp controversy.[34] In an article in *Australian Society*, Professor Brian Martin of the University of Wollongong denounced the manipulations that had led the commission to "acquit Agent Orange."[35] Indeed, in a display of staggering optimism, the report found, according to Martin, that "no veteran had suffered due to exposure to chemicals in Vietnam." The commission concluded: "This is good news and it is the commission's fervent hope that it will be shouted from the rooftops."

In his article, Martin recounted that the expert witnesses called by the Vietnam Veterans Association had been "attacked strongly by counsel for Monsanto Australia." Martin went on to say: "The commission's report evaluated the expert witnesses in similar terms to Monsanto. Those who did not rule out the possibility of the chemicals having harmful effects had their scientific contributions denigrated and their reputations belittled. By contrast, expert witnesses exonerating the chemicals were uniformly lauded by the commission." The authors of the report did not hesitate to copy almost verbatim two hundred pages provided by Monsanto to denigrate the results of the studies by Hardell and Olav Axelson.[36] "The effect of the copying is to present the views of the Monsanto submission as the commission's own," Martin observed. For example, in the vital volume dealing with the carcinogenic effects of 2,4-D and 2,4,5-T, "the Monsanto submission's phrase 'it is submitted that' has been replaced in the commission's report by the phrase, 'the commission concludes,' in the midst of pages and pages of almost verbatim copying."

Very sharply challenged by the report, which insinuated that he had manipulated the data in his studies, Hardell in turn closely examined the document. He was surprised to discover that "the views taken by the commission . . . were supported by Professor Richard Doll in a 1985 letter to Honourable Mr. Justice Phillip Evatt, the commissioner," in which Doll stated that "[Hardell's] conclusions cannot be sustained and in my opinion, his work should no longer be cited as scientific evidence. It is clear, too, . . . that there is no reason to suppose that 2,4-D and 2,4,5-T are carcinogenic in laboratory animals and that even TCDD [dioxin], which has been postulated to be a dangerous contaminant of the herbicides, is at the most, only weakly carcinogenic in animal experiments."[37]

Richard Doll was not just anybody. Long considered as one of the greatest cancer specialists in the world (he died in 2005), this British epidemiologist, knighted by the queen, had made his name by demonstrating the association between smoking and the onset of lung cancer. Having dared to denounce the lies of the cigarette industry, he had the reputation of being incorruptible. In 1981, Doll had published a much-cited article on the epidemiology of cancer in which he asserted that environmental causes played a very limited role in the development of the disease.[38] However, the legend was shattered in 2006 when *The Guardian* revealed that Doll had secretly worked for Monsanto for twenty years.[39] In the archives he had donated to the library of the Wellcome Trust in 2002 was a letter dated April 29, 1986, on Monsanto letterhead. Written by William Gaffey, one of the authors of the controversial studies of dioxin, it confirmed the renewal of a contract providing payment of $1,500 per day. The question of a link between Doll and Monsanto had been raised initially by Hardell and his colleagues in a very enlightening article in the *American Journal of Industrial Medicine* entitled "Secret Ties to Industry and Conflicting Interests in Cancer Research."[40]

But my investigation of the concealment of the harm caused by Agent Orange had more surprises in store—and an element of pure horror, which I found in Vietnam.

The Damned of Vietnam

The nurse in the sky-blue uniform took a bunch of keys out of her pocket and opened the door without saying a word. We went into a room lined with shelves holding dozens of jars straight out of a horror movie. They were fetuses preserved in formaldehyde, monstrous, a cemetery of babies deformed by dioxin: a penis in the middle of a forehead, Siamese twins with one gigantic head, a body with two heads, a shapeless mass attached to a tiny body without limbs. "Anencephaly, 1979," said one label; "Microcephaly," said another; "Hydrocephaly," yet another. Most of the jars had no label because the deformities were so bizarre that they still lacked a medical name.

We were in Tu Du Hospital in Ho Chi Minh City (formerly Saigon) in December 2006. The "museum of the horrors of dioxin," as the Vietnamese called it, had been set up in the late 1970s by Dr. Nguyen Thi Ngoc Phuong, an obstetrician who had long headed the hospital's maternity ward, the largest in the country, and who had recently retired. This renowned dioxin specialist still spends time in the "Peace Village" on the second floor of the hospital, one of the twelve centers set up by Vietnam to care for handicapped children, the victims of Agent Orange. Phuong, a short, slender woman in an immaculate white coat, was conducting her weekly visit to the little patients who occupied five scrupulously clean rooms. Some were confined to bed because they had been born without arms or legs; others were romping on the tile floor under the watchful eye of a nurse sitting in the midst of plastic toys. I was deeply touched by the serenity that emanated from these crippled children, proof that they were receiving high-quality medical and emotional care. "Most of them suffer from neurological problems and severe organic anomalies," the doctor said, taking a little boy born without eyes onto her knees. I could not tear my eyes away from the fetal head attached to a child's body nestled against the doctor's shoulder.

Phuong was still a student when she witnessed for the first time the birth of a deformed baby in the maternity ward of Tu Du Hospital. "It was in 1965," she explained in respectable French. "At the time, I had never heard of dioxin. In the succeeding years we saw a significant increase in the number of stillbirths with severe deformities and of children born with serious handicaps. And it's still going on: in 2005, we recorded more than 800 children born with deformities in this hospital alone, which is significantly above the international average."

"The spraying of defoliants stopped nearly forty years ago. How could dioxin have affected these children?" I asked.

"We know that dioxin accumulates in the food chain and that it is lipophilic, that is, it combines with fats. The mothers of these children may have been contaminated by food or by breast milk from their own mothers. We also know that dioxin can produce chromosomal anomalies, which may also explain the transmission from one generation to the next."

"Have you verified that the parents of these children have dioxin in their bodies?"

"According to admission records, 70 percent of the children treated here have parents living in areas that were sprayed by defoliants. Unfortunately, tests to detect dioxin are very expensive—about €1,000—and there are no laboratories in Vietnam that are able to perform them. The only time we did have a test done was on the mother of Viet and Duc, Siamese twins born with three legs, one pelvis, one penis, and one anus, whom we successfully separated. We found a rather high level of dioxin in her fatty tissue. My country's medical authorities estimate that 150,000 children today have deformities due to Agent Orange and that 800,000 people are ill."

"Are their birth defects characteristic of dioxin?"

"No, but dioxin acts inside cells like a hormone favoring the growth of deformities and diseases that exist otherwise."

"How do you explain that a company like Monsanto and even some American scientists continue to deny the existence of a link between dioxin exposure and genetic deformities?"

"It's history repeating itself. First they denied the link with cancer, and now, to avoid responsibility, they deny the link with birth defects."

Among the thirteen diseases currently acknowledged by the United States to be linked to dioxin, only one is a birth defect, spina bifida.* "The problem," explained Professor Arnold Schechter, who was in Ho Chi Minh City at the time of my visit, "is that we lack scientific data. Only animal studies have been done: they show that when a female is exposed to dioxin the probability that she will have offspring with severe handicaps or deformities, including cerebral deformities, increases considerably." A professor at the University of Texas, Schechter is one of the world's most eminent dioxin specialists. In the early 1980s, he defied the American embargo on Vietnam and joined with scientists in Hanoi to conduct long-term research on the dissemination of dioxin in the environment.

*Latin for "spine split in two." Spina bifida is a defect in the spinal column in which one or more vertebrae are not correctly formed during gestation, forming a space that allows the spinal cord to stick out.

One of these scientists was Professor Hoang Trong Quynh, a former colonel in the Vietnamese army who participated in "the two wars of liberation, first against France, and then against the United States," he explained to me in impeccable French. For thirty years, the two researchers combed the Vietnamese countryside collecting blood samples and fatty tissue from people and animals in order to analyze dioxin levels. Their work has led to many publications, the latest concerning forty-three residents of the South Vietnamese city Bien Hoa, located near a former air base used for Agent Orange spraying missions.[41] The tests showed high blood levels of dioxin, over 5 parts per trillion (ppt), with peaks as high as 413 ppt, even in young children.* In addition, some soil and sediment samples taken in the Bien Hoa region, particularly near Bien Hung Lake, showed extremely high levels of TCDD, over 1 million ppt.

"In Vietnam," Schechter explained, "the urgent task is to decontaminate what we call hot spots, places with high dioxin levels, like the former Bien Hoa air base, because while dioxin does not accumulate in plants, it does penetrate the soil, where its half-life can be as long as one hundred years. Leached by rain, it gets into the water table, lakes, and rivers. There it stays attached to sediments, contaminating phytoplankton, zooplankton, fish, birds, and humans through the food chain. Once it's in the blood, it is distributed to the cells, where it attaches to fatty tissue. Its half-life averages seven years in the human body. It can be eliminated only by weight loss or in breast milk. The problem is that it then contaminates the infant."

On this December day in 2006, the two octogenarians were traveling to Binh Duong Province, about one hundred miles from Ho Chi Minh City, which was one of the regions most heavily sprayed with Agent Orange. They were to meet with a family whose three children in their twenties were mentally handicapped. The father had lived in Bien Hoa from 1962 to 1975. The mother had never left Binh Duong Province.

"Did you see Agent Orange being sprayed?" Schechter asked.

"Yes," said the father. "It smelled like ripe guava."

"In this family," Schechter commented, "if the parents' dioxin levels turned out to be high, one could say that there is a strong probability that the children's handicaps are linked to Agent Orange. If not, we don't know. No

*Unlike other toxic substances, dioxin is usually measured in parts per trillion. In Western countries, the average level of dioxin in humans is 2 ppt.

epidemiological study of the links between dioxin and birth defects has ever been conducted."

"That's not true," interrupted Quynh. "Studies by Vietnamese colleagues have been published showing that in villages sprayed with Agent Orange, rates of miscarriages and birth defects are much higher than in villages that were not sprayed. But because these studies were not headed by Westerners, American scientists pay no attention to them.[42]

"How do you explain that?" I asked, aware that the conversation had entered sensitive ground.

"Dioxin has become a highly political subject," Schechter said, obviously embarrassed. "That's a shame, because in the end we're all concerned: we all have dioxin in our bodies and it's important to know precisely what its effects are on human organisms. Unfortunately, scientists are prisoners of interests that are outside their control."

Meanwhile, one thing is certain: on March 20, 2005, the Bush administration announced the cancellation of a binational research program that had been initiated two years earlier by an agreement between the United States and Vietnam.[43] With a budget of several million dollars, the study was supposed to be headed by Professor David Carpenter of the University of Albany, whom I met in the course of my investigation of PCBs. "This study was to focus on Vietnamese populations, primarily on the link between dioxin exposure and birth defects," he explained to me. "Officially, it was canceled because of lack of cooperation from the Vietnamese government. It's true that the government can be criticized for bureaucratic delays, but I think the decision was very convenient for the manufacturers of Agent Orange, against whom new complaints had been filed."

In fact, on June 9, 2003, the U.S. Supreme Court had decided in favor of Daniel Stephenson and Joe Isaacson, two Vietnam veterans suffering respectively from bone marrow cancer and non-Hodgkin's lymphoma diagnosed in the late 1990s. Considering themselves unaffected by the 1983 settlement, they had decided to sue Monsanto and the other companies. The companies had appealed, but the Supreme Court found against them, opening the way to a new class action that included Alan Gibson as a plaintiff and Gerson Smoger as an attorney. Four years later, the trial had still not begun.

In February 2004, the Vietnam Association of Agent Orange Victims filed a complaint in federal district court in New York. But it was dismissed in

March 2005 by Judge Jack B. Weinstein, the same judge who had presided over the 1983 settlement, on the grounds that the military use of herbicides was not prohibited by any international law and could therefore not be considered a war crime. Citing a 1925 treaty that banned the wartime use of "gases deployed for their asphyxiating or toxic effects on man," the eighty-year-old judge found that this treaty did not concern "herbicides designed to affect plants that may have unintended side-effects on people. If supplying contaminated herbicide had been a war crime the chemical companies could have refused to supply it. We are a nation of free men and women," Judge Weinstein wrote, "habituated to standing up to government when it exceeds its authority."[44]

The language must have delighted Monsanto, which, for its part, had never varied its defense. "We are sympathetic with people who believe they have been injured and understand their concern to find the cause," declared Jill Montgomery, a Monsanto spokesperson, in 2004, "but reliable scientific evidence indicates that Agent Orange is not the cause of serious long-term health effects."[45]

Never-ending denial—which is also a characteristic of the company's current position on Roundup, the herbicide it introduced on the market when in the mid-1970s 2,4,5,-T was finally banned in the United States and subsequently in the rest of the world.

4

Roundup: A Massive Brainwashing Operation

Glyphosate is less toxic to rats than table salt following acute oral ingestion.

—Monsanto advertisement

"If, like Rex, you hate weeds in your garden, here's Roundup, the first biodegradable herbicide. It kills weeds from the inside down to the roots and pollutes neither the soil nor Rex's bones. Roundup, the herbicide that makes you want to kill weeds!" Many television viewers in France will remember this lighthearted ad showing a dog happily spraying Roundup on weeds in the middle of a lawn and then digging up the bone it had buried in the very spot where the herbicide had shriveled the roots of the weeds. Rex's enthusiastic barking suggests that the dog will gnaw on that bone with complete assurance because Roundup is completely harmless. One might almost imagine the nice dog washing down its feast with the dregs from the container of "biodegradable herbicide."

The Most Widely Sold Herbicide in the World

Broadcast 381 times on the major French television channels between March 20 and May 28, 2000, this publicity campaign cost Monsanto the tidy sum of 20 million francs. Similar spots were being broadcast at the same time almost everywhere in the world, because the company was at a critical juncture; its patent on Roundup was set to expire later that year, which would end its monopoly on the most widely sold herbicide in the world and open the door to the production of generic varieties and hence competition. This put Monsanto in a very tight spot, because it was betting its future on the development of transgenic crops that it was calling "Roundup-ready" because they had been genetically manipulated specifically to withstand being sprayed with Roundup. In short, the stakes for the multinational were huge, and it would defend its leading product tooth and nail.

Roundup is the trade name Monsanto had given to glyphosate, a herbicide the company's chemists had discovered in the late 1960s. The distinctive feature of this "unselective" weed killer—unlike 2,4-D and 2,4,5-T—is that it destroys all forms of vegetation because of the way it works: it is absorbed by the plant through the leaves and quickly carried by the sap to the roots and rhizomes. There it inhibits an enzyme essential for the synthesizing of aromatic amino acids, which leads to a decrease in the activity of chlorophyll as well as of certain enzymes. This causes the necrosis of tissue, leading to the death of the plant.

As soon as it was put on the market in 1974, first in the United States and then in Europe, Roundup had "spectacular success," in the words of a promotional Web site for Monsanto and the Scotts Company, which distributed the product in France.[1] In fact, while it was entangled in the ecological and health scandal of 2,4,5-T, the company was advertising the virtues of this innovation on its packaging: "Respects the environment," "100% biodegradable," and "Leaves no residues in the soil."

"The active ingredient in Roundup is inactivated when it touches the soil, which preserves surrounding plants and permits seeding or replanting one week after application," according to the Web site. These enticing promises explained why glyphosate became a farmers' favorite, and they used it in huge quantities to clear their fields of weeds before sowing their next crops. With its ecological aura, Roundup also became the idol of managers of public spaces (parks, golf courses, highways, and so on). In spring, technicians in astronaut suits—airtight clothing covering them from head to toe, along with gas masks and protective boots—were commonly seen patrolling the streets of France with tanks of Roundup on their backs.

One day in May 2006, in an area south of Paris, I went along with one of these teams charged with eradicating these "adventitious" growths—the term professionals use to designate weeds. I had been struck by the unappetizing greenish color of the workers' boots, and the weeders explained that they had to change them "every two months" because the "rubber is eaten away by Roundup." "I'm very careful about my workers' equipment," agreed the head of the company, who asked to remain anonymous. "I also require that they scrupulously follow the dosage recommended by the manufacturer, which unfortunately we haven't always done." He added knowingly, "It seems that the product isn't as safe as we've been led to believe." He said nothing further, and I recalled the repeated ads on television showing kids playing on a lawn while their father in shorts and sandals attacks weeds with a spray bottle of Roundup Patio and Garden.

"In 1988," the Web site explains, "Monsanto established its Garden division to extend access to Roundup to the amateur gardener. A new range of Roundup products for the public came on the market." Roundup thus made its entry into all the gardens in France, where it was used abundantly (and with no protection) before the planting of vegetables for home consumption. "We all use it," I was told by a man renting a garden near the Stade de France in Saint-Denis, north of Paris. In his garden shed, the recently retired man was preparing the mixture that he would spray on his patch of land to prepare his seedbeds. "Look," he said, pointing to his Roundup container, with its soft green color and bird logo. The product description read: "Used according to directions, Roundup poses no risk to people, animals, or the environment."

In the United States, the infatuation with the likable herbicide was such that in 1993 fifteen cities agreed to participate in a "city beautification program" sponsored by Monsanto. Volunteers recruited by the firm found themselves in Spontaneous Weed Attack Teams (SWAT) that patrolled the streets to kill weeds. "The idea is to develop a phobia against weeds and to position Roundup as a socially responsible brand," explained Tracy Frish, one of the leaders of a New York coalition favoring an alternative to pesticides, which was conducting a campaign attacking Monsanto's deceptive advertising.[2]

A Double Fraud

Serious suspicions fell very early on the company's new favorite product. And Monsanto once again succeeded in slipping through the cracks thanks to the negligence of the apparently incorrigible EPA. The EPA's "constancy" was, to tell the truth, not at all surprising: all the facts that I have presented in this book—whether they have to do with PCBs, dioxin, or Roundup—cover the same period, roughly from 1975 to 1995. So, regardless of which product was at issue, the same protective blindness obviously prevailed.

I have already described the trial that made headlines in the early 1980s involving Industrial Bio-Test Labs (IBT) of Northbrook, a private laboratory, one of whose directors was Paul Wright, a toxicologist from Monsanto recruited to supervise a study of the health effects of PCBs. The EPA was quite familiar with IBT, because it was one of the chief laboratories in North America conducting tests of pesticides for chemical companies so that they could obtain regulatory approval for their products. By combing through the laboratory's archives, EPA agents had discovered that dozens of studies had been faked—had "serious deficiencies and improprieties," in the agency's cautious language. In particular, they had discovered a "routine falsification of data" designed to conceal "countless deaths of rats and mice" involved in the tests.[3]

Among the offending studies were thirty tests of glyphosate.[4] It was "hard to believe the scientific integrity of the studies," noted an EPA toxicologist in 1978, particularly "when they said they took specimens of the *uterus* from male rabbits."[5]

In 1991, Craven Laboratories was accused of having falsified studies that were supposed to measure pesticide residues, including Roundup, on plums, potatoes, grapes, and sugar beets, as well as in soil and water.[6] "The E.P.A. said the studies were important in determining the levels of a pesticide that should be allowed in fresh and processed foods," said the *New York Times*. Referring to the earlier IBT fraud, the article reported: "As a result of the falsification . . . the E.P.A. declared pesticides safe when they had never been shown to be."[7] The widespread fraud had resulted in the indictment and conviction of three IBT executives, but Monsanto and the other chemical companies that had benefited from the falsified studies suffered no consequences. With respect to Craven Laboratories, it appeared that the EPA was again sticking its head in the sand: "We don't think there is an environmental or health problem," said Linda Fisher, assistant administrator for pesticides and toxic substances. "First of all, we're dealing with allegations. Right now we're moving out to take preventive measures. . . . [It's a] big deal to me."[8] The following year, the owner of Craven Laboratories and three employees were indicted on twenty felony counts. The owner was sentenced to five years in prison and Craven Labs was heavily fined.

It was indeed a big deal: after ten years at the EPA, Linda Fisher was hired by Monsanto in 1995 to head the company's Washington office, responsible for lobbying political decision makers; she then returned to the EPA in May 2001 as deputy administrator—a typical instance of the revolving door system that exemplifies the collusion between large companies and government authorities.

In the meantime, Monsanto was cognizant of the impact that these two instances of fraud could have on its image. In June 2005, fourteen years after Craven Laboratories had been accused, the company published a note in which it stated with its usual self-assurance: "The damage caused to Monsanto's reputation by discussion of this issue by the media, and then further use by activists to question the integrity of Monsanto's data, cannot be calculated. All affected residues studies have been repeated and the data are sound, up-to-date and accepted by EPA."[9]

After the two scandals, the EPA had indeed required that the questioned studies be repeated. But as Caroline Cox pointed out in the *Journal of Pesticide Reform* in 1998, "This fraud casts shadows on the entire pesticide registration process."[10] On the other hand, these "shadows" had no effect on Monsanto, which continued as though nothing had happened, its advertising campaign promoting Roundup as a pesticide that was "biodegradable and good for the environment."

"False Advertising"

In 1996, complaints filed with the Consumer Frauds and Protection Bureau of New York had compelled the company to negotiate a settlement with the State Attorney General, who had opened an investigation of "false advertising by Monsanto regarding the safety of Roundup herbicide (glyphosate)." In a very detailed statement of findings, the bureau reviewed the numerous Monsanto newspaper and television ads, including "Glyphosate is less toxic to rats than table salt following acute oral ingestion" and "Roundup can be used where kids and pets'll play and breaks down into natural material."[11]

This was "false and misleading advertising," the attorney general found, and he barred Monsanto, under penalty of a fine, from declaring that its herbicide was "safe, nontoxic, harmless or free from risk." Nor could Monsanto claim that Roundup is good for the environment or "known for [its] environmental characteristics." Two years later, the company was forced to pay $75,000 for suggesting in a new advertisement featuring a California horticulturist that the herbicide could be sprayed in and around water.[12]

Oddly enough, these American legal decisions never troubled the European Commission, much less the French authorities, who tolerated unquestioningly the advertising campaign Monsanto launched in the spring of 2000. But the image of the lovable Rex about to gnaw on a bone soaked in Roundup stirred the wrath of the association Eau et Rivières de Bretagne, which in January 2001 sued the French subsidiary of the American giant for false advertising.

"Scientific studies have shown that huge quantities of glyphosate have been found in rivers and streams in Brittany," Gilles Huet, representative of the Breton association, told me in a telephone conversation in the spring of 2006, referring to a report published in January 2001 by the Observatoire Régional de Santé de Bretagne.[13] In fact, analyses conducted of streams in Brittany in 1998 showed that 95 percent of the samples had a level of glyphosate above the legal threshold of 0.1 ppb, with peaks of 3.4 ppb in the Seiche, a tributary of the Vilaine. Huet pointed out that "in 2001 the European Commission, which reauthorized glyphosate, classified it as 'toxic for aquatic organisms' and 'possibly causing long-term harmful effects on the environment.' We are asking for a minimum of consistency: a 'biodegradable' product that is 'respectful of the environment' cannot end up being 'toxic and harmful' in Brittany's rivers."

On November 4, 2004, the criminal court in Lyon, where the headquarters of the French subsidiary of Monsanto was located, began proceedings in which the company was charged with "false and misleading advertising." Until 2003, taking advantage of delays in investigating the complaint by the Breton association, the agrochemical company had been able to continue its advertising campaign. And on the occasion of the Lyon trial, it even gained a further two-year delay by simply not appearing. Company representatives claimed never to have received notification by mail because, according to the prosecution, they had no address in France, and so the prosecution decided to put off the trial until June 2005. "Administrative error or maneuver by the company to avoid an ignominious judgment in terms of its brand image?" wondered the consumer association UFC–Que Choisir, which had joined the suit of Eau et Rivières de Bretagne in 2001. Gossip had it that the delay enabled the company to save the spring weed-killing campaign, crucial for its revenues: in 2004, Monsanto France had 60 percent of the glyphosate market, which amounted to annual sales of 3,200 tons of Roundup—use of the herbicide had doubled between 1997 and 2002.

The hearing was finally held at the Lyon criminal court on January 26, 2007, exactly six years after the complaint had been filed. The heads of Scotts France and Monsanto were fined €15,000, a penalty worth a few dilatory maneuvers. The court found that "the combined use on labels and packaging [of herbicides in the Roundup product range] of the terms and expressions 'biodegradable' and 'leaves the soil clean' . . . could lead the consumer to believe erroneously in the complete and immediate harmlessness of these products following quick biological degradation after use . . . whereas they can on the contrary remain durably in the soil, and even spread into the water table."

Even more bothersome for Monsanto, which appealed, the court determined that the company knew "prior to publication of the challenged advertisements that the products concerned were ecotoxic in nature," because, "according to studies conducted by Monsanto itself, a level of biological degradation of merely two percent may be reached at the end of 28 days." Once again, the company possessed data contradicting what it was publicly claiming, but it carefully refrained from revealing them. Indeed, why should it have done so? As Ken Cook, head of the Environmental Working Group in Washington, said with reference to PCBs, "It pays to keep secrets, because in the end the penalties are very light."

The Very Problematic Process of Pesticide Registration

"We wish to point out that all the statements made on our labels are based on published scientific studies or studies communicated to the regulatory authorities of the Ministry of Agriculture in charge of delivering authorizations to market," a Monsanto France executive wrote on June 8, 2000, to the French government agency in charge of competition policy and consumer protection. It must be conceded that the representative of the firm was right on that point. But by offering a defense of this kind, he was putting his finger on the heart of the problem: namely, that the process for registering chemical products in France (as in a number of developed countries) opens the door to all sorts of abuse and fraud, much to the dismay of consumers.

To be more precise, I would even say that the registration process is in fact a sham: contrary to what the regulatory authorities would have us believe, it is in reality based entirely on the goodwill of the chemical companies, which provide data from studies they are supposed to have conducted to prove the harmlessness of their products. These data are then examined by "experts" who vary widely in their competence, courageousness, and independence. All one need do is read the book *Trust Us, We're Experts* by the American writers Sheldon Rampton and John Stauber,[14] or *Pesticides: Révélations sur un Scandale Français* by the French writers Fabrice Nicolino and François Veillerette,[15] to realize that many toxic products have had a long career after being duly approved by the experts, whose names are concealed by opaque and not very democratic bureaucratic procedures.

In this sense, the history of Monsanto is a paradigm of the aberrations in which industrial society has become mired, forced to manage as best it can—that is, badly—the proliferation of toxic chemical substances that have invaded the planet since the end of World War II. The reasonable solution would be to ban outright any molecule that presents the slightest danger to people and the environment. But instead, to satisfy the interests of the major chemical companies—and, some would say, the interests of modern consumers—every effort is made to regulate dangerous substances only as much as is necessary to limit the most obvious or immediate damage. For the rest, *après nous le déluge*.

The history of pesticides constitutes a perfect illustration of this very twisted mechanism, whose workings it is important to understand, even if that involves entering into rather dry detail—the better to grasp its absurdity. As biologist Julie Marc points out in the doctoral dissertation she defended in 2004 at the University of Rennes, "the use of pesticides goes back to antiquity," but until the twentieth century, the pest killers were of natural origin: peasants and gardeners used mineral derivatives, such as the copper in the old *bouillie bordelaise* (Bordeaux mixture), to treat plants affected by certain diseases or parasites.[16] The development of industrial agriculture was accompanied by the massive use of chemical pesticides belonging to the family of organochlorides, the first of which was DDT. Called "phytosanitary products"—a clever rhetorical trick, replacing the notion of a "killer" with the euphemistic "medicine"—they covered three categories: fungicides, insecticides, and herbicides.

Every pesticide is made up of an active ingredient—glyphosate in the case of Roundup—and numerous inert ingredients, such as solvents, carriers, emulsifiers, and surfactants, the purpose of which is to intensify the physiochemical properties and biological effectiveness of the active ingredients, and which have no pesticidal effect of their own. Hence, the various products in the Roundup range contain between 14.5 and 75 percent glyphosate salts, with the rest of the formulation including a dozen principal additives, whose "composition is often kept secret," as Marc observes. The role of these additives is to enable glyphosate to penetrate into the plant, as in the instance of polyethoxylated tallowamine (POEA), a detergent favoring the spread of spray droplets on the leaves.

In France, the third-largest user of pesticides in the world (after the United States and Japan), with 100,000 tons sold every year—40 percent herbicides, 30 percent fungicides, and 30 percent insecticides—it is estimated that 550 active ingredients and 2,700 commercial formulations are now registered. Following the practice of the rest of the world, notably other countries in Europe, every new phytosanitary product must be registered before it is put on the market, which means an authorization of sale for ten years, granted by the Ministry of Agriculture. To secure an authorization, the company must demonstrate that its formulation is effective and harmless, through a technical dossier containing laboratory tests of the chemical, physical, and biological properties of the product, as well as tests of its possible toxicity for people, animals, and the environment. When one reads the list set out by Julie Marc of the tests supposed to constitute this toxicological dossier, one may think that all is for the best in the best of all industrial worlds and that theoretically there is nothing to worry about.

The tests required by regulatory authorities in France, as in the rest of the European Union, are numerous. They begin by assessing the effects of the substance on rats (and sometimes other animals) when it is absorbed orally, through the skin, or by inhalation. Researchers measure in particular the absorption, distribution, metabolism, and elimination of the molecule by the organism, and calculate what is known as the "lethal dose," that is, the quantity or concentration of the product necessary to bring about the death of 50 percent of a group of test animals ("LD50" or "LC50"), with a view toward avoiding serious accidents during use. They then evaluate what is called "subchronic toxicity," the effects of repeated absorption of the product on bodily organs, chiefly the liver and kidneys. Usually conducted over a period of ninety days or one year (even two years if a problem appears), these tests make it possible to establish what experts call the NOAEL (no observable adverse effect level), that is, the maximum quantity of the substance whose daily absorption has no effects on the test animals. The NOAEL is expressed either in milligrams of active ingredient per kilogram of body weight of the animal tested per day or in milligrams of the substance per kilogram of food (expressed in ppm) if a food ingredient is involved. Finally, tests must verify whether the product is potentially oncogenic (causing cancer), teratogenic (causing birth defects), or mutagenic (causing permanent and inheritable changes in the DNA of subjects exposed).

Taken together, the toxicological data as a whole enable the establishment of regulatory categories, such as the lowest acceptable daily dose (LADD), which designates the quantity of the substance that the user or consumer is supposed to be able to absorb daily over a lifetime with no health effects. In other words, to make the absurdity of the process clear: it is known that a substance is toxic for mammals and one calculates the dose that can be inflicted on them daily before they fall ill, or even die. Then, the data are extrapolated to people. But how do we know that the dose calculated for a rat or a rabbit will protect us effectively from being poisoned? A mystery. And what of the accumulation and interaction among the various toxic substances that we ingest every day, because the LADD (the daily acceptable dose of poison) refers not only to many pesticides, but also to food additives, such as coloring agents and preservatives? The question is ignored. In any event, it is troubling to realize that the calculation of this disquieting LADD is based on tests conducted by the manufacturers, whose purpose is primarily to sell their products.

In addition to toxicity tests intended to assess the danger that a new molecule may constitute for people, there are tests that consider its performance in the environment (including its persistence, mobility, absorption into the food chain, and biodegradability), as well as its ecotoxic potential (for birds, bees, fish, and aquatic plants).

Finally, the toxicological file is examined by the Commission d'étude de la toxicité des produits antiparasitaires à usage agricole, which submits an opinion to the Ministry of Agriculture. This usually conforms to decisions made for the EU by the Permanent Phytosanitary Committee, in charge of preparing an evolving list of authorized active substances by classifying them according to their degree of toxicity (irritant, corrosive, harmful, toxic, and very toxic), along with labeling requirements. "According to French and international bodies, glyphosate is considered an irritant that may produce severe eye lesions, and [is] toxic for aquatic organisms," Marc reports. "According to the World Health Organization, the U.S. EPA, and the European Community, the use of glyphosate in accordance with the manufacturers' instructions therefore poses no human health risks. . . . However, several epidemiological studies have demonstrated a correlation between exposure to glyphosate and cancer."

Roundup Triggers the First Stage of the Development of Cancer

In fact, while regulatory agencies have continued to classify glyphosate-based herbicides as "not a human carcinogen," a series of epidemiological studies tend to demonstrate exactly the opposite. For example, a Canadian study published in 2001 by the University of Saskatchewan showed that men exposed to glyphosate more than two days a year had twice the risk of developing non-Hodgkin's lymphoma of men never exposed.[17] These results were confirmed by a Swedish study published in 2002 by Lennart Hardell (the dioxin specialist) and his colleagues, which compared the health of 442 users of glyphosate-based herbicides with a control group of 741 non-users,[18] as well as by an epidemiological investigation of farmers in the American Midwest by the National Cancer Institute.[19] Another epidemiological study in Iowa and North Carolina of more than 54,315 private and professional users of pesticides suggested an association between the use of glyphosate and multiple myeloma.[20]

In France, a team under Professor Robert Bellé of the Station Biologique de Roscoff, which is under the authority of the CNRS [Centre Nationale de la Recherche Scientifique] and the Université Pierre et Marie Curie, studied the impact of glyphosate formulations on sea urchin cells. "The early development of sea urchins is one of the recognized models for the study of cell cycles," explains Marc, who wrote her doctoral dissertation on the work of the laboratory in Brittany. In fact, the discovery of the sea urchin model, which is crucial for the early phases of carcinogenesis, earned the Nobel Prize for Physiology or Medicine for the Britons Tim Hunt and Paul Nurse and the American Leland Hartwell.

In the early 2000s, Professor Bellé decided to use it to test the health effects of pesticides. His concern at the time had been raised by the level of pollution observed in French waterways as well as food: "The data concerning underground water in France revealed contamination considered suspect in 35 percent of cases," says Julie Marc, who consulted all the available studies. "Ocean waters also showed widespread and perpetual contamination by herbicides. . . . The ingestion of fruits and vegetables also contributes to human pesticide levels. The figures in this area are disturbing, since 8.3 percent of the samples of vegetable foods of French origin analyzed contained pesticide residues above the maximum limits and 49.5 percent had some residues."[21]

In this not very reassuring overview, the Brittany region displayed a record level of contamination, particularly affecting water intended for human consumption, according to Marc: "In 75 percent of cases, the regulatory standard for the combination of substances was exceeded and more than ten were sometimes detected in a single sample, with respective concentrations exceeding the regulatory standard of 0.1 ppb. This pollution originated from agricultural practices, but also from the use of pesticides in non-cultivated areas." She also points to one of the aberrations in the regulatory system: it had set the acceptable level of residues in water as 0.1 ppb, but this referred to a single herbicide, and said nothing about the cumulative effect of different pesticides—a very common occurrence—nor of their interaction.

In the early 2000s, Bellé proposed to the regional council of Brittany that he conduct a study to assess the impact of herbicides on cell division. "The irony of the story," he told me in his Roscoff laboratory on September 28, 2006, "is that we had decided to take Roundup as a control in the experiments, because we were persuaded that the product was completely harmless, as the advertisement of the dog with its bone suggested. And obviously, the huge surprise was that that herbicide produced much more significant effects than the other products we were testing. That's why we changed the focus of our research to concentrate entirely on the effects of Roundup."

"How did you proceed?" I asked.

"Concretely, we had sea urchins lay eggs; characteristically they produce large numbers of ova. We placed those ovocytes in proximity to sperm and put the fertilized eggs in a diluted solution of Roundup. I might emphasize that the concentration was well below that generally used in agriculture. We then observed the effects of the product on millions of cell divisions. We very soon realized that Roundup affected a key point in cell division—not the mechanisms of cell division itself, but those that control it. To understand the importance of this discovery, you have to recall the mechanism of cell division. When a cell divides into two daughter cells, the making of two copies of the genetic inheritance, in the form of DNA, gives rise to many errors, as many as fifty thousand per cell. Normally, a process of repair or the natural death of the defective cell, known as "apoptosis," is automatically initiated. But a cell sometimes avoids the alternative [death or repair], because the point that controls damage to the DNA is affected. It is precisely this checkpoint that is damaged by Roundup. And that's why we say that Roundup induces the early stages leading to cancer. In fact, by avoiding the repair mechanisms, the affected cell will be able to perpetuate itself in a genetically unstable form, and we now know that it can be the origin of a cancer that will develop thirty or forty years later."

"Were you able to determine what it was in Roundup that affected cell division?"

"That is a crucial question. Indeed, we also conducted the experiment with pure glyphosate, that is, with none of the additives that go into Roundup, and we did not observe any effects, so it's Roundup that is toxic, not its active ingredient. But when we looked at the tests that were used for the registration of Roundup, we were surprised to find that they had been conducted with glyphosate alone. In fact, pure glyphosate has no use, not even as a herbicide, because by itself it is unable to penetrate into cells and thereby affect them. That's why I think there is a real problem with the registration process of Roundup and it would be necessary to look more closely at the numerous additives that go into it and at their interaction."

Among the suspect additives is polyethoxylated tallowamine, whose acute toxicity has been confirmed by many studies. Roundup also contains inert ingredients about which nothing can be said because their identity has not been disclosed by the manufacturer, under a claim of "trade secrets."[22] Another factor to consider is the principal product of the biodegradation of glyphosate, α-amino-3-hydroxy-5-methyl-4-isoxazolepropionic acid (AMPA), which has a long half-life.

In the face of these obvious malfunctions of the registration process, some courageous scientists such as Dr. Mae-Wan Ho in the United Kingdom and Professor Joe Cummins in Canada, members of the Institute of Science in Society, have called for an urgent revision of regulations relating to the most widely used herbicide in the world.[23] I say "courageous" because Bellé's story proves, if any proof were needed, that one does not touch the leading product of a company like Monsanto with impunity.

"Obviously, we immediately understood the significance our results might have for the use of Roundup," he explained, since the concentration of the herbicide responsible for the cellular malfunctions was 2,500 times less than that recommended for spraying. In fact, a droplet was enough to affect the process of cell division. Concretely, that means that to use the herbicide without risk, you would have not only to wear protective clothing and a mask, but also make sure that there was no one in a five hundred yard radius. A little naively, we assumed that Monsanto must not be aware of this, or else these recommendations would be in the directions for use, and we sent them our results even before publishing the study.[24] I have to say we were very surprised by their reaction. Instead of seriously considering our results, they responded a little aggressively that all the regulatory agencies had concluded that the product was not a human carcinogen and that in any event, cancer in sea urchins was of no interest to anyone. That is anything but a scientific argument. You would think they didn't even know that the reason the sea urchin model had earned a Nobel Prize for its discoverers was precisely because we know that effects measured in a sea urchin cell are completely transposable to humans."

"And how did your supervisory bodies, the CNRS [Centre Nationale de la Recherche Scientifique] and the Université Pierre et Marie Curie, react?"

"To tell the truth, their reaction was even more surprising," Professor Bellé answered after a silence. "Some representatives traveled to Roscoff to urge us not to communicate with the mass media, on the pretext that it would produce a panic."

"How do you explain it?"

"The question has troubled me for a long time. Now I think they didn't want to make waves so as not to hinder the development of GMOs, which, as you know, have been modified to resist Roundup."

"Aren't you worried about your career?"

"I'm no longer afraid of anything," he said quietly. "I will soon retire and I'm no longer heading the laboratory. That's why I can let myself talk today."

"A Killer of Embryos"

"Not interfering with the development of GMOs" is also the only excuse that Gilles-Éric Séralini was able to find to explain the inertia of the authorities in the face of Roundup's toxicity. A professor at the University of Caen, this biochemist is a member of the French Biomolecular Engineering Commission, charged with assessing the potential risks involved in the deliberate release of GMOs, as well as of the CRII-GEN, the Committee for Independent Research and Information on Genetic Engineering, which has continually called for more exhaustive studies of the health effects of GMOs.

Séralini has conducted several studies to assess the impact of Roundup and its effects on human health, as he explained when I met him on November 10, 2006, at his laboratory in Caen. "The reason I became interested in Roundup is because, with GMOs that have been modified to be able to resist it, it has become a food product, because residues are found in transgenic soybeans and corn kernels. In addition, I had read epidemiological studies done in Canada showing that there were more miscarriages and premature births among farm families using Roundup than in the general population."

In fact, a study of Ontario farm families conducted by Carleton University and Canadian government researchers showed that the use of glyphosate in the three months preceding conception of a child was associated with an elevated risk of late spontaneous abortions (between the twelfth and nineteenth weeks).[25] It is noteworthy that according to another study of North American farm families 70 percent of farmers had detectable levels of glyphosate on the day they used Roundup in the field, with a mean concentration of 3 ppb and a maximum of 233 ppb.[26]

A Texas Tech University laboratory likewise established that exposure of the Leydig cells in the testicles, which play a major role in the male reproductive system, to Roundup reduced their production of sex hormones by 94 percent.[27] Finally, Brazilian researchers found that female rats that were pregnant when exposed to Roundup were more likely to produce offspring with skeletal malformations.[28]

All these results were confirmed by the two studies conducted by Séralini and his team, which measured the toxic effect of Roundup first on human placental cells and then on embryonic cells—"taken," he was careful to point out, "from a line of embryonic kidney cells grown in the laboratory, which involved no destruction of embryos."[29]

"How did you proceed?" I asked.

"We put the cells in solutions of Roundup, varying concentrations of the product from the most minute, 0.001 percent, up to concentrations used in farming, that is, Roundup diluted to 1 or 2 percent. We also varied exposure times to determine the point at which the herbicide had an effect on what we call 'cellular respiration,' which conditions their production of sex hormones. We found that at levels within regulatory limits, such as residual levels accepted in food products like transgenic plants, Roundup literally killed human placental cells within a few hours, and human embryonic cells were even more sensitive."

Séralini opened his laptop to show me the photographs his team had taken of the tests. At first, one could see a string of distinct and transparent cells with a dark spot in the center of each one, the nucleus. After a day of exposure to Roundup, they had dissolved into a formless dark mass, a "kind of purée," in Séralini's words. "In fact," he explained, "under the effects of the product, the cells begin to contract; then, no longer able to breathe properly, they die asphyxiated. And I emphasize that this result occurs at doses well below those used in farming: for example, in this photograph, the concentration was 0.05 percent. That's why I say Roundup is a killer of embryos. When you use an even weaker concentration—by diluting the product bought in a store ten thousand or even a hundred thousand times—you find it no longer kills cells but blocks their production of sex hormones, which is also very serious, because those hormones enable the fetus to develop its bones and form its future reproductive system. It can therefore be concluded that Roundup is also an endocrine disruptor."

"Did you compare the effects of Roundup to those of glyphosate alone?"

"Of course! And we found that Roundup is much more toxic than glyphosate, whereas the tests that were the basis for the registration of Roundup were done with the active ingredient alone. So we contacted the European Agriculture Commissioner, who recognized that it was a problem, but nothing has happened since then."

"What did the French authorities say?"

"Well, first you should know that it is impossible to get institutional financing for this kind of research. In France, as in most industrialized countries, there is no interest and thus no money for laboratories to conduct epidemiological studies or scientific reassessments of the toxicity of the chemical products that have invaded our daily life. But it seems to me that from a public health perspective there is real urgency, because our bodies have become veritable sponges for pollutants. One finds attached to the entire genome of human fetuses, as I have been able to observe, several hundred toxic substances, such as hydrocarbons, dioxins, pesticides, residues of plastic and of glue. These products, which were designed not to be soluble in water, accumulate and concentrate in our fatty tissue, and no one knows what their long-term effects may be. The problem is that the governmental authorities don't really want to know. They are prepared to finance a study to improve the straws used for in vitro fertilization of pigs, but not one on the toxic effects of the most widely sold herbicide in the world. In my case, I found private financing, notably from the Fondation pour une Terre Humaine, but how many young scientists will embark on that kind of venture knowing that they will antagonize their institutional sponsors?"

By the way, the day I went to Caen to interview and film Séralini, his laboratories were strangely empty: "None of my doctoral students wants to appear next to me," he explained. "They're afraid of being associated with my statements and putting their careers in danger." Welcome to the kingdom of independent science.

"The KGB atmosphere" that William Sanjour denounced at the EPA obviously extended beyond that U.S. government agency. Further evidence was provided by the reaction provoked in the French National Assembly by the publication of Séralini's article in *Environmental Health Perspectives* in 2005. It was harshly criticized by the rapporteur of a study commission on the challenges posed by the testing and use of GMOs, Christian Ménard, a doctor and a deputy from Finistère, in a report published in April 2005: "Referring more specifically to the toxic character of glyphosate and glyphosate-based products, the conclusions of the recent study conducted by Professor Gilles-Éric Séralini need to be qualified. . . . The procedure followed and the conclusions of this study are very controversial. . . . The very concept of endocrine disruptor is particularly vague, and the international scientific community is currently in agreement in finding a lack of experimental evidence to establish a causal link between certain molecules suspected of being endocrine disruptors and the appearance of effects in people."[30]

Though Ménard also "emphasizes . . . the necessity of conducting epidemiological studies, particularly on the use of herbicides, in order to establish comparisons among different herbicides," this is something like a snake swallowing its tail. Governmental authorities do not encourage laboratories to conduct studies of the toxic effects of suspect molecules, so there is in fact little experimental evidence, and when it exists it is challenged; this makes it possible to conclude that there's no problem.

Colombia's Agent Orange

In the meantime, thanks to the unwavering collusion between politicians, the giants of the chemical industry, and the international scientific community, the use of pesticides is increasing almost everywhere in the world. It is estimated that 2.5 million tons of phytosanitary products are sprayed every year on the planet's crops and that only "0.3 percent make contact with the target organisms, which means that 99.7 percent of the substances discharged go 'elsewhere,' into the environment, into the soil and water," says Julie Marc.[31] Hence, the contamination of rivers and water sources by the most widely used herbicide in the world might be behind the collapse of frog populations, as revealed by a 2005 study by Rick Relyea, a researcher at the University of Pittsburgh.[32] He observed the effects of two insecticides (Sevin and Malathion) and two herbicides (Roundup and 2,4-D) on a population of twenty-five animal species from a pond (snails, tadpoles, crustaceans, and insects), that were placed in four tanks containing water from their pond. In each tank, he added a dose of pesticide, following the concentrations recommended by the manufacturer. The results were spectacular: "We added Roundup, and the next day we looked in the tanks and there were dead tadpoles all over the bottom," said Relyea. "The most shocking insight coming out of this was that Roundup, something designed to kill plants, was extremely lethal to amphibians."[33] It should be noted that 2,4-D and the two insecticides produced no negative effects on tadpoles.

But animals are not the only ones to suffer from the consequences of pollution due to phytosanitary products. "The number of accidental poisonings by pesticides is estimated at more than a million per year around the world, 20,000 of which are fatal," according to Marc. "If one adds cases of suicide, the figure of 3 million poisonings is reached, 220,000 of which are fatal." In this dark picture, Roundup holds a choice position, because it is the favorite herbicide of would-be suicides by poisoning. According to a study conducted in Taiwan of 131 cases of suicide by swallowing Roundup, the majority of the victims had died after terrible suffering, evidenced by edema coupled with respiratory distress and violent vomiting and diarrhea.[34] A similar study conducted in Japan made it possible to identify the lethal dose of the herbicide for an adult human: approximately 200 milliliters, about three-quarters of a cup.

More generally, Roundup is the most common cause of complaints of poisoning registered in Great Britain and in California, as the news service *Pesticides News* reported in 1996.[35] Corroborating sources indicate that the symptoms of poisoning are always the same: eye irritation, vision problems, headaches, skin lesions and irritation, nausea, dry throat, asthma, respiratory difficulties, nosebleeds, and dizziness.

Writing these lines, I cannot help thinking of the torment suffered every day by the Indian and peasant communities of Colombia, subjected to what Washington strategists call "Plan Colombia." Developed in June 2000 with the active cooperation of the Bogotá government, this program is aimed at eradicating the coca plantations that supply the international cocaine market and are used, in part, to finance guerilla movements. The principal means of eradication is airborne spraying of Roundup. It is estimated that from 2000 to 2006, nearly 750,000 acres had been sprayed, mainly in the departments of Cauca, Nariño, and Putumayo (which extend to the border with Ecuador), whose populations have also been affected by what some call "Colombia's Agent Orange." In the department of Putumayo alone, which is the home of several Indian communities, 300,000 people have been poisoned.

The situation was so dramatic that in January 2002, a United States NGO, Earthjustice Legal Defense Fund, complained to the Human Rights Commission and the UN Economic and Social Council. In its statement, it drew up a list of all the harms it had been able to observe in the field: "gastrointestinal disorders (e.g. severe bleeding, nausea, and vomiting), testicular inflammation, high fevers, dizziness, respiratory ailments, skin rashes, and severe eye irritation. The spraying may also have caused birth defects and miscarriages."[36] Moreover, "the spraying has destroyed more than 1,500 hectares of legal food crops (e.g. yucca, corn, plantains, tomatoes, sugar cane, grass for livestock grazing) and fruit trees and has resulted in the death of livestock (e.g. cows, chickens). . . . In sum, the situation provides a clear example of the link between the environment and human rights—severe damage to the air, water, land and biodiversity caused by the spraying is violating various human rights."

The statement indicated that the herbicide used was Roundup Ultra, to which two surfactants had been added, Cosmo-Flux 411F and Cosmo-InD, whose function was to multiply the "effectiveness" of the product by four. In addition, the concentrations used in the mixtures prepared by the Colombian army under the guidance of North American colleagues were "more than five times greater than levels for aerial application recognized as safe by the US Environmental Protection Agency." Finally, "application methods used contradict the manufacturer's recommendations and exacerbate human and environmental harms. For example, the Roundup Ultra manufacturer recommends that aerial application not occur more than 10 feet above the top of the largest plants unless a greater height is required for aircraft safety. However, according to the Colombian antinarcotics police, the aerial herbicide application program flies aircraft 10–15 meters [approximately 25–30 feet] above the tops of the largest plants." This, of course, meant the spread of the herbicide over a large area.

What can be said in the face of yet another scandal profitable to Monsanto? Nothing except to point to the directions for use that now appear on containers of Roundup Ultra sold in the United States: "Roundup will kill almost any green plant that is actively growing. Roundup should not be applied to bodies of water such as ponds, lakes, or streams as Roundup can be harmful to certain aquatic organisms. After an area has been sprayed with Roundup, people and pets (such as cats and dogs) should stay out of the area until it is thoroughly dry. We recommend that grazing animals such as horses, cattle, sheep, goats, rabbits, tortoises, and fowl remain out of the treated area for two weeks. If Roundup is used to control undesirable plants around fruit or nut trees, or grapevines, allow twenty-one days before eating the fruits or nuts."

Extracted through the vigilance of North American consumer organizations, these warnings obviously did not apply to the poor peasants and Indians of Colombia. One might conclude, a little hastily, that Monsanto had learned the lessons of its inglorious past and that it was now more cautious when it came to the health of its fellow citizens. But, as the history of bovine growth hormone will show, it meant nothing of the kind.

5

The Bovine Growth Hormone Affair, Part One: The Food and Drug Administration Under the Influence

> Because the chemical composition of the milk is not altered as a result of POSILAC, the manufacturing and taste properties do not change.
>
> —Monsanto Web site

"This affair was a real descent into hell. . . . I had joined the Food and Drug Administration thinking I would be working for the good of my fellow citizens, and I discovered that the agency had betrayed its role as guardian of the public health to become the protector of the interests of big business." When I met him in New York on July 21, 2006, nearly twenty years after the events he was describing, Dr. Richard Burroughs still found it hard to talk about. "It's too painful," he said tensely every time I mentioned it. "It feels as though the ground is giving way beneath me all over again and I am going to be swallowed up. It's still hard today to recognize that I was fired from the FDA because I opposed the marketing of a product I considered dangerous. But that was my job."

Witnessing Burroughs's distress made me think of Cate Jenkins, William Sanjour, and all those who will appear later in this book: Shiv Chopra of Health Canada, Arpad Pusztai of the Rowett Institute, Ignacio Chapela of the University of California at Berkeley, and the journalists Jane Akre and Steve Wilson. All had the same choked voice as soon as they began to talk about their experience as whistle-blowers. The story of Richard Burroughs is a classic example.

Fired for "Incompetence"

Burroughs is a veterinarian with a degree from Cornell who started out with a private practice in New York, where his parents had a dairy herd. "I adore cows," he said with a smile that lit up his sixty-year-old face. "That's why I chose this profession." In 1979, he was hired by the FDA, which offered him training in toxicology. "I agreed to leave my hometown for Washington, because I felt it was essential." Like almost everyone, he believed that any product that had been approved in the United States would pose no problems. And the FDA was the agency that did the approving.

Officially named the FDA in 1930 (although it has existed under different names since early in the century), the agency is charged with approving the marketing and sale of food and pharmaceutical products intended for human or animal consumption. Its bible is the Food, Drug, and Cosmetic Act, signed by President Franklin Delano Roosevelt in 1938. This restrictive law, from which the FDA derives its authority, was intended as a response to a national tragedy. A year earlier one hundred people had died after taking Elixir Sulfanilamide, a medication made with a solvent that turned out to be fatal. The new law required that in the future any product containing new substances be tested by the manufacturer and submitted to the FDA for approval before being put on the market. In 1958, the "Delaney amendment" was added, providing that if a product presented the slightest carcinogenic risk, it could not be approved.* It is important to note that the agency itself does not conduct toxicological studies, such as animal testing, but merely examines the data supplied by manufacturers.

So it was that Burroughs, who was working at the FDA's Center for Veterinary Medicine (CVM), was given the task of analyzing the request to approve for sale a bovine growth hormone, bovine somatotropin (BST), manufactured by Monsanto through genetic manipulation and designed to be injected in cows twice a month to increase their milk production by at least 15 percent.† "For the CVM, it was a completely revolutionary prod-

*Named after a Democratic representative from New York, James Delaney (1901–87), who would certainly be spinning in his grave if he could read this.

†In the 1970s, three other companies manufactured the transgenic hormone: Elanco, a subsidiary of Eli Lilly; Upjohn; and American Cyanamid. But finally only Monsanto stayed in the running.

uct," Burroughs explained, "because it was the first transgenic medication that we had had to study."

Somatotropin is a natural hormone abundantly secreted by the pituitary glands of cows after calving, which stimulates lactation by enabling the animal's body to call on reserve energy through its action on the tissues. Ever since its function was described by Soviet scientists in 1936, laboratories tied to agribusiness had tried to reproduce it to increase herd yields. They had no success: twenty cows a day had to be sacrificed to produce from their pituitaries the daily dose of growth hormone required for a single animal. In the late 1970s, researchers financed by Monsanto succeeded in isolating the gene that produces the hormone. They used genetic manipulation to introduce it into a bacterium, *Escherichia coli* (*E. coli*, commonly found in the lower intestine of mammals, including humans), thereby making possible its large-scale manufacture. Monsanto named this transgenic hormone recombinant bovine somatotropin (rBST) or recombinant bovine growth hormone (rBGH).* In the early 1980s, the company organized tests on its own experimental farms or in collaboration with universities such as the University of Vermont and Cornell.

The file Monsanto provided "is as tall as I am," said Burroughs, who is five foot nine. "FDA rules require that we analyze data within ninety days. This is in fact a way for companies to discourage detailed examination: they send tons of paper, hoping that you will merely skim through it. I very quickly understood that the data were intended only to demonstrate that rBGH effectively boosted milk production. The scientists working for Monsanto had paid no attention to crucial questions: What did it mean physiologically for cows to produce milk beyond their natural capacity? How must they be fed so they would survive the exploit? What diseases might it cause? They hadn't even thought that the cows were certainly going to develop mastitis, an inflammation of the udders, a common pathology in herds with high production levels."

"And mastitis is also a problem for the consumer?"

"Of course, because it results in an increase of white blood cells, which means there's pus in the milk! The cows have to be treated with antibiotics,

*The two terms are used today, but not by the same people: anxious to erase the fact that its product is an artificial hormone, Monsanto speaks exclusively of "rBST," while opponents use "rBGH."

which can leave residues in the milk. So it's all very serious. In addition, you have to understand that the transgenic hormone upsets the cow's natural cycle. Normally, a cow produces somatotropin after calving, which enables it to feed its calf. As the calf grows, the secretion of the hormone slows and finally stops. To restart milk production, the cow therefore has to have another calf. rBGH makes it possible to artificially maintain milk production beyond the natural cycle. This is why it can create reproductive problems for the cow, and thus cause financial loss to the breeder. When I saw that all these data were lacking, I asked Monsanto to go back to the drawing board, which took two to three years, because for the study to be valid, it had to follow changes in cows through at least three cycles."

"And what were the results of the new studies?"

"First, I have to say they were of very poor scientific quality. For example, if you want to measure the impact of the transgenic hormone on mastitis, you have to select in each herd a group treated with the hormone and an untreated control group raised in strictly identical conditions. But Monsanto had dispersed the treated and untreated cows among different experimental sites and subsequently mixed all the results together. I was once again forced to make adjustments. I also remember an unannounced visit I made to one of their laboratories that was supposed to be analyzing the effect of the hormone on cows' tissues and organs: I discovered that kidneys had disappeared. In spite of all these technical defects, it came out clearly from the studies that the frequency of mastitis was much higher."

"Did you inform your superiors at the FDA?"

"Yes, and at first they reacted properly."

In fact, a document dated March 4, 1988, attests that Richard P. Lehmann, director of the Office of New Animal Drug Evaluation at the CVM, conveyed Burroughs's concerns to Terrence Harvey of Monsanto:* "We have completed our review and find your application incomplete. The data are incomplete and lack full reports of adequate tests. . . . You have not clearly identified the incidence of clinical and subclinical mastitis in the herds tested. . . . You should clarify your regimen of treatment for mastitis. . . . You should address the use of gentamicin and tetracycline which are not approved for the treatment of mastitis in dairy cattle. . . . You have

*Harvey had spent his entire career at the FDA, where he headed the CVM, before moving to Monsanto as director of regulatory affairs.

compromised the usefulness of your reproductive data by the use of prostaglandins and progesterone essays. It is not possible to evaluate the effect of bovine somatotropin on reproduction if concurrent use of reproductive hormones and diagnostic tests masks or otherwise alters the effect of the drug." Finally, with respect to the toxicological study conducted on rats, the CVM official was stinging; the rats studied were too few (seven), the duration of the study too short (seven days), and the doses ingested by the test animals too small.

This letter marked the beginning of Burroughs's descent into hell. "Suddenly I was put on the sidelines. I was denied access to the data I myself had asked for, and finally the file was completely taken away from me. And then, on November 3, 1989, my boss showed me the door; it was all over for me."

"You were fired?"

"Yes, for incompetence," he said quietly.

The veterinarian sued the FDA for unlawful dismissal and won at trial. The agency filed an appeal, but, according to Burroughs, it was finally obliged to rehire its employee. "I was shifted to the swine division. I didn't know anything about pigs! At any moment I could have made a serious mistake, so I decided to resign. It was a very dark period. I didn't understand what was happening to me. I was ruined, because the litigation had cost a lot and I didn't have a job. Fortunately, there were my wife and two children."

"Were you threatened?"

"You mean physically? I'd rather not talk about it. Psychologically, yes. During the appeal, Monsanto's lawyers threatened to go after me if I revealed confidential information about rBGH. That's typical of Monsanto."

"Do you think the FDA was deceived by Monsanto?"

"'Deceived' is not the right word; that would mean it happened without their knowledge. No, the agency closed its eyes to disturbing data, because it wanted to protect the company's interests by encouraging the marketing of the transgenic hormone as quickly as possible."

The Secret Data of Monsanto and the FDA

As Burroughs was battling for his career, a scientist known for his iconoclastic courage incidentally came upon a subject that was to become one of the fights of his life. Samuel Epstein, now emeritus professor of environmental medicine at the University of Illinois at Chicago, had authored many notable articles and books, particularly on cancer—the upsurge of which, he claimed, was linked to environmental pollution.* In the spring of 1989, he received a phone call from a farmer who had agreed to test rBGH on his herd.

"He got angry when he understood that I had never heard of the transgenic hormone," Epstein told me in his Chicago office on October 4, 2006. "He said to me: 'You ought to look at it, that's your job! This hormone is making my cows sick, and I'm afraid the people who drink my milk will get sick too.'" This led Professor Epstein to comb through the 1987 and 1988 issues of the *Journal of Dairy Science*, where he found many "promotional articles" published by American researchers, and Europeans as well, who had tested rBGH for Monsanto.† "All these publications claimed that the hormone posed no major health problems," Epstein recalled, "but there were very few serious data backing up that claim. The tests had been conducted on small groups of ten cows, which reduced their statistical validity, and most important, they had been of very short duration. Despite these biases, they revealed a significant increase in mastitis and a decrease in the fertility of the cows treated, as well as major changes in the nutritional quality and the composition of the milk."

Epstein then discovered that milk and meat from the American experimental herds had been placed in the food chain even though the hormone was not yet officially approved. On July 19, 1989, he wrote a letter to FDA commissioner Frank Young‡ expressing his concerns, which he revealed a short time later in an article published in the *Los Angeles Times*.[1] On August 11, 1989, the agency published a reply, signed by Gerald B. Guest, director of the CVM and Burroughs's boss: "The pivotal human food safety information led us to make the determination that food derived from BST-treated cows is safe," it said in typical bureaucratic prose. "Our scientists at

*In 1994, Professor Epstein established the Coalition for the Prevention of Cancer. He received the Right Livelihood Award (the "alternative Nobel Prize") in 1998, in 2000 the Project Censored Award, and in 2005 the Albert Schweitzer Golden Grand Medal for his "international contributions to cancer prevention."

†The hormone was tested in France at the Institut Technique de l'Élevage Bovin (located in Le Rheu, near Rennes) as well as at breeding centers of the Institut National de la Recherche Agronomique. (*Le Monde*, December 30, 1988, and August 30, 1990).

‡Frank Young held the post from August 1984 to December 1989. He was replaced by David Kessler (1990–97), who put in place the nonregulation of GMOs.

the FDA are among the most knowledgeable and capable in the world today regarding the evaluation of animal drugs. We are dedicated to the people we serve—consumers of meat, milk, and eggs, and farmers who may use the new animal drug—as well as the animals which may be treated."

Some weeks later, "toward the end of October," Epstein received a "gift from heaven": an unidentified person working at the FDA sent him a box containing all the veterinary data recorded on a Monsanto experimental farm near company headquarters in St. Louis. "It wasn't the first time that had happened to me," he said, smiling broadly. "Over the last thirty years, I've frequently received internal records from regulatory agencies or private companies sent by employees who preferred to remain anonymous for fear of reprisals. But this was really a godsend."

The cancer specialist immediately contacted Pete Hardin, editor of *The Milkweed*, a monthly devoted to milk production that was highly respected for its rigor and editorial independence. Together, they spent hours studying, comparing, and interpreting hundreds of pages filled with figures and raw data. "It was breathtaking to be able to work with original Monsanto documents," Hardin said, still enthusiastic when I met him at his home in Brooklyn, Wisconsin, on October 6, 2006. "Look, most of them are marked 'Company Confidential: this document belongs to the Monsanto Company. . . . It contains confidential information which may not be reproduced, revealed to unauthorized persons or sent outside the company without proper authorization.' This is exactly what an investigative journalist loves to publish!"

The two men together produced an article, published in *The Milkweed* in January 1990, that caused quite a stir.[2] It noted that the Monsanto study involved eighty-two cows followed through one lactation period, that is, forty weeks. The herd was divided into four groups: an untreated control group, a second injected every two weeks with a normal dose of the hormone, a third with three times the normal dose, and the fourth with five times the normal dose. At the end of the experiment, half the cows were slaughtered and their organs and tissue analyzed. The results were edifying:

- The organs and glands of the injected animals (thyroid, liver, heart, kidneys, ovaries, and so on) were much larger than in the control group, whereas the total weight of these cows when slaughtered was significantly lower. For example, the right ovary was on average 44 percent heavier in the group having received five times the dose than in the control group.

- The treated cows had significant reproductive problems: whereas 93 percent of the control group were successfully inseminated during the period, only 52 percent of the injected groups were.
- From one injected animal to another, the hormone level in the blood varied considerably, the highest being a thousand times higher than that recorded in the control group.

Oddly, the Monsanto scientists provided no statistical data on mastitis. On the other hand, it emerged from the documents that the injected cows had been treated with antibiotics much more often than those in the control group; some of the antibiotics used were not authorized by the FDA for use in dairy herds.* One unfortunate animal (no. 85704) had received 120 different treatments. Finally, the milk produced in the experimental groups had been sold to the St. Louis distribution network. No profits are too small.

"Look," said Hardin, taking a document from a thick file folder. "These are photos of cow carcasses after they were skinned. You can see blackish areas: that's dead tissue located at the injection sites. It's a very potent product. I remember reporting from a slaughterhouse where they were worried about making these unappetizing portions into hamburger meat."

A Manipulated Article in the Journal *Science*

Epstein did not let go of the issue. He contacted John Conyers, chairman of the House Government Operations Committee, who in 1979 had asked him to testify in connection with a bill on white-collar crime. "I had mentioned the case of Monsanto, which had hidden health data on nitrilotriacetic acid, which it was manufacturing as a replacement for phosphate-based detergents," he recalled. "The committee had finally defined two categories of company guilty of white-collar crime: those that deliberately concealed data, such as the carcinogenetic effects of a product, and continued to sell it as though nothing was the matter, and those that hid or destroyed information

*The following antibiotics were mentioned: Banamine, Ditrim, Gentamicin, Ivomec, Piperacillin, Rompun, and Vetislud.

and in addition continued to advertise the product as safe.[3] Monsanto fit into both categories: the first for PCBs, and the second for dioxin and rBGH."

On May 8, 1990, Conyers officially asked Richard Kusserow, inspector general of the Department of Health and Human Services, to open an investigation of bovine growth hormone, alleging that "Monsanto and the FDA suppressed and manipulated data from veterinary tests to secure approval for the commercial use of rBGH." The request led to an investigation by the General Accounting Office (GAO), the investigative arm of Congress. Among the witnesses testifying before the GAO were Richard Burroughs, whose case had recently been reported in the *New York Times*, and Samuel Epstein.[4]

But the FDA and Monsanto very quickly organized a counterattack. In August 1990, the agency decided to breach its statutorily mandated duty of confidentiality. For the first time in history, it took a public position in favor of a product it had not yet authorized by publishing an article in the prestigious journal *Science*, in which it asserted that milk from cows treated with rBGH was "safe for human consumption."[5] Officially, the article was written by two FDA scientists, Judith Juskevich and Greg Guyer, who were careful to note at the outset: "The FDA requires that the pharmaceutical companies demonstrate that food products from treated animals are safe for human consumption. . . . The companies also submit the raw data from all safety studies that will form the basis for approval of the product." Specifically, the authors referred to two toxicological studies conducted by Monsanto: in the first, rats received injections of the transgenic hormone over a period of twenty-eight days; in the second study, which lasted ninety days, the test animals ingested rBGH, in order to test its particular effects on the gastrointestinal system. In both cases, the same conclusion was reached: "No significant change."

"This publication was pure and simple manipulation," Dr. Michael Hansen told me when I met him in New York in July 2006. Hansen is an expert who works for the Consumer Policy Institute and who, along with Samuel Epstein, has become one of Monsanto's major gadflies.* "First," he explained,

*The Consumer Policy Institute is a division of Consumers Union, established in 1936, which publishes *Consumer Reports*, the second most widely circulated American consumer magazine (4.5 million subscribers).

"the main reviewer of the article was Professor Dale Baumann of Cornell University, who had been paid by Monsanto to test rBGH on cows. It is clear that this was a real conflict of interest that *Science* never should have allowed."

For a neophyte, the question of the reviewer may seem trivial, but it is essential. All reputable scientific journals operate in the same way. When a researcher submits an article for publication, the editorial board appoints a panel of reviewers (at least two) selected for their scientific expertise, who are asked to evaluate the quality of the paper. These unpaid reviewers may ask to consult the raw data on which the research relies if they think it necessary. If their opinion is positive, the editor gives the green light for publication. It is important to note that both the identity of the reviewers and the content of the article are kept confidential until the date of publication, to avoid pressure of any kind, a principle that, as will be seen, is not always respected. In any event, the label "peer-reviewed" constitutes a guarantee of quality and independence.*

"The second point," Hansen went on, "is that the FDA truncated the results of the ninety-day study. Contrary to what it claimed, the absorption of rBGH by rats did have a significant effect, since 20 to 30 percent of them produced antibodies, which means that their immune system had been mobilized to detect and neutralize pathogenic agents." Made public in 1998 thanks to disclosures in Canada that will be discussed in the next chapter, this information forced an agency representative, John Scheid of the CVM, to acknowledge that the agency had never had access to the study's raw data, but instead had relied on a summary provided by Monsanto.[6] "Those results should have induced further long-term studies of the effects of the growth hormone, and especially of IGF-1, on the quality and composition of milk from treated cows," said Hansen, "but the FDA preferred to shut its eyes."

IGF-1 (insulin-like growth factor 1, also known as tissue growth factor) is at the heart of the dispute about rBGH. This hormonal substance is produced by the liver under the stimulus of growth hormone in all mammals. In humans, it is present in large quantities in colostrum, making infant growth possible. Its production reaches a peak at puberty and declines with age.

*The week following the publication of the article in *Science*, Dr. Jean-Yves Nau, editor of the science page in *Le Monde*, devoted an article to it: "Les chercheurs de la FDA estiment que l'utilisation de cette hormone ne présente pas de danger pour le consommateur," *Le Monde*, August 30, 1990.

The IGF-1 produced by human growth hormone is identical to that produced by bovine growth hormone, whereas the two hormones are noticeably different. Therein lies the problem and also the origin of the scientific sleight of hand used by the promoters of rBGH to substantiate its harmlessness. Said another way, the pituitary of the cow and of the human each produce a specific growth hormone, both of which, however, induce the production of the same substance, IGF-1, the function of which is to stimulate the proliferation of cells, thereby causing organisms to grow. When, for example, Jean-Yves Nau writes in *Le Monde*, "This hormone is specific to the animal species from which it comes and thus can have no effect on human metabolism, whether it is in milk or meat that is consumed," he is mistaken.[7]

The "detail" is all the more important because there is one fact about which everyone is in agreement: the level of IGF-1 is distinctly higher in the milk of cows treated with transgenic growth hormone than in natural milk. According to the *Science* article, the increase can be as high as 75 percent![8] But the FDA added that "rBGH is biologically and orally inactive in humans" because it "cannot be absorbed in the blood" and when ingested "would be expected to be degraded in the human gastrointestinal tract in the same manner as other proteins." "This is completely false," according to Epstein and Hansen. "Several studies have confirmed that IGF-1 is not destroyed in digestion, because it is protected by casein, the principal protein in milk."[9]

Clearly fully aware of the stakes, the FDA scientists hazarded a final argument: "Furthermore, 90% of rBGH is destroyed upon pasteurization. . . . The use of rBGH in dairy cattle presents no increased health risk to consumers." "This is the height of bad faith," according to Hardin, who showed me the study on which this assertion is supposedly based. It was conducted by Paul Groenewegen, a Canadian doctoral student, on an assignment from Monsanto. He heated milk of transgenic origin to 162° Fahrenheit (approximately 90° Celsius) for thirty minutes. "The normal time for pasteurization is fifteen seconds," Hardin pointed out. "Milk pasteurized in these conditions no longer has any nutritional value, and yet 10 percent of the IGF-1 was not destroyed."

A Serious Public Health Problem

There should be no mistake: the squabble over IGF-1 was more than a battle of experts, particularly in the United States, the third-largest consumer of milk in the world.* When you know in addition that the greatest milk drinkers are children, then you can understand the concerns of the opponents of rBGH. Beyond that, this affair was symptomatic of an evolution of the American administration that leaves one breathless and whose repercussions also call into question European practices: how many dubious products have ended up on the market in the Old World through an approval process that is as opaque as it is expeditious?

The case of rBGH is simply devastating. "We have known for several decades," Epstein explained to me, "that an elevated level of IGF-1 in the body can cause a disorder called acromegaly or giantism. Those affected have a very short life expectancy and generally die of cancer when they're about thirty. There is nothing surprising about that: IGF-1 is a growth factor that stimulates the proliferation of all cells, good and bad. That's why rBGH represents a real public health danger. About sixty studies have demonstrated that an elevated level of IGF-1 substantially increases the risks for breast, colon, and prostate cancer."

He showed me the mass of publications carefully organized on the shelves of his library. The oldest studies were from the 1960s; the FDA could not have been unaware of them and ought, at a minimum, to have applied the precautionary principle required by the Delaney amendment. The most recent were published in the 1990s. One of them, conducted by a team of researchers at Harvard, followed a cohort of fifteen thousand men and concluded that an elevated level of IGF-1 in the blood multiplied the risk of prostate cancer by four.[10] Another study published in *The Lancet* revealed that premenopausal women younger than fifty with an elevated level of IGF-1 had seven times more likelihood of developing breast cancer than those with normal levels.[11]

"Look," Hardin told me, "I recently published two studies that confirm our concerns. The first was conducted by Paris Reidhead, who went through

*In 2004, annual consumption was 89.1 liters per person. If other dairy products are added (yogurt, ice cream, cheese, and so on), the figure rises to 270 liters per year.

the national statistics and found that the rate of breast cancer among American women older than fifty has increased by 55.3 percent between 1994, the year rBGH was put on the market, and 2002.[12] Similarly, a study conducted by Dr. Gary Steinmann of Albert Einstein Medical College in New York showed that American women who consume dairy products every day are five times more likely to give birth to twins than those who don't, and that the rate of twin pregnancies increased by 31.9 percent between 1992 and 2002. All of that is the work of IGF-1."[13]

"All right," I said, rather taken aback, "but the GAO conducted an investigation. What did that produce?"

"Not much. Without exaggerating, I would say that it capitulated in the face of lack of cooperation from the FDA and especially Monsanto."

In fact, at the initiative of the GAO, Congress asked the National Institutes of Health (NIH) to review the scientific findings on rBGH. From December 5 to December 7, 1990, the NIH held a special conference, which came to very cautious conclusions but nonetheless recommended that "more research be conducted" to determine the acute and chronic actions of IGF-1, if any, in the gastrointestinal tract."[14] Three months later, it was the turn of the American Medical Association (AMA) to publish an article: "Further studies will be required to determine whether ingestion of higher than normal concentrations of bovine insulin like growth factor is safe for children, adolescents, and adults," wrote the association's Council on Scientific Affairs.[15]

"The FDA and Monsanto completely disregarded these recommendations," Epstein said angrily. "That's why I say their attitude is criminal. Worse: with respect to the problem of residues of antibiotics, everything was done to obstruct the search for truth." Recall that one of Richard Burroughs's major worries had to do with the effect of the transgenic hormone on the frequency of mastitis. Mastitis is an infection treated with antibiotics, which pass into the milk in the form of residues. The milk drinker swallows these residues, which are in turn absorbed by the bacteria populating his or her intestinal flora. If you add to that the fact that this same milk drinker may be prescribed antibiotics for conditions such as colds, which are not caused by bacteria and thus cannot be treated by antibiotics, or for very minor bacterial infections, you can understand why many bacteria that were thought to have been conquered since Arthur Fleming discovered penicillin in 1928 have become resistant to antibiotics. The result, of course, has been new outbreaks of diseases that the medical community had thought were eradicated. Hence, as early as 1983, three hundred renowned scientists submitted a petition to the FDA asking that it ban the use of antibiotics in animal feeds.[16]

Since then, publications on the damaging effects of resistance to antibiotics have proliferated. In 1992, when the argument over rBGH was raging, two researchers wrote in *Science*: "After a century of decline in the United States, tuberculosis is increasing. . . . One third of the cases tested in a New York City survey in 1991 were resistant to one or more drugs."[17] In the same year, the Centers for Disease Control (CDC) found that 13,300 hospital patients in the country had died from infections caused by bacteria resistant to the antibiotics doctors had prescribed for them.[18]

Extreme Pressure

This is the reason the GAO took the mastitis problem very seriously.[19] Having learned of a study by the University of Vermont, which had been paid by Monsanto to test the transgenic hormone on forty-six cows, the GAO asked for the results, but according to them the scientists refused. Publicized by the Vermont Public Interest Research Group, a consumer and environmental advocacy organization, the affair caused a stir, forcing the FDA to break its silence. It was then discovered that 40 percent of the cows treated had required treatment for mastitis, compared to less than 10 percent in the control group.

At the same time, Monsanto was involved in a dispute with five British scientists, including Professor Erik Millstone of the University of Sussex, who, unlike others, did not give up. The episode is worth looking at in some detail because it clearly illustrates Monsanto's attitude toward independent research. The seriousness of mastitis is measured by what is called somatic cell count (SCC): a high number of leukocytes (white cells) in a cow's blood means that there is inflammation of the udder and therefore there is pus in the milk. It is also important to know that as the FDA was considering authorizing the marketing of rBGH, Monsanto had submitted a similar request to European countries. In this connection, the company had associated with twenty-six international research centers to conduct tests on the hormone.

On October 4, 1989, Millstone met Neil Craven, Monsanto's Brussels representative, who agreed to send him the raw data from the tests conducted in eight centers located in the United States, Holland, Great Britain, Germany, and France (where the test was conducted by the Institut Technique de l'Élevage Bovin). A week later, a letter—which would soon become the heart of the matter—specified the framework of the agreement: "We would be interested to hear your views on the data when you have had chance to assess it," Craven wrote. "As you know, we request that the raw data be kept confidential. We hope that you will discuss any interpretation of the data with us before disclosing it to third parties."

Millstone dissected the data sent by the eight international centers, which concerned 620 cows, 309 of which had been injected with the hormone. He discovered that a certain number of animals had been "prematurely withdrawn from the statistics," which of course distorted the results. For example, in Dardenne, Missouri, this was the case for cow number 321, found dead on March 28, 1986, and removed from the trial. Number 391 was withdrawn because of mastitis. In Arizona, number 4320 died of peritonitis; in Utah, number 5886 succumbed to lymphosarcoma; in Holland, number 701 was eliminated because of acute anemia caused by the rupture of blood vessels in the mammary glands; and so on. By doing a meta-analysis of the data, Millstone found that the SCC was on average 19 percent higher in treated cows than in the control groups. Knowing that the Veterinary Products Committee of the British Ministry of Agriculture, Fisheries, and Food was in the process of studying the request to authorize marketing of rBGH, he sent them a summary of his findings, emphasizing that "some of Monsanto's published figures did not coincide with those provided directly to us" and that "in commercial use, rBST might be responsible for a decrease in milk quality."

Then on December 5, 1991, he contacted Doug Hard, Monsanto's new Brussels representative, to ask him for authorization to publish an article in a scientific journal, and he attached a draft, in accordance with the prior agreement. "As the raw data are confidential, all subsequent analyses are as well." Hard answered a month later, although he was conciliatory: he conditioned eventual publication on the prior appearance of a paper by Monsanto consultants in the *Journal of Dairy Science*, whose publication was imminent, he said. Two years passed with no further news. Millstone then wrote to the *Journal of Dairy Science* to suggest joint publication of his paper and Monsanto's. He discovered that no article had ever been submitted by the company on rBGH and SCC.*

Worn out by these maneuvers, Millstone embarked on a long struggle that says a lot about the purported independence of scientific journals. He contacted the *Veterinary Record*, which agreed to publish his article provided he get consent from Monsanto. Erik Millstone then turned to the *British Food Journal*, which initially gave a green light without prior consent from Monsanto but eventually reversed itself because the company threatened to attack the journal for "plagiarism." "When you take someone else's data and you submit it without putting their names on it . . . it's called plagiarism," argued Robert Collier, Monsanto's point man for rBGH. Finally Millstone's article, co-signed by Eric Brunner of University College London, was published on October 20, 1994, in *Nature*, the British counterpart of *Science*, in which he revealed the whole affair.[20] The publication was accompanied by a note in which Millstone explained that Monsanto had no rights over the analyses conducted by his laboratory, only over the raw data, which he had kept confidential in accordance with his promise. It is interesting to note that in order to block the distribution of an article challenging its findings, Monsanto did not hesitate to brandish its intellectual property rights on data that raised questions about the health of consumers.

Welcome to the World of the Revolving Door

Back in the United States, on March 2, 1993, the GAO wrote to Donna Shalala, secretary of health and human services: "The approval of rBGH products should not be forthcoming until the antibiotic risk is validly assessed. The Department's response suggests that our recommendations have not been seriously addressed." And they never would be. On November 5, 1993, the FDA gave a green light to the marketing of Posilac, the trade name of rBGH. The only minor "restriction" was that the directions for use had to indicate that the product could produce twenty-two side effects in cows. Among them were reduced fertility, ovarian cysts and uterine disor-

*When the affair came to light, Monsanto finally published an article in the *Journal of Dairy Science* in the summer of 1994.

der, a reduction in gestation time and in the weight of calves, a higher rate of twins, an increase in mastitis and SCC, and abscesses of one to two inches and sometimes as much as four inches, at injection sites. Hardly of minor importance.

Michael Hansen of the Consumer Policy Institute explained that rBGH "is the most controversial product ever authorized by the FDA." "You have to understand that the transgenic hormone is not a drug designed to treat any cattle disease, but a product with a strictly economic purpose which has no benefit for either animals or consumers. The agency should therefore have required that it be totally harmless before approving its sale. Instead, it acknowledged that it might pose countless health problems by creating a new criterion, which violated the Food, Drug, and Cosmetic Act: 'manageable risk.'" An internal FDA document reveals that at a meeting on March 31, 1993, the CVM concluded that the risks the transgenic hormone posed for human and animal health were "manageable" and that the agency should therefore proceed to its approval. According to Hansen, "the agency surreptitiously changed its regulatory criteria to satisfy the needs of Monsanto, which had been able to maneuver very cleverly by placing some of its representatives in key positions in the agency." This was a perfect illustration of the revolving door, the hiring of private industry employees by government agencies and vice versa. Later on, I will discuss more thoroughly this national pastime, at which Monsanto is unquestionably a master, even considering only the case of rBGH. For example, it turns out that one of the hidden authors of the controversial article published by the FDA in *Science* was Susan Sechen, a former student of Dale Baumann (the principal reviewer of the article), who, you will recall, had been paid by Monsanto to test the transgenic hormone at Cornell University. After writing her dissertation on rBGH, Sechen had been hired by the CVM to evaluate the data provided by the company. Her supervisor was Margaret Miller, who had worked for Monsanto from 1985 to 1989 before becoming assistant to Dr. Robert Livingston, head of the CVM's Office of New Animal Drug Evaluation.

Miller's presence in such a strategic position had in fact created some controversy. On March 16, 1994—the date Posilac came on the market—CVM employees wrote an anonymous letter to David Kessler, the FDA commissioner, with copies to the GAO and to Consumers Union: "We are afraid to speak openly about the situation because of retribution from our director, Dr. Robert Livingston, who openly harasses anyone who states an opinion in opposition to his," the whistle-blowers wrote. "The basis of our concern is that Dr. Livingston had Dr. Miller write a policy on use of antimicrobials in milk. She picked an arbitrary and scientifically unsupported number of 1 ppm as being the allowable amount of antimicrobial in milk permitted without any consumer safety testing. This is for an antimicrobial. A cow could be treated with several antibiotics and each one would be permitted to be in milk at a level of 1 ppm without additional consumer safety testing. Effects of the different antibiotics could be addititive and this is not taken into account."

"As soon as we learned of this letter, we began to hope again," recounted Jeremy Rifkin, the media-savvy president of the Foundation on Economic Trends, whom I met in his Bethesda office in July 2006. Author of the best-seller *The Biotech Century*, this economist was unquestionably the first American intellectual to recognize what was at stake with rBGH, the initial GMO product put on the market by Monsanto.[21] He launched a national campaign in February 1994, which he called the "Pure Food Campaign." In television archives he can be seen pouring containers of milk into the gutters of New York as a young activist accosts passers-by. "Transgenic growth hormone is a test to persuade us to accept GMOs," she shouts through a megaphone, brandishing a sign reading "No to transgenic milk!" Relying on the anonymous letter from the CVM whistle-blowers, Jeremy Rifkin managed to convince three members of Congress to ask the GAO to open an investigation. The investigative arm of Congress, which had just pathetically buried its first investigation of the health risks of rBGH, opened a second one, this time on a possible conflict of interest affecting FDA handling of the question.* Under scrutiny were Susan Sechen, Margaret Miller, and a man named Michael Taylor.

Taylor perfectly embodies the revolving door system and, beyond that, the links between Monsanto and U.S. regulatory agencies. According to his résumé, this lawyer, born in 1949, worked first at the FDA (from 1976 to 1980), where he helped draft documents concerning food safety for the *Federal Register*. In 1981 he joined the prestigious firm of King and Spalding in Atlanta, whose clients included Coca-Cola and Monsanto. On July 17, 1991, he was appointed deputy commissioner for policy of the FDA, a posi-

*On April 15, 1994, the three members of Congress wrote to the GAO explaining that the first investigation failed "because of the refusal by the Monsanto company to communicate all available clinical data on rBGH."

tion tailor-made for him. He remained there for three years, long enough to supervise the drafting of basic documents concerning the regulation of rBGH and beyond that of GMOs, then spending a brief period at the Department of Agriculture, and was then hired in the late 1990s as a vice president of Monsanto.

When I finally tracked him down in July 2006, he was a senior fellow at Resources for the Future (RFF) and director of its Center for Risk Management. RFF is a "nonpartisan organization" based in Washington that "conducts independent research . . . on environmental, energy, and natural resource issues." Taylor never agreed to meet me, with or without a camera. But, oddly, he granted me a telephone interview, and I was able to record our conversation. I remember that, in a fit of paranoia, I assumed that he had done the same thing. At the time I was waiting for an answer from Monsanto about interviews in St. Louis, and I knew that the company was investigating me, conscientiously weighing the pros and cons.

"The fact that you worked for seven years as an attorney for Monsanto and that you later supervised the approval process for one of its most controversial products never raised an ethical issue for you?" I asked cautiously.

"No, no. There are rules and I respected them."

"You don't think there was a conflict of interest?"

"Absolutely not. Besides, the GAO conducted a very detailed investigation and it totally cleared me."

In fact, much to the dismay of Jeremy Rifkin, the GAO investigation found no conflict of interest. "Welcome to Washington!" Rifkin said ironically. "At their hearing, Michael Taylor and Margaret Miller claimed that they had voluntarily withdrawn from all meetings having to do with rBGH. So move on, there's nothing here."

"The GAO investigation was a charade," according to Samuel Epstein. "How could they accept that alibi at face value when it was Michael Taylor who signed an FDA directive recommending not labeling natural milk as 'rBGH free' or 'hormone free'? You understand what that means? The agency in charge of food safety published an absolutely unprecedented document that prevented consumers from choosing the milk they wanted to drink and, most important, allowed Monsanto to sue all the sellers of dairy products that publicly rejected milk with hormones. Don't you think the country has lost its mind?"[22]

6

The Bovine Growth Hormone Affair, Part Two: The Art of Silencing Dissenting Voices

Posilac is the single most tested new product in history.

—Monsanto promotional film

Samuel Epstein was right: the further you go in this unbelievable story, the more you have to pinch yourself to make sure you're not dreaming. It feels as though it's straight out of a science fiction novel. On November 5, 1993, the FDA granted authorization to put Posilac on the market. Ninety days later, at the expiration of the legal waiting period, precisely on February 4, 1994, Federal Express trucks traveled to every corner of the American countryside delivering the first doses of the transgenic hormone. In fact—and this is another peculiarity—rBGH is not purchased in veterinary pharmacies but is ordered directly from Monsanto on a toll-free number.

Labeling Prohibited under Threat of Legal Action

Six days later, the *Federal Register* published a document entitled "Interim Guidance on the Voluntary Labeling of Milk and Milk Products from Cows That Have Not Been Treated with Recombinant Bovine Somatotropin," the purpose of which was to "prevent false or misleading claims regarding rBST."[1] It is one thing to state, as the document does in the first part, that "the agency found that there was no significant difference between milk from treated and untreated cows, and therefore, concluded that . . . [it] did not have the authority . . . to require special labeling for milk from rBST-treated cows." It is another for the FDA not to require milk producers who use the transgenic growth hormone to inform cooperatives or distributors of dairy products, meaning that milk from cows injected with rBGH will be mixed with natural milk, with no particular notice. And what about people who are absolutely determined to drink natural milk? Well, their suppliers will not have the right to use the simple label "rBST free." The argument presented by the FDA is rather surprising: "Because of the presence of natural BST in milk, no milk is 'BST-free,' and a 'BST-free' labeling statement would be false. Also, FDA is concerned that the term 'rBST free' may imply . . . that milk from untreated cows is safer or of higher quality than milk from treated cows. Such an implication would be false and misleading."

To be sure, this guidance had no legal force and the agency did not formally prohibit the label "rBST free," but it strongly suggested that it should be accompanied by a short statement intended to "inform the consumer," which it called a "contextual statement": "No significant difference has been shown between milk derived from rBST-treated and non-rBST-treated cows."

And who signed the guidance? Michael Taylor.[2] "Of course I'm the one who signed the document—it was my role to sign all the documents the FDA published—but I didn't write it," Taylor told me over the phone, sounding embarrassed. "And why come back to this old story fifteen years later?" Why? Because it sheds light on the way GMOs would in the end be imposed on the entire planet under the influence of a multinational company that had planned everything with implacable logic. That is why I was interested in small details—because the company had left nothing to chance.

To be precise, it is true that Michael Taylor himself had not written the guidance. And that is understandable: as number two at the FDA, he had other things to do. As he acknowledged in the course of our conversation, his task was to "supervise the regulatory process." The person who drafted the document was Margaret Miller, the former Monsanto employee who had become a deputy director in the CVM. This is what the CVM whistle-blowers claimed in their 1994 anonymous letter: "The basis of our concern is that Dr. Margaret Miller, Dr. Livingston's assistant and, from all indications, extremely 'close friend,' wrote the FDA's opinion on why milk from BST treated cows should not be labeled. However, before coming to FDA, Dr. Miller was working for the Monsanto company as a researcher on BST.

At the time she wrote the FDA opinion on labeling, she was still publishing papers with Monsanto scientists on BST. It appears to us that this is a direct conflict of interest. As you know, if milk is labeled as being from BST-treated cows, consumers will not buy it and Monsanto stands to lose a great deal of money."

Although Michael Taylor did not draft the guidance, his former law firm inspired its content. The guidance apparently drew on a confidential document sent to the FDA on April 28, 1993, by King and Spalding. Recall that Michael Taylor had served for seven years as counsel for Monsanto, working, says his CV, on food labeling, particularly of transgenic origin. Entitled "Mandatory Labeling of Milk and Other Foods Derived from Dairy Cows Supplemented with Bovine Somatotropin Would Be Unlawful and Unwise," the memorandum from King and Spalding, which was "submitted on behalf of Monsanto Company for the Food and Drug Administration," presents arguments quoted directly in the FDA guidance: "In addition to being unlawful, such a requirement would be unwise. Consumers would be misled into believing that there is some difference between milk and other foods derived from BST-supplemented dairy cattle and foods from untreated animals. In fact, there is no significant difference."[3]

"This FDA guidance takes the cake," said Michael Hansen, the Consumer Policy Institute expert who sent a detailed critique to the agency on March 14, 1994. "First—and the FDA knew this very well—milk from treated cows is not identical to natural milk; second, it has long since authorized labels such as 'organic product,' 'cheese from Wisconsin,' 'produced by Amish,' or 'Angus beef,' and it never thought that might mislead consumers by implying a difference in terms of quality or food safety. Why would it be different for milk labeled 'rBST free'? Once again, the document was tailor made for Monsanto, which knew very well that if milk was labeled, consumers would do everything to avoid products from the transgenic hormone." He referred to eleven surveys conducted in the 1990s, all of which confirmed that the vast majority of consumers preferred to buy milk without rBGH if they had the choice.*

In the meantime, the guidance had thoroughly benefited Monsanto,

*These surveys were conducted by the Department of Agriculture, Cornell University, the University of Wisconsin, *Dairy Today*, etc.

which brandished it to sue anyone who dared to use the label "rBGH free." The first victim, in 1994, was Swiss Valley Farms, a dairy cooperative in Davenport, Iowa, which informed its 2,500 members that it would not buy their milk if they used rBGH. "If things like that were repeated, it would cause irreparable harm to Monsanto," company spokesman Tom McDermott said in justifying the suit.[4] The case ended in a settlement authorizing the cooperative to label its milk, provided that it add the brief "contextual statement" highly "recommended" by the FDA guidance: "The FDA has found no significant difference between milk from cows treated with rBST and non-treated cows." "All dairy professionals are terrified," a director of a cooperative in the Northeast said shortly afterward, demanding anonymity for fear of reprisals.[5]

In 2003, it was the turn of Oakhurst Dairy Inc., the largest dairy company in northern New England, to find itself in court. This family business had sharply increased sales ($85 million) by labeling its products with the statement: "Our Farmers' Pledge: No Artificial Growth Hormones Used." In return, it paid its producers a bonus. Monsanto sued on the grounds that the label constituted a "disparagement of the use of growth hormones in dairy herds." "We don't feel we need to remove that label," declared Stanley T. Bennett, president of Oakhurst Dairy. "We ought to have the right to let people know what is and is not in our milk."[6] Like its Davenport counterpart, the company, however, settled by adding the celebrated brief statement.[7]

In February 2005, Tillamook County Creamery Association, one of America's largest cheese producers, was the target of Monsanto's thunderbolts. In the face of growing demand from its customers to supply natural milk, the dairy cooperative had asked its 147 members to stop using the transgenic hormone. Monsanto had immediately dispatched a lawyer from King and Spalding to Portland, Oregon, to persuade some of the members of the board of directors to reconsider the decision. In a press release, the cooperative expressed surprise at these "intrusive tactics" aimed at "sowing dissension " among its members.[8]

It was indeed hard to see why the company should have abstained from such practices, since it could boast of always having received unfailing support from the FDA. Evidence of this is provided by a letter from Dr. Leslie Crawford, deputy administrator of the agency, sent in 2003 to Brian Lowery—long in charge of rBST matters at Monsanto, and later director of the company's human rights policy—which the International Dairy Food Association (IDFA), a powerful pro-rBGH dairy lobbying group,* hastened to put on its Web site: "You stated that these deceptive practices mislead consumers about the quality, safety, or value of milk and milk products from cows supplemented with rBST. . . . We share your concerns. . . . FDA . . . is in the process of exploring ways to document current labeling practices for certain milk and milk products to determine if these products are labeled in a manner that is false and misleading."

Illegal Propaganda

"Look to your left—that's one of the largest dairy farms in the region, and it is certain that it uses the transgenic growth hormone," said John Peck, executive director of Family Farm Defenders. "If you want to film it, be discreet—you never know." The young farmer carefully pulled over to the side of the road. We had picked up his anxiety, and we did the three shots as quickly as possible. In front of us was a huge dairy establishment holding several hundred cows penned in straight rows. The animals never went outside and were fed entirely with food supplements—genetically modified (GM) soy and meal. Dark-skinned workers moved around the site. "They're undocumented Mexicans," Peck explained. "This kind of business operates like a factory that employs a low-paid workforce that is easily exploitable."

It was October 2006, and we were in Wisconsin, long the largest dairy producer in the United States until it was outstripped by California, where farms like the one in front of us had proliferated in the last ten years, thanks to rBGH. "Today," Peck said, "Wisconsin farms have an average of 50 cows compared to 400 in California, but we are the largest producer of organic milk."

We got back on the road and drove through hilly green country dotted with tidy farms, many displaying the sign "Amish Products." Wisconsin harbors the fourth-largest Amish community in the country, who continue to abide by the rules set by the old order, unchanged since the sect from Switzerland settled in the United States in the late seventeenth century:

*On the site www.idfa.org, one finds the following: "Monsanto is a supplier of agricultural products that increase farm productivity and food quality. The company manufactures and markets Posilac, a technology that has demonstrated its profitability by enabling dairy farmers to produce 8 to 12 more gallons per cow per day."

beards for the men and bonnets for the women, all wearing traditional clothing, and the rejection of all techniques arising from "progress," starting with electricity. The Amish light their buildings with candles, travel in horse-drawn buggies, and work their land with teams of oxen.

"Amish agricultural products are having great success these days, because they are necessarily organic," according to John Kinsman, president of Family Farm Defenders, whom we had just met at his home. "They sell their milk directly from the farm, which allows them to avoid attacks from Monsanto." Around sixty and very talkative, Kinsman is one of the leading opponents of the transgenic hormone. He became active against it very early on, first for economic and social reasons. "rBGH is a real aberration," he said, laying out a thick file on his kitchen table. "When Monsanto submitted it to the FDA, the American government was paying farmers to slaughter their cows, because we had been overproducing milk for a quarter century." In fact, in 1985, to deal with milk surpluses that were annually costing the federal budget the tidy sum of $2 billion, Congress passed the Food Security Act, intended to reduce the cost of the price support program by reducing the number of dairy farms. Some fourteen thousand farmers agreed to accept subsidies to send more than 1.5 million cows to the slaughterhouse (this program cost $1.8 billion). "Growth hormone is part of the system of industrial agriculture which drives toward the concentration of production and consequently the disappearance of many agricultural units unable to handle the expenses incurred by the intensive farming model," Kinsman explained. "We think this model is contrary to sustainable development and the production of quality food, which only family and organic farming can provide."

We went back on the road and met a farmer who had used rBGH for a while but gave it up because of serious veterinary problems. "It's very hard to find a farmer who will testify about his difficulties," Kinsman commented. "First, because most of them are ashamed of having subjected their animals to such abuse and at the same time threatening the health of their customers; and then, to be able to get the hormone, you have to sign a contract that includes a confidentiality clause in case there is a problem. I've met farmers who have been sued by Monsanto because they had spoken publicly," Kinsman said.

The farmer we met was named Terry, and he had a herd of about forty Holsteins that were peacefully grazing not far from his house. The black-and-white animals were guarded by two Peruvian llamas. "They're excellent, better than dogs," he explained, amused by my surprise. Then, suddenly turning serious, he warned at the outset: "I've agreed to meet with you, because John persuaded me that you were on the level and someone has to speak out so companies like Monsanto will stop spreading their control over agriculture in this country. My story is unfortunately quite common. One day, in 1992 or 1993, my veterinarian told me about a miracle drug that was about to come on the market and that, he said, would considerably increase my revenues. Since we're in a business where we're often financially on edge, I agreed to try it as soon as it was available."

"Did veterinarians commonly promote rBGH?" I asked, a bit surprised.

"Yes," said Kinsman. "Monsanto was constantly propagandizing for its product even before it was approved. The company offered a $300 bonus to every veterinarian who persuaded a farmer to use it. It also organized promotional banquets in all the dairy states, at which it distributed a video touting the virtues of the growth hormone."

I saw this for myself when I got a copy of the video produced by Monsanto, showing a gentleman with a very professorial air strolling around a dairy farm and talking about the advantages of rBGH, "the most studied product in the history of the FDA." "The drug has been tested for years, and it works," he asserts, as a man nearby injects a row of astonishingly docile cows. Monsanto started distributing this video to farmers in the late 1980s, provoking the anger of the FDA. On January 9, 1991, Gerald B. Guest, director of the CVM, sent a letter to David Kowalczyk of Monsanto: "Over the past several years your company has developed a large number of items, including brochures, video tapes and sponsored meetings, which in part promote bovine somatotropin (BST) as being safe and/or effective for increasing milk production in dairy cattle even though all BST products are still under investigation to determine whether they can be legally marketed in the United States. [Federal regulations do] not permit the sponsor, or anyone acting on behalf of the sponsor, to represent that a new animal drug is safe or effective for the purposes for which it is under investigation." The letter goes on to note that among the events were "cocktails and dinners" organized for veterinarians and "CVM personnel," who always refused to participate. Honor was preserved. Meanwhile, the FDA official politely ordered Monsanto to halt these illegal promotional practices or face penalties.

Massacre on the Farm

"Look," Kinsman said to me, "I kept a promotional leaflet from Monsanto singing the praises of Posilac." He read: "Cows treated with Posilac are in very good health. . . . The performance of calves born from treated cows is excellent."

"That's a lie," Terry protested. "I used the growth hormone on twelve cows in my herd. I very soon noticed that they were losing a huge amount of weight. I kept on increasing the feed rations, but nothing could be done and they grew thinner right before your eyes. At the end of the lactation period, I wanted to have them inseminated, and I tried four or five times, but it never worked. None of the cows I had injected gave me a calf. In the end, I sold them for slaughter. Fortunately, I saved the rest of the herd, or else I would have lost everything."

"That's what happened to many farmers in Wisconsin," said Kinsman, who referred me to a 1995 study by Mark Kastel, an independent consultant working at the time for the Wisconsin Farmers Union.[9] In late summer 1994—that is, six months after Posilac had been put on the market—the farmers' organization, in cooperation with the National Farmers Union, based in Denver, set up a toll-free number for users of the hormone. The first farmer who would allow his name to be used was John Shumway of New York, who had given an interview about his difficulties to a local weekly.[10] After barely two months of injections, he had had to sacrifice one-quarter of his herd, about fifty cows, because of acute problems with mastitis. Recontacted a year later, in September 1995, Shumway said that he had replaced 135 of his original herd of 200, and that his losses came to about $100,000, from a combination of the decline in milk production and the purchase of new animals.

The toll-free number was soon swamped with calls from dairy farmers around the United States. For example, Melvin Van Heel—70 cows in Minnesota—reported that he no longer knew how to treat his animals, which were suffering from mastitis and huge abscesses at injection sites. Al Core—150 cows in Florida—noticed that his cows could no longer walk because of the great weight of their udders and that they limped because of wounds on their legs and hooves; in addition, three treated cows had given birth to deformed calves (legs above their heads or external stomachs); Jay Livingston—

200 cows in New York State—reported that he had had to replace 50 animals, some of which had died suddenly, and that after stopping injections, he had had the rest of the herd inseminated: 35 cows gave birth to twins, most of them with very weak constitutions, "good for absolutely nothing."

Reading this apocalyptic report, I recalled the emotional reaction of Richard Burroughs, the veterinarian fired by the FDA. "It's terrible what they're doing to cows," he had said. "To be able to turn themselves into milk factories, they are forced to draw constantly on their reserves, which weakens their bones. Encumbered with monstrous udders, they limp and can hardly stay standing."

All the farmers in Mark Kastel's survey had sent reports to Monsanto, as provided in the contract they had signed, but the company had not responded. Worse, although Monsanto was legally obligated to report the secondary effects that its product caused in the field, it had, according to Kastel, improperly delayed transmitting some of the reports to the FDA. And in any event, what good would that have done? On March 15, 1995, although he was inundated with alarming reports, Stephen Sundlof, the new director of the CVM, coolly noted: "Based on these reports, the FDA does not find any cause for concern."[11]

Today, whereas major food distribution companies are trying to obtain milk that if not organic is at least natural, to satisfy increasing consumer demand, no official assessment has ever been made of the use of the transgenic hormone.* "The FDA has kept its head in the sand," said Kinsman, "but inadvertently, its irresponsible conduct has in fact fostered the growth of organic farming. By attempting at all costs to avoid milk from cows treated with rBGH, consumers have fallen back on organic milk producers, and as a result they've started to wonder about the quality of their food. I don't think any official decision will ever ban the use of the hormone, but in the long run consumers will make it disappear from our farms. And that will be a massacre."

"Why a massacre?" I was taken aback.

"Because rBGH is a real drug," said the experienced activist. "When cows stop being injected, they experience withdrawal and they literally collapse. It's been called 'crack for cows.' The day when large dairy farms are forced to

*On June 5, 2006, *Dairy and Food Market Analyst* reported that chains such as Dean Foods, Wal-Mart, and Kroger, although not very inclined to support organic farming, were promising to sell only milk that was rBST free.

stop injections, because no one wants their milk, they will have to send their herds for slaughter, which will represent, according to our estimates, one-third of the dairy cows in the country."

"That's alarming, but how did such madness happen?"

"The power of blind money. Monsanto was able to rely on a real war machine to silence all dissonant voices."

Lobbying and Control of the Press

"This is just a note to say thanks for the efforts of you (and the Animal Health Institute) to keep me and the AMA generally informed on the progress of BST and the public climate for it. The amount of advance communicating Monsanto and the other BST companies have been doing on this product is impressive. . . . I see no reason for the medical community to be anything but comfortable with the safety of this product for people and milk. Let's stay in touch on BST's progress and continued good luck at Monsanto." This smooth letter was sent on June 30, 1989, by Roy Schwartz, a vice president of the American Medical Association, to Dr. Virginia Meldon, vice president of Monsanto for scientific affairs. It is a perfect illustration of the "war machine," composed of charm and influence, that Monsanto put together to nip in the bud any criticism of its products.

Established in 1847, the AMA counts 250,000 doctors as members, one-third of all practitioners in the country. In support of "Helping Doctors Help Patients"—its official motto—it publishes the *Journal of the American Medical Association*, the most widely read medical journal in the world. "The AMA has always argued in favor of rBGH," Samuel Epstein explained, "just like the American Cancer Society* and the American Dietetic Association, which is one of the alibis of the Dairy Coalition, a powerful dairy lobby, set up as if by chance in 1993, at the time the FDA approved Posilac, which brings together representatives of the dairy industry, large food distribution chains, the association of Agriculture Secretaries of the fifty states, scientists sponsored by Monsanto, et cetera." Epstein continued, "Relying on these networks, the Dairy Coalition flooded the press with deceptive infor-

*It was precisely to denounce the collusion between the American Cancer Society and the pharmaceutical multinationals that Epstein established the Coalition against Cancer.

mation about rBGH and organized defamation campaigns against anyone who, like me, never stopped warning about the dangers of the transgenic hormone."

"What about the press?" I asked. "How did it behave?"

"Ah, the press. It didn't get involved very much, either because it didn't understand anything about this transgenic hormone business or because it was blinded by the aura of respectability around the FDA. Indeed, how could anyone imagine that the agency would betray its duty on this issue? Finally, the few reporters who really did their work were severely punished, like, for example, Jane Akre and Steve Wilson."

Now a symbol of press censorship in the United States, this husband-and-wife team of reporters was hired on November 18, 1996, by Channel 13, WTVT, in Tampa, which was owned by New World Communications, to work on an investigative magazine program launched with much fanfare. "The investigators. They uncover the truth! They protect you!" promised the repeatedly broadcast promo. Jane Akre and Steve Wilson are well-known investigative reporters who between them have won a number of prestigious awards, including three Emmys and a National Press Club award.[12]

"We were delighted to be able to work together on a magazine that gave us carte blanche to investigate subjects of our choosing," Akre explained in July 2006 in their Jacksonville, Florida, home. "The first subject we proposed had to do with rBGH, because we had heard about the disputes surrounding the product. I was in charge of the investigative reporting and Steve of production. I will always remember the first report I did. I'd managed to film a farmer as he was injecting his cows; they shook violently every time the nine-inch needle was plunged in their flanks." Jane showed me the pictures, copies of which she kept in a box in her basement: they showed the farmer squeezing the huge udder of a cow; a thick, brownish liquid spurts into his hand. "You see these little lumps?" he says, extending his palm toward the camera. "That's what you call mastitis." A few minutes later, a long tracking shot sweeps along a shelf piled with all kinds of antibiotics.

Jane Akre filmed for a month. She met defenders of the transgenic hormone, such as a scientist from the University of Florida, and Robert Collier, a Monsanto representative, but also opponents such as Samuel Epstein and Michael Hansen. She interviewed the representative of a small dairy sued by the company for having labeled its milk "BST free." But the FDA refused to grant her an interview. "At the time, I was still very naive," she said with a smile, "and this refusal surprised me, so convinced was I that the agency must have had good reasons for approving a drug that seemed very dangerous, so much so that Steve and I decided to give only organic milk to our daughter Alix."

In the meantime, New World Communications of Tampa, and along with it Channel 13, had been bought by Fox News, owned by Rupert Murdoch, the Australian American press baron noted for his very commercial and conservative conception of journalism.

When the editing was completed, the couple showed the report to Daniel Webster, the news director, who was enthusiastic. He decided to broadcast it in four parts and to promote it with an expensive radio advertising campaign. The first broadcast was set for Monday, February 24, 1997, in prime time.

"The Friday before the broadcast we were called into Daniel Webster's office, and he handed us a letter that had been faxed," Akre said. "It was signed by John Walsh, a partner in the prominent New York law firm Cadwalader Wickersham & Taft, and addressed to Roger Ailes, the CEO of Fox News": "I write to bring to your attention a situation of great concern to Monsanto involving your recently acquired, owned and operated station in Tampa, Florida," the letter began, even though Walsh had never seen the report. "Serious questions arise about [the] objectivity [of your reporters] and [their] capacity for reporting on this highly complex scientific subject. . . . The fact is that every scientific, medical, or regulatory body in the world which has reviewed and approved this product has come to the same conclusion: milk from rBST treated cows poses no risk to human health. . . . There is a lot at stake in what is going on in Florida, not only for Monsanto, but also for Fox News and its owner, as well as for the American people and a world population that can benefit significantly from the use of rBST and other products of agricultural biotechnology." Then, aware of the recipient's sore points, Monsanto's lawyer pointed out that the conduct of the two Tampa journalists was all the more regrettable because it was occurring "shortly after the verdict in the Food Lion case." The subtext: *Be careful, because the same thing could happen to you.**

*In 1992, the program *Prime Time* on ABC News had broadcast a report showing employees of the Food Lion chain, filmed on hidden camera, mixing ground beef past its expiration date with fresh meat. Following the broadcast, the price of Food Lion shares had collapsed, and almost one hundred stores had been forced to close. The company sued ABC News and at trial won $5.5 million in damages. The verdict had caused serious worries in the country's newsrooms (the damages were reduced on appeal to $2).

Bob Franklin, the general manager of the station, asked to see the report: "He found it very good," Jane Akre recalled, "and together we agreed to offer Monsanto another interview. The company asked us to send the list of questions beforehand, which we did, but finally it refused to meet with me."

A few days later, another letter reached Fox News headquarters. This time the tone was openly threatening: "I find it nothing short of amazing that one week after my detailed letter to you about the concerns of my client Monsanto Company . . . I should be writing to you again to advise you that the situation has not improved, but clearly worsened in terms of the irresponsible approach being taken by WTVT's news correspondent Jane Akre," the lawyer wrote. He inveighed against the eight questions submitted by the reporter, particularly one on "crack for cows." "Indeed some of the points clearly contain the elements of defamatory statements," he continued, "which, if repeated in a broadcast, could lead to serious damage to Monsanto and dire consequences for Fox News."

"What could Fox News be afraid of?" I asked after carefully reading the two letters.

"Of losing advertising," Jane answered. "Monsanto is a major advertiser, particularly for Roundup and NutraSweet, its two leading products, which represents a substantial budget."

"That's how you and Steve became whistle-blowers?"

"Yes. We never would have imagined going through an experience like that in a country that prides itself on being the greatest democracy in the world."

Open warfare had now been declared. In Tampa, it was conducted by Dave Boylan, who had just been appointed general manager of Channel 13 after it was purchased by Fox. He asked the two journalists to start from scratch and prepare a new version of the report, which had been canceled until further notice. "We rewrote the script eighty-three times!" Jane said with some amusement, adding that she had kept drafts of every version. "But it was never suitable. For example, we couldn't use the word 'carcinogenic,' but had to replace it with 'possible health implications.' Or we had to minimize the scientific competence of Dr. Samuel Epstein, and so on. We later discovered that the Dairy Coalition had flooded Fox News with documents supposedly demonstrating the harmlessness of rBGH. Every version was carefully reviewed by Carolyn Forrest, a lawyer for Fox News, who one day in exasperation said: 'Don't you understand? It's not the truth of the facts that's important. I don't think this story is worth going to court and to trial spending a couple of hundred thousand dollars to fight Monsanto.' "

According to Jane Akre (when I met her in July of 2006), on April 16, 1997, Dave Boylan allegedly threatened to fire the two reporters for "insubordination" if they refused to reedit the report following to the letter the "recommendations" of Fox News: "We paid $3 billion to buy these stations," he was reported to have said. "We're the ones who decide what the content of the news should be." Steve Wilson replied that if the report was broadcast without their consent, they would file a complaint with the Federal Communications Commission for violation of the Communications Act of 1934.

On May 6, the new manager of Channel 13 reportedly changed tactics: he offered to pay the reporters a full year's salary, including benefits (about $200,000), and to appoint them to fictitious positions as consultants. In return for this "golden handshake," they had to promise never to reveal how Fox had censored the report or what they had discovered about rBGH. "Put your offer in writing, and we'll look at it," Steve answered, to Jane's great surprise. But she soon understood.[13]

The invaluable document was an exhibit in the complaint they filed after they had been fired "without cause" on December 2, 1997. To establish their claim, Jane and Steve relied on a recent Florida law on whistle-blowers, pointing to the fact that the various lies that their employer had wanted to force them to include in their report were contrary to the public interest and violated the regulations of the Federal Communications Commission.* This was the first time that reporters had used this law and Fox News took the case very seriously, hiring a dozen lawyers, including some from the firm of Williams and Connolly, which represented Bill Clinton in the Monica Lewinsky affair.

For two years, they filed countless motions to dismiss in order to avoid a trial. Jane and Steve were forced to sell their house to pay for their legal costs, but they won a preliminary victory: the case would be heard in a Tampa court in July 2000. After five weeks of trial, the jury had to answer one question: "Do you find that the plaintiff, Jane Akre, has demonstrated by a preponderance of the evidence that the defendant . . . terminated her em-

*According to the terms of this law, a whistle-blower is an employee who is the victim of retaliatory measures for having refused to participate in an illegal activity carried out by his company or for having threatened to denounce that activity to the authorities.

ployment contract because she threatened to reveal under oath and in writing to the Federal Communications Commission the broadcast of a falsified, distorted, or tendentious news report?" The jury answered in the affirmative and Jane won damages of $425,000.*

"Were you supported by the press?" The question obviously saddened Jane, who replied: "No. The major national media ignored the trial. The CBS news magazine *60 Minutes* and the *New York Times* promised to do something, but we never heard from them again. There were even incredible manipulations. For example, we had a long meeting with a reporter from the *St. Petersburg Times*, a very respected Florida newspaper. She had assiduously followed the trial. When we read her article, we came down to earth. There was one sentence that said: 'The jury did not believe the couple's claim that the station bowed to pressure from Monsanto to alter the news report.' In fact, that sentence had been added by the editor in chief without the reporter's knowledge. It was then repeated word for word on CNN, which never granted us a right to reply. But the worst thing was that our troubles weren't over."

Indeed, Fox appealed. On February 13, 2003, a Florida appeals court reversed the decision. The judges considered that no law prohibited a television network or a newspaper company from lying to the public. To be sure, the rules established by the Federal Communications Commission prohibited it, but they did not have the force of law. As a consequence, the court found that the law on whistle-blowers could not apply in the case of Jane and Steve. At the conclusion of a very technical opinion, which did not consider the underlying question—namely, the dishonesty of Fox News toward its viewers—the two reporters were required to reimburse the network's attorneys' fees, which amounted to at least $2 million.

"In fact," Jane insisted, "the court adopted the arguments of the company's lawyers, who felt no shame in declaring that no law prohibited the distortion of the news. We appealed, and finally the Florida Supreme Court threw out Fox News's claim for reimbursement of legal fees. But after what happened to us, you can understand that investigative journalism is dead in this country, and that no reporter will try to stand in Monsanto's way."†

*Steve had decided to handle his case on his own, which he did with the spirit of an experienced lawyer, but the jury thought that the principal victim was Jane.

†Since then, Jane Akre and Steve Wilson have won many prestigious awards: the First Amendment Award of the Society of Professional Journalists; the Joe Callaway Award for Civic Courage; a Special Award for Heroism in Journalism from the Alliance for Democracy; and the Goldman Environmental Prize for North America.

An Attempt at Corruption in Canada

I left Florida shaken by my colleague's story. I naively thought I had explored every one of the "special" methods Monsanto had no hesitation in using to impose its products on the market. But I had not had my last surprise. As my plane took off for Ottawa, I dove into the file of press clippings I had put together on the approval process for rBGH in Canada. "Health Canada Researchers Accuse Firm of Bribery in Bid to OK 'Questionable' Product" was the headline in the *Ottawa Citizen* on October 23, 1998. "The scientists' testimony before a Senate committee was like a scene from the conspiratorial television show *The X Files*," said the *Globe and Mail* on November 18.

I discovered that Monsanto had filed a request for authorization to put its transgenic hormone on the market with Health Canada, the Canadian counterpart to the FDA, in 1985. Generally, Health Canada models its decisions on those of the U.S. agency, but this time, even though the machine had been well oiled, it jammed. Three scientists from the Bureau of Veterinary Drugs (BVD) took on the uncomfortable role of whistle-blowers by publicly denouncing the imminent authorization of rBGH. In June 1998, they were called to testify before a Senate committee that met over a period of several months before publishing a report recommending that Monsanto's product not be approved for sale in Canada. I obtained a transcript and a video recording of the committee's hearings—the atmosphere does indeed recall *The X Files*.

The opening session immediately took on a solemn tone when the three whistle-blowers asked to take an oath on the Bible or the Canadian constitution. They were Drs. Shiv Chopra, Gérard Lambert, and Margaret Haydon, who had been working at Health Canada for thirty, twenty-five, and fifteen years, respectively. One after the other they rose, stretched out their hands, and swore to tell "the truth, the whole truth, and nothing but the truth."

There was a lengthy silence among the rather stiff and formal members of the audience, who seemed both embarrassed and surprised, until Senator Terry Stratton spoke. "I have two lines of questioning. The first goes back to the fact that you swore on oath," he said. "Are you now satisfied that your personal professional life will be protected? In other words, do you believe that you no longer have any worries about actions taken against you? . . . The minister has sent a letter to the committee stating that you are, as a group, to testify honestly and directly without fear of reprisal. Do you feel comfortable with that letter?"

"If I have sworn an oath in the presence of God, then I am supposed to tell the truth, the whole truth and nothing but the truth," answered Chopra. "Then my question to myself was what truth am I going to tell—the one I know or what the minister is telling me to tell? That is my conflict. . . . It has been said that there is a guarantee that there will be no repercussions, but . . . that remains to be seen because I am still under a complete gag order, to the extent that I cannot even attend meetings. If I said something at a dinner meeting and somebody heard and reported to the [Human Safety] department [of the BVD], I could be in trouble."

"Finally, it does not appear to me that you have a lot of confidence in the process, obviously," answered Stratton. "One thing I would like to impress upon you, if you do have a problem with respect to grievances or with respect to threats from management, this committee would be delighted to hear from you."

"The department is saying all over the place that the client—and this is in writing—the client now is the industry and we have to serve the client," said Chopra. "The conflict was that our concern at the BVD, particularly in the Human Safety Division, has been that we have been pressured and coerced to pass drugs of questionable safety, including rBST. . . . I stuck my neck out and I wrote to the previous Minister of Health and the current Minister of Health as well as all the way to the Deputy Minister complaining about these very serious problems of secrecy, conspiracy . . . and saying that something needs to be done. I specifically wrote to Minister Dingwall. I urged him to intervene to safeguard the public interest. I never received a reply. In November 1997, we met with Dr. Paterson, one executive of Health Canada, and we told him we wanted a scientific gaps analysis of the records to be done. . . . When we got them, there were no raw data, only a summary sent by the FDA. . . . [The raw data] are locked up. They are kept in the personal custody of Dr. Ian Alexander, who is called the file manager of rBST. No one else is supposed to look into it."

Next to be questioned was Margaret Haydon, who had been given the task of examining the request for approval from 1985 to 1994, until it was taken away from her. "My files were stolen in May of 1994 from my locked cabinets," she testified in a delicate voice. "I discovered that there appeared to be a lot of things missing, so I was quite shocked. . . . It amounted to most of my work over the ten previous years dealing with rBST reviews. So I decided that I would document this and send a memo to my chief. I came back after the weekend and again looked in the file cabinet and there were some additional files back in. . . . Then an investigation began with the security group in Health Canada. After interviewing me, Sergeant Fiegenwald took some of the files that were back for fingerprinting, and he also asked me to write statements on anything I could recall or that I thought might have caused these documents to be removed. I provided him with the original, and I kept a copy for myself. A few months later, in November of 1994, when I was away on sick leave, a member of Health Canada Security Department called me and demanded that I provide the copy of my statements. She came to my home and demanded these. I have never seen those since that time, so that was kind of a surprise."

"Would you say the files that you had were pro-rBST or anti-rBST?" asked Senator Nicholas Taylor.

"There were a lot of questions that I had asked, and there were numerous what we call 'additional data letters' asking questions of the company to provide additional information. So at that point I was not recommending that the drug be approved from a point of view of safety in the intended species or efficacy."

"I gather if, after your research, people offered you rBST-treated milk, you would not want to drink it?"

"Personally, I would probably decline."

"I do not believe I am in Canada, when I hear you that your files have been stolen!" said Senator Eugene Whelan. "What in the hell kind of system have we got? Do not forget that I was minister for eleven years, and research was my favorite topic. . . . So I have every reason to be skeptical when we become more dependent all the time on Monsanto and these companies doing the research for us. . . . I have strong reservations about less and less public research and big grants by Monsanto to Agriculture Canada for this, and big grants to somebody else from another company for that. . . . I would ask each one of you, has any one of you been lobbied by Monsanto?"

"I will describe the situation," said Haydon. "I am not sure 'lobby' is the correct word, but I did attend a meeting back approximately in 1989 to 1990, and Monsanto representatives had met with myself and my supervisor, Dr. Drennan, and my director, Mr. Messier. At that meeting, an offer of one to two million dollars was made by the company. I do not know any more about what became of that, but my director indicated after the meeting that he was going to report it to his superiors."

"Dr. Haydon talked about the one to two million dollars from Monsanto at a meeting," Shiv Chopra added. "The Fifth Estate channel network did a program. They checked with Dr. Drennan, who is now retired. They asked him: 'Was that offered?' He said: 'Yes.' They said: 'Did you consider that to be a bribe?' He said: 'I would say so.' They said: 'Well what did you do after?' He said: 'Well, I laughed.' They said: 'After you laughed, what did you do? Did you report?' He said: 'I did.' They said: 'Then what happened?' He said: 'I don't know.'"

Tension in the room was at its peak. The members of the committee maintained a long silence, finally broken by Senator Mira Spivak, who put her finger on an extremely important point: "In The United States, the FDA approved the product based on some summary which turned out to be incorrect, because the raw data were not available or they didn't get access to them. Now the JECFA, which is the joint committee of the WHO and the FAO, said there is nothing wrong with rBST as of this year, also based on summaries which had nothing to do with raw data. May we trust the JECFA?"

To understand the importance of this question, you should know that the Joint Expert Committee on Food Additives (JECFA) is a scientific consultative committee established in 1955 by the World Health Organization (WHO) and the Food and Agriculture Organization (FAO), two UN bodies. The committee meets regularly to examine requests to put new food products on the market. With that in mind, it relies on appropriate experts supposed to be selected by member countries for their competence and impartiality. The decisions of the JECFA are transmitted to the Codex Alimentarius Commission, also dependent on the WHO and the FAO, established in 1963 to develop uniform standards for food products and issue international recommendations regarding the health and safety of technological practices related to food. Documents published by the commission have an aura of international scientific expertise and the imprimatur of the United Nations.

The work of the Canadian Senate committee was very informative about the way in which the JECFA and the Codex Alimentarius Commission operate, providing confirmation of what some had suspected, namely, that their work had been controlled by Monsanto. On December 7, 1998, the

Senate committee heard testimony from Michael Hansen, the expert from the Consumer Policy Institute, who was familiar with the ins and outs of the UN organizations since he had participated in several meetings as a representative of consumer organizations. He revealed that the first panel of scientists assembled by the JECFA in 1992 to evaluate the transgenic growth hormone included six representatives from the FDA, among them Margaret Miller—the former Monsanto employee—and Drs. Greg Guyer and Judith Juskevich, the authors of the controversial *Science* article. The rapporteur of the second panel, in 1998, was none other than Margaret Miller. Under these circumstances, it is not hard to see why the JECFA issued a favorable decision on rBGH, as Ray Mowling, Monsanto vice president for government and public affairs, was eager to point out: "The UN report reaffirmed that treating cows with BST to increase milk production is safe . . . there are no food safety or heath concerns related to BST residues in products such as meat and milk from treated animals."

A Testing Ground for GMOs

That same afternoon, December 7, 1998, the senators heard testimony from David Kowalczyk, Monsanto's director of regulatory affairs, who was caught lying openly. "In the memos, you are suggesting to Health Canada who should be on the JECFA panel," declared Senator Spivak, looking him in the eye. "Do you think that suggesting who Canada's representative on the JECFA panel should be might be overstepping the boundaries in your relationship with Health Canada?"

"This is the first time I have heard that. I have never made any recommendation on who should be on JECFA."

"There are memos and minutes of meetings that report conversations that you have had with Mr. [Ian] Alexander, who controlled the data provided by your company. He denied it as well."

After hearing from Health Canada's three whistle-blowers, Senator Whelan concluded: "I still maintain we should be doing more research. Now, for instance, Monsanto is paying $600,000 to Agriculture Canada to find a wheat that is immune to Roundup. I wrote to ten universities. One of them told me to mind my business when I wanted to find out what strings are attached to the research grants they are getting. A couple of them called me and said: 'You are on the right trail, but we cannot give you any information.' They are scared to death. I am so proud of you people because you are not scared to death. If they ever do anything to you, let us know."

Whelan, a former Minister of Agriculture, was prophetic. After the act of national catharsis produced by the senate committee, everything went back to normal. Canada definitively banned rBGH from its territory, leading to the same action by the European Commission, although the latter had been on the verge of following the advice of the JECFA and lifting the moratorium that had been in force since 1990.* Australia and New Zealand followed suit. But at Health Canada, old habits reasserted themselves: in July 2004, Shiv Chopra, Margaret Haydon, and Gérard Lambert were fired for insubordination. "After our testimony to the committee, we were harassed and marginalized," Chopra explained when he met with me in July 2006 in his home outside Ottawa. "Everything we were afraid of happened, and no one lifted a little finger. We took the case to court, but there is no law protecting whistle-blowers in Canada. This country is corrupt to the core, and that's the title of the book I'm working on."

"Do you think Monsanto played a role in your firing?"

"I have to be very careful about my answer," he said, smiling. "Let's say our testimony came at a very bad moment for the company, which was just launching its GMOs in Canada. It's clear that the growth hormone was a testing ground, which turned out badly in part, but it enabled the company to work out its techniques for conquering the market."

*On March 10, 1999, the Scientific Committee on Animal Health and Animal Welfare of the European Commission issued a ninety-one-page report recommending that "rBST not be used in dairy herds." At no point was there any mention of risks that the hormone might pose for human health. The hormone has been officially banned in the European Union since January 1, 2000.

PART II

GMOs: The Great Conspiracy

7

The Invention of GMOs

> The health and safety of biotechnology products is not an issue: the food, feed and environmental safety of the products must be demonstrated before the products enter the agricultural production system and supply chain.
>
> —Monsanto, *Pledge Report*, 2005

"The cow hormone drug was simply the first major application of biotechnology to food production and Monsanto is a very powerful corporation with many, many linkages to top level persons in government. I think the prevailing ethic at the federal government was 'Biotechnology is so important that we can't let a few little questions about cow safety or human safety get in the way.' The drug got approved, regardless of its demerits," Michael Taylor told me.

Indeed, by the time rBGH was approved by the Food and Drug Administration, dozens of GMOs were in development in the laboratories of biotechnology companies, chiefly Monsanto, which had just filed an application for the marketing of Roundup Ready soybeans, genetically modified to resist the spraying of Roundup. The connection between the company's maneuvering to secure approval of the controversial hormone at any cost and its plan to position itself in the market as the "Microsoft of biotechnology" was confirmed, unexpectedly, by Taylor, who, it will be recalled, worked as counsel for Monsanto, was appointed deputy commissioner of the FDA in 1991, and a few years later became a Monsanto vice president.

"I think in terms of public acceptance, it's been one blunder after another," he confessed in our telephone conversation. "If you're trying to have a strategy for having the public understand and accept a new technology, having the first application of it be related to milk, which we already have more than we need, it helped create a climate of . . ."

"Suspicion?" I suggested, completely astounded by what I was hearing.

"Suspicion, yes," he answered. "I think that Congress should change the law. It should create a mandatory notification system that ensures that every product is looked at by FDA and the FDA makes a safety judgment about every product."

I still find it hard to understand why Taylor made this surprising confession. Was it belated remorse, or an attempt to exculpate himself for the role he played in supervising the writing of U.S. regulations of GMOs, which influenced all governments and international organizations, including the European Community? The answer is a mystery.

The Scramble for Genes

Before recounting in detail the genesis of what can be considered one of the greatest conspiracies in the history of the food industry, it is appropriate to outline in broad terms the saga of genetic engineering. And just this once I must admit that Monsanto's tenacity and enthusiasm were impressive—it overcame all its many competitors to become the unchallenged leader in this advanced field.

It is generally accepted that the story began in 1953 when the American James Watson and the Briton Francis Crick discovered the double helix structure of DNA (deoxyribonucleic acid), the molecule that contains the genetic code for every living organism. The discovery won the two geneticists and biochemists a Nobel Prize in 1962 and signaled the birth of a new discipline: molecular biology. As Hervé Kempf has noted in *La Guerre secrète des OGM*, it also led to the emergence of a "doctrine" according to which "the organism is a machine" entirely dependent on genes alone, the key to the understanding of the mechanisms of life. This "doctrine"—not to call it a "dogma"—was clearly summarized by the 1958 Nobel Prize winner Edward Tatum: "(1) All biochemical processes in all organisms are under genetic control. (2) These overall biochemical processes are resolvable into a series of individual stepwise reactions. (3) Each single reaction is controlled in a primary fashion by a single gene. . . . The underlying hypothesis, which in a number of cases has been supported by direct experimental evidence, is that each gene controls the production, function, and specificity of a particular enzyme."[1]

In other words, every biological reaction that characterizes the functioning of a living organism is governed by *one* gene that expresses a function by triggering the production of a specific protein. This exclusive idea, which some call "all gene," is the source of one of the greatest misunderstandings underlying the development of biotechnology, one that persists today. "In reality," as Arnaud Apotheker, holder of a doctorate in biology and spokesman on GMO issues for Greenpeace France, pointed out in 1999, "every day phenomena turn out to be more complex: a single gene may code for proteins having very different primary structures and biological properties depending on the tissues of an organism or the organism itself. The molecular machinery of living things is of a complexity that we are barely beginning to glimpse."[2] We now know, for example, that some genes interact with others and that it is not a simple matter to extract them from one organism and introduce them into another in order for them to express the protein and hence the function that has been selected. Rather, transferring genes this way may cause unexpected biological reactions in the host organism.

Beginning in the early 1960s, molecular biologists set to work to develop techniques that would enable them to manipulate genetic material to create chimerical organisms that nature never would have been able to produce on its own. To do so, they strove to divide and put together fragments of DNA, to copy and multiply genes with the aim of transferring them from one species to another. This genetic tinkering was often justified by a generous humanitarian vision, expressed, for example, in 1962 by Caroll Hochwalt, Monsanto's vice president for research, in a commencement speech at Washington University in St. Louis: "It is entirely conceivable that, through the manipulation of the genetic information at the molecular level, a crop such as rice could be 'taught' to build a high protein content into itself, literally working a miracle of alleviating hunger and malnutrition."[3] It should be pointed out that at the time the secrets of DNA were of little concern to Monsanto, which was busy making its fortune in the jungles of Vietnam.

So it was at Stanford University, not in St. Louis, that the first genetic manipulations took place. In 1972, as Monsanto was preparing to launch

Roundup, Paul Berg succeeded in "recombining" DNA—that is, putting together two fragments of DNA from different species into a hybrid molecule. A little later, his colleague Stanley Cohen announced that he had succeeded in transferring a frog gene into the DNA of a bacterium able to reproduce the intruder in large quantities. These discoveries, which broke a law that had been considered inviolable, the impossibility of crossing what was known as the "species barrier," created great excitement, along with deep concern, in the international scientific community. The worries turned into an uproar when Paul Berg announced his intention to insert a carcinogenic virus, SV-40, from a monkey into an *E. coli* cell, a bacterium that colonizes the human digestive tract. Some scientific authorities, such as Robert Pollack, a cancer virus specialist, worried: "What will happen if the manipulated organism inadvertently escapes from the laboratory?"[4] The general outcry led to a temporary moratorium on genetic manipulation and, on February 25, 1975, the first international conference on recombinant DNA. For two days at Asilomar, a Pacific seaside resort in California, leading figures in the rising discipline considered the risks of genetic engineering, focusing the debate on experimental safety and the formulation of rules, such as measures to contain manipulated organisms. But at no point did they broach ethical questions, which were excluded from the outset. It was as though the biologists had already decided to "limit the involvement of the public and the government in their affairs to the minimum."[5] The message was soon received loud and clear by the future world leader in biotechnology.

After the Asilomar conference, genetic engineering experiments proliferated in the United States—the National Institutes of Health recorded more than three hundred in 1977. While attempts to place legal restrictions on these extremely hazardous new scientific activities were buried one after the other—in 1977 and 1978, sixteen bills were proposed in Congress, but none passed—start-ups and risk capital companies were flourishing, particularly in California, where another promising technology had just given birth to Silicon Valley. Companies such as Calgene and Plant Genetics Systems were established by biologists who had previously worked in universities and who, carried away by an extraordinary burst of research activity and the prospect of huge financial rewards, plunged into the economic arena, raising millions of dollars on the New York Stock Exchange or taking shares in and joining the boards of private companies.

This veritable "race for genes" brought about an unprecedented association between science and industry, which radically transformed research practices, as the sociologist Susan Wright explains in her standard work on the history of biotechnology, published in 1994: "As genetic engineering became seen as a promising investment prospect, a turn from traditional scientific norms and practices toward a corporate standard took place. The dawn of synthetic biology coincided with the emergence of a new ethos, one radically shaped by commerce."[6] This development was very markedly stimulated by Monsanto through the patent system that controlled research and the products derived from it.

The Triumph of Genetic Tinkering

While start-ups were making news on the stock market, one man in St. Louis was conducting a solitary battle. His name was Ernest Jaworski, and he had joined Monsanto in 1952. This researcher, who was an expert on glyphosate and had worked out the details of its manner of acting on plant cells, had an idea that seemed completely preposterous to his colleagues in the old chemical company: instead of trying to manufacture new herbicides, why not create selective plants by manipulating their genetic makeup precisely so they could survive the spraying of herbicides?

Encouraged by John Hanley, who became CEO of Monsanto in 1972 and was also convinced that biology represented the future of chemistry, Jaworski initiated himself into the cultivation of plant cells in a Canadian laboratory and then supervised the work of thirty researchers, including such rising stars of molecular biology as Robert Fraley, Robert Horsch, and Stephen Rogers. "These young genetic engineers did believe that their work would be good for the planet, possibly making it easier to grow food or reducing agriculture's dependence on chemicals," according to Daniel Charles, author of *Lords of the Harvest*, who was able to interview the pioneers of biotechnology before they decided to sink into stubborn silence. "Some of them, working inside chemical companies, often saw themselves as 'green' revolutionaries fighting against the entrenched power of the chemists, whom they dismissed as 'nozzleheads.'"[7]

Meeting on the fourth floor of U Building at Monsanto's Creve Coeur location, a suburb of St. Louis to which the company had recently moved, the team was nicknamed "Uphoria" by company skeptics, who saw this group of excited young men as economically irresponsible oddballs. At the same time, the "Kremlin," as the company management, located in D Building, was called, had broken with company habits and for the first time in its history plunged headlong into basic research without knowing what applications it would lead to. "Scientific excellence was the priority," according to Rob Horsch. "There was no pressure to produce a product. For example, we were working on petunias. No one came and said to us: 'Petunias? What do you think we are? A university?' In fact, we were a kind of entrepreneurial unit protected by the management."[8]

Following the lead of laboratories in California, Belgium, and Germany, the Uphoria researchers developed a three-stage research program: first, to manipulate DNA to extract genes that might be useful, known as "genes of interest"; next, to transfer those genes into plant cells; and finally, to develop tissue cultures in order to reproduce and encourage the growth of these manipulated embryonic cells. The first stage was worked out thanks to the discovery of restriction enzymes, which functioned like scissors, enabling molecular biologists to cut DNA to extract genes of interest.

But the second stage was another story. Contrary to the argument often put forth by promoters of biotechnology, the techniques of genetic manipulation have absolutely nothing to do with the genealogical selection that has been practiced by breeders since the work of Louis de Vilmorin in the mid-nineteenth century. Seed companies have merely rationalized and systematized the ancestral practices of farmers who, since the advent of agriculture in Mesopotamia ten thousand years ago, have endeavored to keep the best grains from their harvests to seed their fields the following year. The contribution of professional breeders is to *cause* the cross-breeding of two plants—the "parents" of the line—selected for complementary agronomic qualities (such as resistance to disease or crop yield), in the hope that their descendants will preserve the same characteristics because of the laws of heredity. The best examples from the second generation are then selected and forced to cross-breed, and so on over several generations. It is clear that genealogical selection is based on natural laws, in this case the sexual reproduction of plant organisms; human action is aimed only at orienting the range of possibilities within a single genetic reservoir, but in the end the "improved" plant might very well have been created by Mother Nature in the fields. I will return to the effects of genealogical selection on biodiversity in Chapter Eleven, but for now, it is important to understand that this agronomic procedure cannot be identified with the techniques of genetic manipulation, which, rather than respecting the natural laws of plant development, attempt instead to break them in any way possible.

Molecular biologists knew very well that plant organisms possess defense mechanisms designed to protect them from the intrusion of foreign bodies, including, of course, genes coming from other living species. From the very beginning, those biologists understood that genetic manipulation could not be carried out without using an intermediary, or a "mule," able to transport the selected gene and make it enter *by force* into the target cell. For this purpose, they turned to a bacterium that is abundant in the soil, *Agrobacterium tumefaciens*, which has the capacity to insert some of its genes into plant cells to cause tumors.* In other words, this bacterium is a pathogen that changes the genetic inheritance of cells by infecting them.

In 1974, a Belgian research team succeeded in identifying the plasmid (a ring of DNA) constituting the vector by which the gene that induces the tumor is transferred from the bacterium to the plant. In St. Louis, as in laboratories around the world at the time, they then attempted to isolate in the plasmid the gene responsible for the tumors and replace it with the gene of interest by adding a gene "promoter," a sequence of DNA that triggers the expression of the gene to be triggered. The gene in question is often 35S, from the cauliflower mosaic virus, which is related to the hepatitis B virus, raising the alarm of some opponents of unrestricted tinkering with genes.

But there was more: if the gene-inducing tumors had been suppressed, how could one know that the plasmid was doing its work and inserting the substitute gene in the plant cell? The only solution the sorcerer's apprentices found was to attach to the genetic construction what they called a "selection marker," in this case a gene resistant to antibiotics, usually kanamycin. To verify that the transfer had actually taken place, the cells were sprayed with an antibiotic solution, and the "chosen" were those that survived this shock treatment. (This gave rise to further health concerns—at a time when resistance to antibiotics was in the process of becoming a serious public health problem, some Cassandras were afraid that the selection marker would be absorbed by bacteria populating the human intestinal tract, reducing medicine's ability to fight infectious agents.)

**Agrobacterium tumefaciens* causes crown gall disease, which attacks the roots of some plants by inducing the growth of a tumor. It was discovered by two American researchers in 1907.

In the meantime, on January 18, 1983, at the symposium on molecular genetics in Miami, representatives of three laboratories—one Belgian and two Americans, one of whom was Rob Horsch of Monsanto—announced that they had succeeded in inserting a genetic construct, a kanamycin resistance gene to be exact, into cells of petunia and tobacco plants (two plants susceptible to *Agrobacterium tumefaciens*). The three laboratories had filed patents on their simultaneous discoveries. For Monsanto, serious work was beginning and the call to battle had sounded.

The "Artificial Cassette" of Roundup Ready Soybeans

"I'll never forget the first time I used the phrase 'We are not in the business of the pursuit of knowledge; we are in the business of the pursuit of products.' You could have heard a pin drop. They were furious."[9] The words are those of Richard Mahoney, who, as soon as he was appointed CEO of Monsanto in 1984—a position he held until 1995—decided to shake up the Uphoria troops. The end had come for lavishly funded research on tinkering with petunias, and the aim was now clear: to create transgenic plants that brought in money. Called by *Fortune* one of "America's toughest bosses," Mahoney was an unselfconscious businessman who bluntly declared: "Forgiveness is out of style, shoulder shrugs are out of fashion. Hit the targets on time without excuses."[10]

Subjected to unprecedented stress, Ernest Jaworski's team understood that the laboratory's success was a question of life or death and that a failure would signal the victory of the pure chemists. From then on, all research was focused on the production of plants resistant to Roundup, which, ten years after its introduction, had become the most widely sold herbicide in the world. Furthermore, the implacable boss reminded everyone that the patent guaranteeing a monopoly on glyphosate derivatives would expire in 2000 and that GMOs soon to be known as "Roundup Ready" would be a good way of pulling the rug out from under manufacturers of generics. This was a concrete objective that delighted Jaworski, because in the end this had been his original idea: to manipulate plants so that they could survive the use of herbicides, which could therefore be sprayed at any time on crops—corn, soybeans, cotton, rapeseed, and why not wheat?—to destroy only weeds.

But they hadn't gotten there yet. In 1985, the Monsanto researchers were obsessed by only one thing: finding the gene that would immunize plant cells against Roundup. This was especially urgent because Calgene, a California start-up, had just announced in a letter published in *Nature* that it had succeeded in making tobacco resistant to glyphosate.[11] Discussions were already under way on an agreement with the French company Rhône-Poulenc to develop crops resistant to glyphosate. At the same time, the German company Hoechst was going all out to find the gene resistant to its herbicide Basta, not to mention DuPont (Glean) and Ciba-Geigy (atrazine). In short, all the chemical giants were pursuing the same goal, because the stakes were primarily economic: companies were already imagining the patents they could file on all the major food crops in the world.

In St. Louis, stress took up permanent residence, because the notorious gene remained elusive. Jaworski's researchers were going around in circles. They had succeeded in identifying the gene responsible for the enzyme that, as I reported in Chapter Four, is blocked by the action of glyphosate molecules, causing tissue necrosis and plant death. The idea was to manipulate it so as to deactivate the reaction to the herbicide, and then introduce it into plant cells, but nothing worked. "It was like the Manhattan Project," said Harry Klee, a member of the research team. "The antithesis of how a scientist usually works. A scientist does an experiment, evaluates it, makes a conclusion and goes on to the next variable. With Roundup resistance we were trying twenty variables at the same time: different mutants, different promoters, multiple plant species. We were trying everything at once."[12]

The search lasted for more than two years, until the day in 1987 when engineers thought of rummaging through the garbage in Monsanto's Luling plant, located 450 miles south of St. Louis. At this site on the banks of the Mississippi, Monsanto produced millions of tons of glyphosate annually. Decontamination pools were supposed to treat production residues, but some of the residues had contaminated nearby land and ponds. Samples were taken to collect thousands of microorganisms in order to detect the ones that had naturally survived glyphosate and identify the gene that gave them that invaluable resistance. It took a further two years for a robot analyzing the molecular structure of the bacteria collected to finally come up with the rare pearl. It was "a great Eureka moment," said Stephen Padgette, one of the "inventors" of Roundup Ready soybeans, now a Monsanto vice president.[13]

But the game was far from over. They now had to find the genetic construct that would enable the gene to function once it was introduced into plant cells, specifically soybeans, the oil-producing plant the team was working with after preliminary trials with tomatoes. The stakes were huge: along with corn, soybeans dominated American agriculture at the time, annually contributing $15 billion to the national economy. Until 1993, when Roundup Ready soybeans were officially launched, Stephen Padgette and his colleagues in the Roundup resistance program divided their time between the laboratory and the greenhouses covering the roof of the Chesterfield Village biotechnology research center that Monsanto had set up in a wealthy suburb of St. Louis. It took "700,000 hours and an $80 million investment" to attain the result: a genetic construct including the gene of interest (CP4 EPSPS), the promoter 35S from the cauliflower mosaic virus, and two other fragments of DNA derived from the petunia intended to control the production of the protein.[14] The "'Roundup tolerant soybean gene cassette' is a completely artificial one that never existed in natural life kingdom nor could have evolved naturally," reported Japanese biologist Masaharu Kawata of Nagoya University.[15]

This was so much so that the Monsanto researchers encountered enormous difficulties in introducing it into soybean cells. They had to give up the "mule," *Agrobacterium tumefaciens*, because they had constantly faced the same problem: whenever they inundated the cells with antibiotic, the ones that had not absorbed the cassette died, but those dead cells poisoned the genetically modified cells in a phenomenon Rob Horsch named "colloperative death," a sinister-sounding neologism indicating death from cooperative collapse.[16]

In the face of this resistance from nature, the team decided to bring out the heavy artillery, a "gene gun" invented by two Cornell University scientists, developed in collaboration with Agracetus, a Wisconsin biotech company that Monsanto acquired in 1996. When John Sanford and his colleague Ted Klein came up with the idea for this last-ditch weapon, they were considered crazy, even though laboratories at the time were prepared to do anything to force the desired DNA to penetrate into the target cells: some researchers were using microscopic needles, while others employed electric charges to make little holes in cell walls to enable the DNA to enter (evidence, if any were needed, that biotechnology has nothing to do with the traditional technique of genealogical selection). But nothing was working.

The gene gun is now the insertion tool most frequently used by the "artillerymen" of genetic engineering. It works by attaching genetic constructs to microscopic gold or tungsten bullets and shooting them into a culture of embryonic cells. A clear picture of the imprecision of the technique can be found in the description Stephen Padgette provided in 2001 to Stephanie Simon of the *Los Angeles Times*: "Trouble was, the gene gun inserted the DNA at random. Sometimes a bundle would splinter before landing in a cell. Or two gene packets would double up. Even worse, the DNA would at times land in a spot that interfered with cellular function. The team had to fire the gun tens of thousands of times to get a few dozen plants that looked promising. After three years of field tests on these promising plants, a single line of transformed soybean shone as superior. It could resist heavy doses of glyphosate, as the greenhouse experiment proved. . . . 'It was bulletproof,' Padgette recalled with pride. In 1993, Monsanto declared it a winner."[17]

But at what cost? As Arnaud Apotheker points out in *Du poisson dans les fraises*: "In their determination to subjugate nature, humans use the technologies of war to force cells to accept genes of other species. For some plants, they use a chemical or bacteriological weapon to infect cells with bacteria or viruses; for others they use only classic weapons, such as gene guns. In both cases, waste is considerable, because on average one cell out of a thousand enters the transgene, survives, and is able to generate a transgenic plant."[18]

In 1994, in any event, Monsanto filed a request for authorization to market Roundup Ready soybeans, the first widely grown GMO. And once again the company had "bulletproofed" everything, as its vice president said.

Maneuvers in the White House

While the team in Chesterfield Village was desperately tracking the glyphosate resistance gene, company management was demonstrating a capacity for foresight that might be surprising if one were unaware of the consequences. As the *New York Times* reported in a very well-informed article in 2001: "In late 1986, four executives of the Monsanto Company, the leader in agricultural biotechnology, paid a visit to Vice President George Bush at the White House to make an unusual pitch."[19]

To fully understand the subtlety of the strategy managed by Leonard Guarraia, then director of regulatory affairs for the company, recall that the Reagan administration's watchword was "deregulation," intended to "liberate market forces" by shrinking the intrusive state. This ideology was aimed at fostering American industry by reducing to the maximum extent possible what White House hard-liners called "bureaucratic obstacles," which is how they saw the health and environmental tests required by regulatory agencies before a new product could be marketed: the FDA for food and drugs, the EPA for pesticides, and the Agriculture Department (USDA) for crops.

The United States at the time was conducting a merciless struggle to impose its superiority in competition with Japan, and to a lesser extent with Europe, particularly in the area of new technologies, but also in agricultural products. In this extremely competitive context, the stakes involved in biotechnology were considerable. For this reason, on June 26, 1986, the White House issued a policy document entitled "Coordinated Framework for the Regulation of Biotechnology," directed primarily at preventing Congress from getting involved in this delicate issue by introducing specific legislation for the regulation of GMOs. Addressed to the three relevant regulatory agencies (FDA, EPA, and USDA), the directive provided that products derived from biotechnology would be regulated within the framework of already existing federal laws, insofar as "recently developed methods are an extension of traditional manipulations" of plants and animals.[20] In other words, GMOs did not require special treatment and would be subject to the same system of approval as non-transgenic products.

But the document did not satisfy Monsanto, which clearly had another idea in mind. "'There were no [GMO] products at the time,' Leonard Guarraia, a former Monsanto executive who attended the Bush meeting, recalled. . . . 'But we bugged Bush for regulation. We told him that we have to be regulated.'"[21] So what was behind what the *New York Times* called an "unusual pitch"?

"In fact," Michael Hansen of the Consumers Union told me in July 2006, "Monsanto wanted an appearance of regulation. The company knew that after the PCB and Agent Orange scandals, when it had lied or concealed data, it would not be believed if all it did was to say that GMO products posed no danger to health or the environment. It wanted federal agencies, primarily the FDA, to be the ones to say that the products were safe. So, whenever a problem arose, it would be able to say: 'The FDA has established that GMOs do not pose any risks.' This was also a way of covering itself in case things turned out badly."

According to the *New York Times* reporter, the Washington meeting bore fruit: "In the weeks and months that followed, the White House complied, working behind the scenes to help Monsanto . . . get the regulations that it wanted. It was an outcome that would be repeated, again and again, through three administrations. What Monsanto wished for from Washington, Monsanto—and, by extension, the biotechnology industry—got."[22]

To understand just how unusual Monsanto's approach was, one has to consider that at the time some high FDA officials were absolutely opposed to the idea of regulating GMOs, even in the form of a document that would be an "appearance of regulation." This was so, for instance, for Henry Miller, the agency spokesman for biotechnology, who had no compunctions about calling GMO opponents "troglodytes" or "intellectual Nazis" and whom the White House would have to fight hard.[23]

But that wasn't all. The *New York Times* was able to get its hands on a draft of a secret document, dated October 13, 1986, in which the company's directors established a veritable battle plan to impose GMOs in the United States. Among the primary objectives were "'creating support for biotechnology at the highest U.S. policy levels,' and working to gain endorsements for the technology in the presidential platforms of both the Republican and Democratic Parties in the 1988 election."[24]

In fact, I found evidence on film of the company's boundless self-confidence: it was capable of expressing thinly veiled threats to George Bush when it felt the administration was resisting it. I was able to see extraordinary archive footage filmed on May 15, 1987, by the Associated Press. It shows Ronald Reagan's vice president, who was then running for president, walking through Monsanto's St. Louis laboratories wearing a white coat. Followed by a pack of reporters, the future president first participates in a class on genetic manipulation.

"What I'd like to do today is show you some of the steps we go through when we're moving genes from one organism to another," explains Stephen Rogers, one of Uphoria's three rising stars, with a test tube in his hand. "We take DNA, cut it apart, mix different pieces together, and then rejoin them. . . . This tube contains DNA that was made from a bacterium. DNA would look the same whether it was from a plant or an animal."

"Oh, I see," says George Bush, his eyes fixed on the test tube. "This will lead you to do what? To have a stronger plant? Or a plant that resists . . ."

"In this case it resists the herbicide," Rogers answers.

"We have a fabulous herbicide," says a voice off-camera.

Then Bush walks through the greenhouses on the Chesterfield Village roof, where a Monsanto executive in suit and tie shows him transgenic tomato plants that turn out to be the real purpose of this self-serving guided tour. Next comes an absolutely astounding conversation: "And we have before USDA right now a request to test this for the first time on a farm in Illinois this year," the executive says.

"We keep hallucinating about it . . . the expense goes up and nothing happens," says Rogers.

"And I would say quite frankly we have no complaint about the way USDA is handling it," the executive goes on. "They're going through an orderly process; they're making sure as they deal with these new things [that] they do them properly, and uh, no, if we're waitin' until September and we don't have our authorization we may say somethin' different!"

"Call me, I'm in the dereg business," says Bush with a great burst of laughter. "I can help."

On June 2, 1987, exactly two weeks after the amazing guided tour, the Monsanto researchers conducted their first field test of transgenic crops in Jerseyville, Illinois. There is a photograph showing Stephen Rogers, Robert Fraley, and Rob Horsch posing in front of a tractor wearing farmer's caps. Facing them are crates containing tomato shoots manipulated through the magical power of the bacterium *Agrobacterium tumefaciens*.

Political Regulation Made to Order

George H.W. Bush assumed the presidency in January 1989. In March, he appointed his vice president, Dan Quayle, to head the Council on Competitiveness, "with responsibility for reducing the regulatory burden on the economy."[25] On May 26, 1992, Vice President Quayle presented American policy on GMOs in front of an audience of business executives, government officials, and reporters. "We are taking this step as part of the President's regulatory relief initiative, now in its second phase," he declared at the outset. "The United States is already the world leader in biotechnology and we want to keep it that way. In 1991 alone, it was a $4 billion industry. It should reach at least $50 billion by the year 2000, as long as we resist the spread of unnecessary regulation."

Three days later, on May 29, Monsanto was victorious: the FDA published in the *Federal Register* its regulatory policy on "foods derived from new plant varieties."[26] It should be noted that the title of this twenty-page document, considered a bible around the world, carefully avoided any reference to biotechnology, presented in the introduction as merely an extension of genealogical selection, following recommendations issued by the White House six years earlier: "Foods . . . derived from plant varieties developed by the new methods of genetic modifications are regulated within the existing framework . . . utilizing an approach identical to that applied to foods developed by traditional plant breeding."

Anyone wanting further information was asked to contact a man named James Maryanski. I went through a long struggle to locate the man who held the key position of Biotechnology Coordinator for Food Safety and Applied Nutrition at the FDA from 1985 to 2006. In 2006, this microbiologist who had joined the agency in 1977 was enjoying an active retirement, working as an "independent consultant" on the "safety of GM foods" for various governments, as the CV he gave me states. An interesting sidelight: as I was about to give up locating him, I asked to interview an FDA representative about the 1992 regulation, explaining that I was producing a documentary on Monsanto, particularly on the approval of Roundup Ready soybeans. On July 7, 2006, I received an e-mail from Mike Herndon, one of the agency's press officers: "I must respectfully decline your request for an on-camera interview. FDA must appear neutral in its relationship with food manufacturers. Being interviewed in a documentary about a company whose products FDA regulates is inappropriate."

The statement is ironic in light of the fact that the 1992 policy statement was developed in close cooperation with Monsanto, which in fact wanted the agency to present an "appearance of regulation," in the words of Michael Hansen. And this task was confided to none other than Maryanski under the supervision of Michael Taylor, who was then deputy commissioner of the FDA. (I have already described Taylor's role in the bovine growth hormone affair; I will come back to his subsequent career as a Monsanto vice president.)

I was finally able to meet the former FDA official one day in July 2006 in New York, on his return from a consultation in Japan. I was surprised to encounter a short, shy man with light-colored eyes and a calm, quiet voice. Later, viewing this filmed three-hour conversation, I was able to recognize his controlled panic, perceptible only in the nervous blinking that seized him on several occasions.

To start with, I questioned him on the instructions transmitted by the White House regarding the drafting of the regulation of transgenic foods. "Basically, the government had taken a decision that it would not create new laws," he explained cautiously. "For the FDA, it felt that the Food, Drug, and Cosmetic Act, which ensures the safety of all foods except meat, poultry and egg products, which are regulated by the United States Department of Agriculture (USDA), had enough authority for the agency to deal with new technologies. And actually what occurred at FDA was that the commissioner, Dr. David Kessler . . . established a group of scientists under my authority and lawyers, who were given the charge to see whether in fact we could regulate foods developed by biotechnology under the existing Food, Drug, and Cosmetic Act."

"But this decision that GMOs should not be submitted to a specific regulatory regime wasn't based on scientific data, it was a political decision?" I asked. The question made him a little tense.

"Yes, it was a political decision. It was a very broad decision that didn't apply to just foods. It applied to all products of biotechnology," he said hesitatingly.

The Amazing Trick of the Principle of Substantial Equivalence

I then proceeded to read a paragraph of the regulation that lies at the heart of the dispute around GMOs: "In most cases, the substances expected to become components of food as a result of genetic modification will be the same as or substantially similar to substances commonly found in food such as proteins, fats and oils, and carbohydrates."[27]

These few apparently anodyne lines pointed to a concept that has been adopted around the world as the theoretical basis for the regulation of GMOs: the "principle of substantial equivalence." Before I dissect why it represents the nub of what I called earlier one of the greatest conspiracies in the history of the food industry, let me give the floor again to James Maryanski, who continued to defend it stubbornly: "What we do know, is that the genes that are being introduced currently, to date, using biotechnology, produce proteins that are *very similar* to proteins that we've consumed for many centuries. . . . Using Roundup Ready soybeans as an example, this is a plant which has a modified enzyme that is *essentially the same* enzyme that's already in the plant: it has a *very small* mutation, so, in terms of safety, *there's no big difference* between that introduced enzyme and the one that already occurs in the plant" (emphasis added).

In other words, GMOs are roughly identical to their natural counterparts. And it is precisely this "roughly"—rather surprising coming from a microbiologist—that makes the concept of substantial equivalence suspect in the eyes of those who denounce its emptiness, such as Jeremy Rifkin of the Foundation on Economic Trends, one of the earliest opponents of biotechnology. "Here, in Washington, if you were to have an evening and go out and get a drink at one of the local haunts where all the lobbyists hang out, everybody would laugh about this. They all know this was a joke, this 'substantial equivalency.' This was simply a way to paper over the need for these companies, especially Monsanto, to move their products into the environment quickly, with the least amount of government interference. And I should say they were very, very good at getting their interests expressed," he said to me in July 2006.

Michael Hansen, the Consumers Union expert, drove the point home when I spoke to him around the same time. "The principle of substantial equivalence is an alibi with no scientific basis created out of thin air to prevent GMOs from being considered at least as food additives, and this enabled biotechnology companies to avoid the toxicological tests provided for in the Food, Drug, and Cosmetic Act and to avoid labeling their products. That's why we say that American regulations of transgenic foods violate federal law." To support his argument, Hansen showed me a document relating to an amendment to the Food, Drug, and Cosmetic Act, passed in 1958, entitled the Food Additive Act. As the name indicates, this amendment was aimed at regulating food additives such as coloring agents, preservatives, or "any substance the intended use of which results or may reasonably be expected to result, directly or indirectly, in its becoming a component or otherwise affecting the characteristics of any food (including any substance intended for use in producing, manufacturing, packing, processing, preparing, treating, packaging, transporting, or holding food)."

Following this definition, many are the substances that might be considered food additives, the safety of which would then have to be rigorously assessed through an obligatory procedure, including toxicological tests that might last, depending on circumstances, from twenty-eight days to two years. Answering to the "precautionary principle," as Congress required, the tests would have to demonstrate that there is a "a reasonable certainty that the substance in the minds of competent scientists is not harmful under its intended conditions of use." Excluded from the category of "food additives," and therefore not subject to toxicological tests, were substances "generally recognized as safe" (GRAS), either because they were "used in food before January 12, 1958," or because "scientific procedures" have shown that they pose no health risk.

I asked Maryanski, "Could you give me an example of substances classified as GRAS?"

"Yeah, those are common food processing enzymes, or salt, pepper, vinegar, things that have been used for many years and that the scientific community has established as safe."

"And how was the FDA able to decide that the gene introduced into a plant by genetic manipulation was GRAS?" I asked, looking him in the eye.

We had reached the heart of the debate between advocates and adversaries of GMOs. Indeed, even though no scientific study had yet been conducted to verify it, the FDA had decided a priori that transgenes did not fit into the category of food additives and that GMOs therefore could be marketed without prior toxicological testing. This is all the more curious because when the agency published its regulation, it had been considering a request that showed how essential it was to wait. The California biotech company Calgene (the one that had given Monsanto a chill by announcing in *Nature* that it had succeeded in producing Roundup-resistant tobacco) had filed a request for the approval of a tomato christened "Flavr Savr," manipulated to slow the ripening process.

There is no need to insist on the significance of a tomato tinkered with so that it can remain firm on supermarket shelves for an extended period. But it is important to know that it contained the kanamycin resistance gene and that its inventors had rightly concluded that the gene should be considered a "food additive." They had therefore asked a laboratory (the International Research and Development Corporation of Michigan) to conduct toxicology tests designed to measure the health effects of transgenic tomatoes on rats. But the FDA did not yet know the results of the study when it published its regulation. It was later found that seven of the forty test animals had died after two weeks for unexplained reasons and that a significant number of them had developed stomach lesions. Even so, adhering to its dogma, the agency had given Calgene the green light on May 18, 1994.

Before coming back to James Maryanski, let us look at the end of this appalling story. The cultivation of the transgenic tomato, which seemed so promising in the laboratory, turned out to be a catastrophe: yields in California were so low that the inventors decided to move production to Florida, where the crop was decimated by diseases. "There are so many things that can kill a plant, and it's all in the details," said a former plant breeder for Calgene.[28]

Flavr Savr was then shifted to Mexico, where the results were far from acceptable. As a 2001 FAO study soberly commented: "Since 1996, Flavr Savr tomatoes have been taken off the fresh produce market in the United States. The manipulation of the ripening gene appeared to have had unintended consequences such as soft skin, strange taste and compositional changes in the tomato. The product was also more expensive than non-modified tomatoes."[29]

In the interim, Calgene had fallen into the pocket of Monsanto, which had definitively buried the doomed tomato.

The L-Tryptophan Affair: A Strange Fatal Epidemic

Had Maryanski understood what I was getting at? In any event, he blinked nervously when I asked him on what scientific data the FDA had based its decision to declare transgenes to be GRAS. "What FDA was saying was: if you introduce a gene into a plant, that gene is DNA . . . and we have a long history of consuming DNA and we can establish that that is GRAS," he said, seeming to search for his words.

"If we come back to the example of Monsanto's soybeans, that means that the agency considers that a gene from a bacterium imparting resistance to a powerful herbicide is by definition less dangerous than a coloring agent?" I insisted.

"Correct," answered the former biotechnology coordinator, blinking even more rapidly.

The FDA's position, supported by Maryanski, infuriated Hansen, who pinpointed the question that Monsanto and its allies had always wanted to evade: "Currently, when you want to add a microscopic amount of a preservative or a chemical agent to a food product, it is considered a food additive and you therefore have to do all kinds of tests to prove that there is a reasonable certainty that it is safe. But when you manipulate a plant genetically, which can create countless differences in the food, you're not asked to do anything. In fact, the whole misunderstanding or confusion comes from the fact that the FDA has always refused to assess the technique of genetic manipulation and not just the final product; it made the assumption that biotechnology was intrinsically neutral, even though it had received a warning sign that should have made it much more cautious."

Hansen then told me the dramatic story of L-tryptophan, which has been thoroughly documented by Jeffrey Smith of the Institute for Responsible Technology, based in Fairfield, Iowa, a rigorous critic of GMOs.[30] L-tryptophan is an amino acid found naturally in turkey, milk, brewers' yeast, and peanut butter, among other things. A precursor to serotonin, it was prescribed in the form of a dietary supplement as a remedy for insomnia, stress, and depression. In the late 1980s, thousands of Americans suffered from a mysterious illness that was called eosinophilia-myalgia syndrome (EMS), because muscular pain (myalgia) was a symptom experienced by all victims. They also suffered from a litany of recurrent ailments: edema, coughs, skin lesions, respiratory difficulties, puckering of the skin, mouth ulcers, nausea, visual and memory problems, hair loss, and paralysis.

The strange epidemic was first reported on November 7, 1989, by Tamar Stieber, a reporter for the *Albuquerque Journal*, who had learned that all the victims had taken L-tryptophan (her reporting won her a Pulitzer Prize in 1990). Four days later, 154 cases were reported to medical authorities, and the FDA requested that the public avoid taking the dietary supplement. But the number of victims grew: a preliminary survey in 1991 counted thirty-seven dead and fifteen hundred permanently disabled.[31] According to later estimates by the Centers for Disease Control, EMS was fatal to one hundred patients and caused illness or paralysis in five thousand to ten thousand people.

As Jeffrey Smith reported, L-tryptophan in the United States was imported from Japan, where six producers shared the market. Investigation by the health services revealed that only the product made by Showa Denko was associated with the epidemic. Investigators then discovered that in 1984 the company had modified its production process by using biotechnology to increase yields: a gene had been introduced into the bacteria from which the substance was extracted after fermentation. The manufacturer gradually changed the genetic construct so that the final strain (Strain V), produced in December 1988, turned out to contain five different transgenes and a large number of impurities.[32]

Then began a strange battle about the origin of the disease, which everything indicated was directed primarily at discrediting the hypothesis that the disease could have been triggered by genetic manipulation. Some researchers argued that the problem could have come from a change in the filter used by Showa Denko to purify the product, but it was later shown that this change had not taken place until January 1989, *after* the outbreak of the epidemic. Others suggested that L-tryptophan itself was the problem, but as the expert Gerald Gleich pointed out, "Tryptophan itself clearly is not the cause of EMS in that individuals who consumed products from other companies than Showa Denko did not develop EMS."[33] Only Showa Denko was sued, and after settlements negotiated in 1992, it paid more than $2 billion in damages to more than two thousand victims.

Nonetheless, the FDA had decided in 1991 to permanently prohibit the sale of L-tryptophan, even if it was produced conventionally, and in subsequent official reports it does not even mention the fact that the strains involved were transgenic.[34] But one man at the FDA had very seriously considered the hypothesis that EMS might have been caused by the technique of genetic manipulation: James Maryanski.

In September 1991, six months before the FDA published its regulation on GMOs, according to a declassified document of which I have kept a copy, Maryanski met GAO representatives "at their request." They wanted "to discuss issues related to food biotechnology for the studies they are conducting on new technologies," he wrote. "They asked about L-Tryptophan and the potential that genetic engineering was involved. I said that we . . . do not yet know the cause of EMS, nor can we rule out the engineering of the organism."[35]

When I met the former FDA official in July 2006, he did not know that I was aware of this document. "The FDA had considered the use of genetic manipulation, but it had no information indicating that the technique itself could create products that would be different in terms of quality or safety," he said with assurance.

"Do you remember what happened with L-tryptophan in 1989?"

"Yes," he mumbled.

"It was a genetically manipulated amino acid. In theory, we know amino acids very well."

"That's right."

"It caused an epidemic of an unknown illness, EMS."

"That's true," he said. His eyes started blinking nervously.

"How many people died?"

"Well, but we have many—"

"At least thirty-seven. And more than one thousand disabled," I said.* "Do you remember?"

"I remember."

"According to a declassified FDA document, you said: 'We do not know the cause of EMS and we cannot rule out the manipulation of the organism.' You did say what I just read?"

"Yes."

But six months after his statement to the GAO representatives, Maryanski did not balk at signing the FDA document approving GMOs, which stated loud and clear: "The agency is not aware of any information showing that foods derived by these new methods differ from other foods in any meaningful or uniform way or that, as a class, foods developed by the new techniques present any different or greater safety concern than foods developed by traditional plant breeding."[36]

Beyond what it reveals about the FDA's blind spots, the L-tryptophan affair is exemplary in more ways than one. As Jeffrey Smith points out in *Genetic Roulette*, "The epidemic took years to identify. It was discovered only because the disease was rare, acute, came on quickly, and had a unique source. If one of these four attributes were not present, the epidemic might have remained undiscovered. Similarly, if common GM food ingredients are creating adverse reactions, the problems and their source may go undetected."[37]

Contrary to James Maryanski's assertions, FDA scientists were perfectly aware of the unknowns and the risks associated with biotechnology and GMOs, but the agency chose to ignore their warnings.

*I did not know at the time that the preliminary estimate of the number of victims was much lower than the reality.

8

Scientists Suppressed

"The Composition of Glyphosate-Tolerant Soybean Seeds Is Equivalent to That of Conventional Soybeans."

—Title of a study published by Monsanto in the *Journal of Nutrition*, April 1996

"When we finished the policy [relating to GMOs], all the scientists agreed with the policy," James Maryanski told me with sudden assurance.

"You mean there was a consensus on the principle of substantial equivalence?"

"All of the different views were taken into account in the agency's final decision about how it would proceed."

No Consensus at the FDA

Maryanski was out of luck. The day before our meeting. I had visited the Web site of the Alliance for Bio-Integrity, an NGO based in Fairfield, Iowa.[1] Headed by a lawyer named Steven Druker, it had sued the FDA for violation of the Food, Drug, and Cosmetic Act.[2] With scientists, clergy members, and consumers as plaintiffs, the complaint was filed in federal court in Washington in May 1998, together with the Center for Food Safety, an NGO established in 1997.[3] As one might have expected, the case was dismissed in October 2000, because the judge determined that the plaintiffs had not proved that the FDA regulation constituted a *deliberate* violation of federal law.[4]

Despite this legal setback, the complaint led to the declassification of some forty thousand pages of internal FDA documents related to GMOs. The least one can say is that this treasure trove of notes, letters, and memoranda presents a not-very-pretty picture of the way the agency handled this delicate issue, in light of its duty to protect the health of American consumers. In a document dated January 1993, FDA representatives acknowledged in plain language that, in accordance with government policy, their aim was to "promote" the biotechnology industry in the United States.[5] But the highlights of this mass of information are the reports written by agency scientists, intended to express their opinions on the draft regulations submitted to them. The Alliance for Bio-Integrity had the excellent idea of putting these documents online.[6] Some of them, of course, were addressed to the Biotechnology Coordinator.

For example, on November 1, 1991, Maryanski received a memorandum from the Division of Food Chemistry and Technology. The document pointed to all the "undesirable effects" that might be produced by the technique of genetic manipulation, such as an "increased levels of known naturally occurring toxicants, appearance of new, not previously identified toxicants, increased capability of concentrating toxic substances from the environment (e.g., pesticides or heavy metals), and undesirable alterations in the levels of nutrients."[7]

And on January 31, 1992, Samuel Shibko of the Toxicology Section of the FDA, wrote: "We cannot assume that all gene products, particularly those encoded by genes from non-food sources, will be digestible. For example, there is evidence that certain types of proteins . . . are resistant to digestion and can be absorbed in biologically active form."[8]

A few days later, it was the turn of Dr. Gerald Guest, director of the Center for Veterinary Medicine, to sound the alarm: "In response to your question on how the agency should regulate genetically modified food plants, I and other scientists at CVM have concluded that there is ample scientific justification to support a pre-market review of these products. . . . The FDA will be confronted with new plant constituents that could be of a toxicological or environmental concern."[9]

Dr. Louis Pribyl of the FDA's microbiology division dismissed out of hand the argument commonly put forth by promoters of biotechnology: "There is a profound difference between the types of unexpected effects from traditional breeding and genetic engineering, which is just glanced over in this document. . . . Multiple copies inserted at one site could become potential sites for rearrangements, especially if used in future gene transfer experiments, and as such may be more hazardous."[10]

I could continue with examples showing that many divisions of the FDA, whatever their specialty, expressed strong concerns about the unknown health effects that might result from the process of genetic manipulation. In contradiction to what Maryanski now claims, there was no consensus on the FDA's proposed regulation of GMOs even a few months before it was issued. Indeed, the former coordinator himself acknowledged this fact in a letter he sent on October 23, 1991, to Dr. Bill Murray, chairman of the Food Directorate, Canada: "There are a number of specific issues . . . for which a scientific consensus does not exist currently [in the FDA], especially the need for specific toxicology tests. . . . I think the question of the potential for some substances to cause allergenic reactions is particularly difficult to predict."[11]

During my meeting with Maryanski, I read to him a memorandum he had been sent on January 8, 1992, by Dr. Linda Kahl, a compliance officer with responsibility for summarizing her colleagues' views on the proposed regulation: "The document is trying to force an ultimate conclusion that there is no difference between foods modified by genetic engineering and foods modified by traditional breeding practices. This is because of the mandate to regulate the product not the process." She went on to note that this mandate resembled a "doctrine": "The processes of genetic engineering and traditional breeding are different, and according to the technical experts in the agency, they lead to *different risks*" (emphasis added).[12]

"What did you answer to Linda Kahl?" I asked Maryanski, who had lost his composure as soon as I began to read the document.

"My job was really to bring together the scientists who would be—provide the expertise to deal with, you know, to identify the issues and understand how to address them. I'm not the decision maker. The decision maker's ultimately the commissioner, Dr. David Kessler."

"Yes, but Dr. Kahl asked you a very specific question: 'Are we asking the scientific experts to generate the basis for this policy statement *in the absence of any data*?' (emphasis added). What was your answer?"

"Well, this is part of the early discussions that were going on."

"Are you sure? Linda Kahl wrote this memorandum to you in January

1992—three months before the FDA published its policy. How could it get scientific data in this very short time?"

"Right, but the policy was designed to provide the guidance to the industry for the kinds of testing they would need to do."

The Myth of Regulation

We had gotten to the point. Indeed, as Maryanski acknowledged, the document published by the FDA in 1992 was in no way a regulation, since its purpose was primarily to provide justifications for *not* regulating GMOs. It was only a statement of policy intended to provide direction to the industry and provide guidance in case of need. This was clearly indicated in the final section of the document, which provided for a mechanism for "voluntary consultation," if companies so desired: "Producers should consult informally with FDA on scientific issues or design of appropriate test protocols when the function of the protein raises concern or is not known, or the protein is reported to be toxic. FDA will determine on a case-by-case basis whether it will review the food additive status of these proteins."[13]

This outraged Joseph Mendelson, legal director of the Center for Food Safety. "In fact," he told me, "the health of American consumers is at the mercy of the goodwill of the biotech companies that are licensed to decide, with no government supervision, whether their GMO products are safe. This is absolutely unprecedented in the history of the United States. The policy was drafted so the biotechnology industry could propagate the myth that GMOs are regulated, which is completely false. In the process, the country has been turned into a huge laboratory where potentially dangerous products have been set loose for the last ten years without the consumer being able to choose, because, in the name of the principle of substantial equivalence, labeling of GMOs is banned, and there is no follow-up."

In March 2000, relying on various surveys indicating that more than 80 percent of Americans favored the labeling of transgenic foods[14] and 60 percent would avoid them if they had the choice,[15] the Center for Food Safety filed a citizen petition with the FDA asking it to review its policy on GMOs and that testing be required before they were sold and labeled.[16] When the agency failed to respond, the Center for Food Safety filed suit in federal court in the spring of 2006. "We won't give up," Mendelson told me, "especially because quite obviously the mechanism for voluntary consultation the FDA had set up wasn't working."

He showed me a study by Dr. Douglas Gurian-Sherman, a former FDA scientist who had worked on assessing transgenic plants before joining the Center for Science in the Public Interest.[17] He had gotten access to fourteen "voluntary consultation" files submitted to the FDA by biotechnology companies between 1994 and 2001 (out of a total of fifty-three), five of which concerned Monsanto. He found that in six cases, the FDA had asked the producer to provide more data so the agency could completely assess the safety of the products. "In three [50 percent] of those cases FDA's requests were either ignored by the developer or the developer affirmatively declined to provide the requested information." Two of these three cases concerned Monsanto's transgenic corn, notably MON 810, to which I will return. Monsanto had never provided the further information the FDA had requested to be able to determine whether GM corn was in fact substantially equivalent to its conventional counterpart. The agency could do nothing, because, as Dr. Gurian-Sherman noted, the policy document—unlike an actual regulation—gave it "no authority to require the developers to submit the desired additional data unless it decided to evaluate the crop as a food additive."

This was a decision the FDA made only once, on the Flavr Savr tomato at Calgene's request. A declassified document shows that that decision had little effect and that, despite the results of toxicological tests, the agency approved the product. On June 16, 1993, Dr. Fred Hines sent a memorandum to Linda Kahl concerning the three toxicological tests conducted on rats fed with transgenic tomatoes for twenty-eight days. "In the second study, gross lesions were described in the stomachs of four out of twenty female rats fed one of the two lines of transgenic tomato. . . . The Sponsor's . . . report concluded that . . . these lesions were incidental in nature. . . . The criteria for qualifying a lesion as incidental were not provided in the sponsor's report."[18] But one year later, the FDA gave its approval to the tomato with the long shelf life.

Dr. Gurian-Sherman also examined the data summaries companies provided to the FDA for their "voluntary consultation" and found that in three cases out of fourteen, they contained "obvious errors" that had not been detected by agency scientists during their review. This point is very important, because it underscores the imperfection (to put it mildly) of the process for approving food or chemical products as it is conducted around the world. Very seldom do companies provide the raw data of the tests they have conducted; they generally merely prepare a summary that reviewers sometimes only skim. As Dr. Gurian-Sherman very persuasively puts it: "The more highly summarized and less detailed those data, the greater the role of the developer in determining the safety of the crop, and conversely the more the FDA must rely on the developer's judgment."

He also analyzed the quality of the tests conducted by the producers, and his conclusions are troubling. He found that some fundamental health considerations were frequently neglected, such as the toxicity or allergenicity of proteins in the transgenic plants.

Finally, he raised a concluding technical point that is of primary importance because it undermines the validity of practically all the toxicological tests conducted on GMOs, particularly by Monsanto. Generally, to measure the toxicity and allergic potential of the proteins produced in the plant by the inserted gene, the companies did not use the proteins as they were expressed *in the manipulated plant*, but those present *in the original bacterium*, that is, before the gene derived from the bacterium was transferred. Officially, they proceeded in this way because it was difficult to remove a sufficient quantity of the pure transgenic protein from the plant but much easier to do so from the bacterium, which could produce as much protein as was needed.

In the view of some scientists, this practice might well represent a manipulation intended to conceal a fact that companies such as Monsanto had always made a point of denying: the inserted genes, and hence the proteins they produced, were not always identical to the original genes and proteins. Indeed, random insertion caused the appearance of unknown proteins. Dr. Gurian-Sherman concluded: "Therefore, bacterially produced protein may not be identical to, and have the same health effects as, the GE protein from the plant."

The Unshakable Team of Maryanski and Taylor

Even as FDA scientists were expressing their disagreement with the policy document, it was published on May 29, 1992. Two months earlier, on March 20, Commissioner David Kessler wrote a very curious memorandum to the secretary of health and human services, urgently requesting authorization to publish the document in the *Federal Register*. "The new technologies give producers powerful, precise tools to introduce improved traits in food crops, opening the door to improvements in foods that will benefit food growers, processors, and consumers. Companies are now ready to commercialize some of these improvements. To do so, however, they need to know how their products will be regulated. This is critical not only to provide them with a predictable guide to government oversight but also to help them with public acceptance of these new products. . . . Furthermore, the Biotechnology Working Group of the Council on Competitiveness wants us to issue a policy statement as soon as possible. . . . The approach and provisions of the policy statement . . . respond to the White House interest in assuring the safe, speedy development of the U.S. biotechnology industry."

The commissioner's memorandum concluded with the mention of "potential controversy," fostered by "environmental defense groups," including Jeremy Rifkin's: "They may challenge our policy as leaving too much decision making in the hands of industry and not adequately informing consumers." Attached to the memorandum was a copy of the policy statement with two very interesting notations: "Drafted: J. Maryanski. Cleared: M. Taylor."

"This document is proof that the FDA policy statement was not written to protect the health of Americans, but to satisfy strictly industrial and commercial aims," asserts Steven Druker of the Alliance for Bio-Integrity. "To reach its goal, the American government has continually lied to its own citizens and to the rest of the world, claiming that the principle of substantial equivalence was supported by a broad consensus in the scientific community and that a good deal of scientific data substantiated it: these two assertions are blatant lies. Decided on at the highest levels, with the active complicity of Monsanto, this huge enterprise of disinformation was carried out by an unshakable team: James Maryanski and Michael Taylor."

"What exactly was Maryanski's role?" I asked, a little shaken by the vehemence of his language.

"His role was to propagate the transgenic gospel inside and outside the agency. I met him several times, and he never deviated from the party line, even when he testified before Congress."

In fact, the complaint filed by the Alliance for Bio-Integrity had created a stir, and Maryanski was called to testify before the Senate Committee on Agriculture, Nutrition, and Forestry on October 7, 1999. After explaining at length the grounds for the FDA policy statement, he concluded his statement with this: "FDA takes seriously its mandate to protect consumers in the United States and to ensure that the United States' food supply continues to be one of the safest in the world. . . . We are confident that our approach is appropriate. It allows us to ensure the safety of new food products and . . . it gives manufacturers the ability to produce better products and provide consumers additional choices."

"Maryanski's other role was to smooth over differences inside the FDA, if necessary by stifling dissident voices with the support of Michael Taylor," Druker went on, showing me another declassified document that his organization had put online. This was a letter dated October 7, 1991, from the biotechnology coordinator to the deputy commissioner for policy, which contained the following: "Suggest that you consider discussing *your goals* for developing our food biotechnology policy by the end of the year with Dr. Guest, CVM. Most crops developed by the new biotechnology that will be used for human foods will also be used as feed for animals. . . . I think CVM would appreciate hearing your thoughts."[19] Maryanski was obviously putting Taylor forward to stifle the rebellion that was brewing in the Center for Veterinary Medicine. The document also shows that Michael Taylor, the former Monsanto lawyer, was the person who determined the purposes of the regulation that was then being drafted.

"Michael Taylor was Monsanto's man at the FDA, which hired him specifically to supervise the regulation of GMOs and created the position for that purpose," said Druker. "The declassified documents reveal that he worked to empty the policy statement of any scientific substance, which caused a good deal of discontent on the staff."

During my lengthy recorded telephone conversation with the former Monsanto vice president, he persistently denied any direct involvement in the preparation of the policy statement: "That's false. I wasn't the author of the policy. I was the deputy commissioner for policy who oversaw the process. But the policy was developed by the FDA's professional career people based on the law and the science."

When I reported these words to Michael Hansen, he literally jumped out of his chair and pulled out a document published in 1990 by the International Food Biotechnology Council (IFBC). This ephemeral body was set up in 1988 by the International Life Sciences Institute (ILSI), well known to all anti-GMO activists. Established in 1978 by major food industry corporations—the Heinz Foundation, Coca-Cola, Pepsi-Cola, General Foods, Kraft (owned by Philip Morris), and Procter and Gamble—ILSI calls itself a "non-governmental organization" and describes itself on its Web site as "a global network of scientists devoted to enhancing the scientific basis for public health decision-making."[20] As the British daily *The Guardian* revealed in 2003, the organization was well connected in the World Health Organization and the Food and Agriculture Organization, two UN bodies it lobbied in favor of GMOs through a document published in 1990 by the IFBC.[21] And it was precisely this document, a statement of principles on the way GMOs should be regulated, entitled "Biotechnologies and Food: Assuring the Safety of Foods Produced by Genetic Modification," that Hansen had just pulled out.[22]

"Remember that Michael Taylor came to the FDA in July 1991," Hansen went on. "Until then he'd been working at the law firm of King and Spalding. His clients included not only Monsanto but also the IFBC, the International Food Biotechnology Council. He wrote this document setting out the way the organization would like GMOs to be regulated. If you compare this proposal Taylor wrote for the IFBC and the policy statement published by the FDA, you can see they are very similar. If he didn't write the statement, then someone took his proposal and changed it slightly before publishing it." The anonymous IFBC document, oddly unavailable on the Web, is in fact the first reference cited in the appendix to the FDA policy statement.[23]

"Again, it's false," Taylor insisted. "I could not possibly have anything to do with it because I'm not a scientist. So, again, this is why you need to be talking to Dr. Maryanski and people who were actually involved in developing the FDA policy." When I subsequently interviewed Maryanski, he found it hard to get rid of this new hot potato. "Mr. Taylor was the deputy commissioner at the time, and he provided leadership for the project and served as the chief, sort of the leader . . . policy person, in terms of making sure the project got done."

"Did you know that he used to work for Monsanto as an attorney?"

"I think I knew that he had, you know, been at Monsanto, but, you know, we often have people come in and they're appointed as commissioner or deputy commissioner."

"What was the role of Monsanto in the FDA?"

"Well, Monsanto was very active, in fact very helpful to FDA in terms of helping us to understand just what does it mean to use genetic engineering in food crops. I remember meetings that we had where the Monsanto scientists met with the FDA scientists and they went through the kinds of modifications that they were making and how those were being done. And basically, what they were also saying to FDA was, 'How will these products be regulated?'"

The Champion of the Revolving Door

"You think it was a plot?" The question I asked Jeffrey Smith when I met him in Fairfield, Iowa in October 2006 made him pause for reflection. He is the executive director of the Institute for Responsible Technology and the author of two very well-informed books on GMOs that I have already referred to.[24] The silence is something I have encountered from most of those I have interviewed who have dared to denounce Monsanto's practices, because the company is so ready to threaten costly litigation. Smith knew this very well: he had been forced to self-publish his books because he could not find a publisher willing to stand up to Monsanto. Monsanto says that it is only trying to protect its patents, but the company has been willing to spend millions of dollars and even lose at trial, as if its real purpose were to bleed its opponents dry. This is why every word had to be weighed before it was launched into the public arena.

"The word 'plot' is a little strong," he finally answered. "But from the company's point of view, let's say it took power without a single misstep, thanks to its savoir-faire and its ability to infiltrate all the decision-making machinery in the country." Among the elements behind its success were financial contributions to the election campaigns of the two major parties. According to figures from the Federal Election Commission, in 1994, Monsanto contributed $268,732, almost equally divided between Democrats (then holding the White House) and Republicans. In 1998, the amount was $198,955, almost two-thirds for the Republicans. Two years later, George W. Bush's party received $953,660 compared to $221,060 for Al Gore's Democrats. Finally, in 2002, as the White House was launching its crusade against "international terrorism," the Republican party collected $1,211,908 compared to $322,028 for the Democrats. At the same time, lobbying expenses for the leading producer of GMOs were officially $21 million between 1998 and 2001, with a record of $7.8 million in 2000, the year of Bush's election.*

Probably more decisive than these political expenses—rather modest in American terms—was the ability to infiltrate, illustrated by a system already glimpsed in the case of bovine growth hormone: the revolving door, at which, according to Smith, "Monsanto is the national champion." "Take the Bush administration," he said, showing me a list covering several pages. "Four important departments are headed by people close to Monsanto, either because they've received contributions from the company or because they have worked directly for it. Attorney General John Ashcroft was backed by Monsanto when he ran for reelection in Missouri, and the company supported Tommy Thompson, the secretary of health and human services [which oversees the FDA], when he ran for governor in Wisconsin. Ann Venneman, the secretary of agriculture, was on the board of directors of Calgene, owned by Monsanto. Secretary of Defense Donald Rumsfeld was the CEO of Searle, a Monsanto subsidiary. And let's not forget Clarence Thomas, who was a lawyer for Monsanto before working for Senator Danforth of Missouri and later being appointed to the Supreme Court."

On Smith's list, which can in part be found on the Web, one discovers that the revolving door moves people in at least four directions.[25] First, consider movements from the White House to Monsanto. For example, Marcia Hale, former assistant to President Bill Clinton for intergovernmental affairs, was appointed director of international government affairs for Monsanto in 1997. Her colleague Josh King, former director of production for White House events, has continued his career as director of global communications in Monsanto's Washington office. Mickey Kantor, U.S. trade representative from 1992 to 1997 and commerce secretary from 1996 to 1997, immediately thereafter joined the company's board of directors, and so on.

The second direction is that taken by former members of Congress and their staffs, who have become registered lobbyists for the company, such as former Democratic congressman Toby Moffett, who became a political strategist for Monsanto, and Ellen Boyle and John Orlando, former congressional staffers later hired as lobbyists.

*For 2000, 2001, and 2002, these expenses also include lobbying by Pharmacia, which acquired Monsanto in 2000 and resold it in 2002. These figures can be consulted on the Web site of Capital Eye: www.capitaleye.org-monsanto.asp. (*Translator's note*: This site is no longer active.)

The revolving door also moves people from the regulatory agencies to Monsanto. We have already seen that Linda Fisher was appointed Monsanto vice president for governmental affairs in 1995 after serving as assistant administrator of the EPA, and William Ruckelshaus, who headed the agency from May 1983 to January 1985, later joined the company's board of directors. Similarly, Michael Friedman, former deputy director of the FDA, was hired by Monsanto's pharmaceutical subsidiary Searle.

But the flow of people is even stronger in the other direction, from Monsanto to governmental or intergovernmental agencies. Recall that in 1989 Margaret Miller moved from the company's labs to the FDA. Her colleague Lidia Watrud joined the EPA. Virginia Meldon, former Monsanto public relations director, was hired by the Clinton administration. More recently, Rufus Yerxa, former chief counsel for Monsanto, was appointed U.S. representative to the World Trade Organization (WTO) in August 2002, and in January 2005, Martha Scott Poindexter was hired by the Senate Committee on Agriculture, Nutrition, and Forestry after serving as director of governmental affairs for Monsanto's Washington office. Finally, Robert Fraley, one of the "discoverers" of Roundup Ready soybeans, who became a Monsanto vice president, was named a technical adviser to USDA.

Dan Glickman: "I Had a Lot of Pressure on Me"

"You know, the revolving door is not just in agriculture. It tends to be in many, many areas, finance areas, health care." These were the words not of an anti-GMO activist but of Dan Glickman, Bill Clinton's secretary of agriculture from March 1995 to January 2001, whom I interviewed in Washington on July 17, 2006. Known for having been a strong advocate of biotechnology, he had been familiar with the USDA long before taking charge of it: he had represented Kansas in Congress for eighteen years and chaired the House Agriculture Committee.

When he arrived at this strategic department, which then had an annual budget of $70 billion and more than 100,000 employees throughout the country, it had changed a good deal since being established in 1862 by Abraham Lincoln, who called it the "people's department," because it was supposed to be at the service of farmers and their families, then 50 percent of the population. One hundred and forty years later, its many detractors call it the "Agribusiness Department" or "USDA, Inc." because it is accused of serving the interests of the companies that control the production, processing, and distribution of food. "These industry-linked appointees have helped to implement policies that undermine the regulatory mission of USDA in favor of the bottom-line interests of a few economically powerful companies," writes Philip Matera in a 2004 article titled "USDA, Inc.: "How Agribusiness Has Hijacked Regulatory Policy at the US Department of Agriculture."[26]

To illustrate his argument, the former journalist, now working at Good Jobs First in Washington, took the example of biotechnology, for which, he said, the USDA had become one of the most fervent promoters. Begun under the first Bush administration, this direction was followed by the Democratic administration of Bill Clinton, whose campaign director was Mickey Kantor, later U.S. trade representative and commerce secretary, and, as I've already noted, later a member of the Monsanto board of directors. In 1999, the intransigent American trade representative became famous for the harsh comments and the threats he made against his European counterparts when they announced their intention to label GMO products. In this area, his greatest ally was Dan Glickman.

The *St. Louis Post-Dispatch* once referred to Glickman as "one of biotechnology's leading boosters, admonishing reluctant Europeans not to stand in the way of progress."[27] Clinton's agriculture secretary firmly believed in the benefits of genetic manipulation: "I believe that biotechnology has enormous potential for consumers, for farmers, and for the millions of hungry and malnourished people in the developing world" was the language he was still using in April 2000 in a speech to the Council for Biotechnology Information.[28] He had already seen the fervor of people on the opposite side of the issue: at the World Food Summit, held under the auspices of the FAO in Rome in November 1996, governments had just committed themselves to cutting the numbers of the malnourished in half by 2015, and the American representative was holding a press conference. Greenpeace activists who had gotten forged press credentials stood up, took off their clothes, and displayed anti-GMO slogans on their naked bodies as they pelted Glickman with Roundup Ready soybeans.

Appointed Secretary of Agriculture just after Monsanto's transgenic soybeans had gone on the market, Dan Glickman was the one who authorized all subsequent GMO crops. When I met him in July 2006, he had completely changed hats: in September 2004 he had been appointed CEO of the Motion Picture Association of America, which brings together the six majors in Hollywood. I had asked to interview him, of course, because of the position he had held in the Clinton administration, but also because he had expressed some regrets in a *Los Angeles Times* article published on July 1, 2001: "Regulators even viewed themselves as cheerleaders for biotechnology. It was viewed as science marching forward, and anyone who wasn't marching forward was a Luddite."

I read him the quotation and asked him why he had said that.

"When I became secretary of agriculture [in 1995], . . . most of the regulatory climate was basically focused on approvals, approvals of the crops, facilitating the transfer of the technology into agriculture in this country and pushing the export regime for these. I found that there was a general feeling in agribusiness and inside our government in the U.S. that if you weren't marching lock-step forward in favor of rapid approvals of GMO crops, then somehow you were anti-science and anti-progress."

"Do you think that the Monsanto soy, for instance, should have received more scrutiny?"

"Well, I think that, frankly, there were a lot of folks in industrial agriculture who didn't want as much analysis as probably we should have had, because they had made a huge amount of investments in the product. And certainly when I became secretary, given the fact that I was in charge of the department regulating agriculture, I had a lot of pressure on me to push the issue too far, so to speak. But I would say even when I opened my mouth in the Clinton administration, I got slapped around a little bit by not only the industry, but also some of the people even in the administration. In fact, I made a speech once where I said we needed to more thoroughly think through the regulatory issues on GMOs. And I had some people within the Clinton administration, particularly in the U.S. trade area, that were very upset with me. They said: 'How could you, in agriculture, be questioning our regulatory regime?'"

Mickey Kantor was probably involved in that pressure. The speech Glickman mentioned did contain some surprises, breaking as it did with the line he had followed until then. Speaking at the National Press Club in Washington on July 13, 1999, the secretary of agriculture began with a stirring tribute to the "promise of biotechnology," speaking of "bananas that may one day deliver vaccines to children in developing countries." (In this vein, I might mention that eight years later we were still waiting for the appearance of these magical GMOs that had been announced back in the 1980s. Except for plants resistant to herbicides or producing insecticides, we have seen nothing.)

"With all that biotechnology has to offer, it is nothing if it's not accepted," Glickman went on in his speech, before speaking the words that so infuriated his colleagues in foreign trade and likely Monsanto. "This boils down to a matter of trust, trust in the science behind the process, but particularly trust in the regulatory process that . . . must stay at arm's length from any entity that has a vested interest in the outcome. At the end of the day many observers, including me, believe *some type of informational labeling* is likely to happen."[29]

The words were cautious, but they were the ones picked up by the press the next day. His conclusion was a real shot across Monsanto's bow: "Industry needs to be guided by a broader map and not just a compass pointing toward the bottom line. Companies need to continue to monitor products, after they've gone to market, for potential danger to the environment and maintain open and comprehensive disclosure of their findings. . . . We don't know what biotechnology has in store for us in the future, good and bad, but . . . we're going to make sure that biotechnology serves society, not the other way around."

Glickman says today that he would not change a word of his 1999 speech. "The Congress never really got into it too much."

"Why?"

"Well, first of all, it's complicated, okay? Any issue that is technical and complicated is very hard for a legislative body to get into. After all, like in Europe, in the United States most members of Congress are not scientists."

Scientists under the Influence

The point may seem simplistic, but I am convinced that it explains in part politicians' lack of interest in the issues raised by biotechnology. For my part, it took me months of intense work before I could claim to have come to a reasoned and reasonable opinion about genetic manipulation. I would even say that Monsanto has been able to gain acceptance for its products so easily precisely because it was able to take advantage of the fact that it was a "complicated subject" that only scientists seemed able to master. To guarantee its domination, the company understood that it had to control the scientists who discussed the subject and to make sure that they spoke in the right places, such as international forums sponsored by UN organizations and renowned journals and universities.

Evidence is provided by an internal Monsanto document marked "company confidential" that arrived mysteriously (clearly from a whistle-blower) at the office of GeneWatch, a British association that keeps a close watch on GMO issues.[30] This ten-page "Monthly Summary," made public on September 6, 2000, details the activities of Monsanto's Regulatory Affairs and Scientific Outreach team during the months of May and June 2000. "The leaked report shows how Monsanto are trying to manipulate the regulation of GM foods across the globe to favour their interests," said Dr. Sue Mayer, GeneWatch UK's director, in a press release. "It seems they are trying to buy influence with key individuals, stack committees with experts who support them, and subvert the scientific agenda around the world."

The document congratulates the team for having been "instrumental in assuring that key internationally recognized scientific experts were nominated to the FAO/WHO expert consultation on food safety which was held in Geneva this past month. The consultation and final report were very supportive of plant biotechnology, including support for the critical role of substantial equivalence in food safety assessments. . . . Information on the benefits and safety of plant biotechnology was provided to key medical experts and students at Harvard. . . . An editorial was drafted by Dr. John Thomas (Emeritus Professor of U. of Texas Medical School in San Antonio) to place in a medical journal as the first in a planned series of outreach efforts to physicians. . . . A meeting was held with Prof. David Khayat, an internationally well known cancer specialist, to collaborate on an article demonstrating the absence of links between GM food and cancers. . . . Monsanto representatives were successful at the recent Codex Food Labeling Committee meeting in maintaining two labeling options for further consideration by the committee." There is much more of the same.

Among the scientists who generously cooperated with the Monsanto team's initiatives, the report also refers to Domingo Chamorro from Spain, Gérard Pascal and Claudine Junien from France, and Nobel Prize winner Jean Daucet from France, who participated in the Forum des Biotechnologies launched by the team.

Reading this document makes it easier to understand why the WHO and the FAO organized a "consultation," like the one described in the report, in Geneva from November 5 to 10, 1990. Titled "Strategies for Assessing the Safety of Foods Produced by Biotechnology," it brought together representatives from international health authorities as well as "experts," including James Maryanski as a member of the secretariat.* Oddly, although no GMO had yet seen the light of day, this "consultation" produced the following peremptory diagnosis: "The DNA from all living organisms is structurally similar. For this reason, the presence of transferred DNA in produce in itself poses no health risk to consumers." The reference cited in the appendix was the article published by Monsanto scientists a short time earlier in *Nature* on the transgenic growth hormone, which, as I have noted, had been strongly challenged.[31]

From then on, it is very clear that Monsanto played a major role in imposing, internationally and with no scientific data, the principle of "substantial equivalence." It appeared, for instance, in 1993, in an OECD document entitled "Safety Evaluation of Foods Derived by Modern Biotechnology: Concepts and Principles." This seventy-one-page document begins with a long argument designed to establish that "biotechnology" has existed ever since humanity learned how to select plants and hence that the techniques of genetic manipulation are only a modern extension of ancestral knowledge. On that basis, it argues: "For foods and food components from organisms developed by the application of modern biotechnology, the most practical approach to the determination of safety is to consider whether they are substantially similar to analogous conventional food product(s), if such exist." To back up this new concept, which came out of nowhere, the report relies on the example of GMOs such as Calgene's long-shelf-life tomato (which was, of course, withdrawn from the market) and Monsanto's Roundup Ready tomato (which remained at the experimental stage).

Among the authors of this founding document was the ubiquitous James Maryanski as well as a representative of the President's Council on Competitiveness. In an appendix, the document lists ten publications to consult, including one from the International Life Sciences Institute (established, it will be recalled, by agribusiness companies), the notorious document from

*According to his CV, James Maryanski served as an expert for WHO and FAO, then as a U.S. delegate to the Codex Alimentarius Committee and the Organisation for Economic Co-operation and Development (OECD).

the International Food Biotechnology Council drafted in part by Michael Taylor, and the report of the WHO/FAO 1990 "consultation." Like the other documents cited as references, none of these publications involves scientific studies conducted to assess the safety of GMOs, for a simple reason: there were none.

A year later, it was the turn of the WHO to carry the torch for this vigorously conducted propaganda campaign. From October 31 to November 4, 1994, it sponsored a workshop with an unambiguous title: "Application of the Principle of Substantial Equivalence to the Safety Evaluation of Foods or Food Components from Plants Derived by Modern Biotechnology." This time the "principle of substantial equivalence" was carved in stone, even though there was still no new scientific evidence. And to prove that their work was indeed serious, the participants in the workshop, including Dr. Roy Fuchs from Monsanto, pointed out that "the comparative approach was first proposed by WHO/FAO, and was further developed by OECD."

The circle was fully completed two years later when the FAO and the WHO hammered home the point—two UN organizations amount to something—by organizing a second joint consultation, from September 30 to October 4, 1996 (in which both James Maryanski and Roy Fuchs participated). The timing was critical: the first shipments of Roundup Ready soybeans were already on their way to Europe. The final report, which is unavailable online (though I managed to get hold of a copy), is frequently cited as the international document of reference for the principle of substantial equivalence. It includes the following "scientific" information: "When substantial equivalence is established for an organism or food product, the food is regarded to be as safe as its conventional counterpart and no further safety consideration is needed. . . . When substantial equivalence cannot be established, it does not necessarily mean that the food product is unsafe. Not all such products will necessarily require extensive safety testing."

A Questionable Study

As Sussex University professor of science policy Erik Millstone pointed out in 1999: "The concept of substantial equivalence has never been properly defined; the degree of difference between a natural food and its GM alternative before its 'substance' ceases to be acceptably 'equivalent' is not defined anywhere, nor has an exact definition been agreed by legislators. It is exactly this vagueness which makes the concept useful to industry but unacceptable to the consumer. Moreover, the reliance by policy makers on the concept of substantial equivalence acts as a barrier to further research into the possible risks of eating GM foods."[32]

Monsanto used and abused the concept, and it had no qualms about rewriting its history to vindicate the safety of its GMOs by referring to the imprimatur of UN organizations, precisely the goal of the series of maneuvers I have just recounted. "A basic principle in the regulation of foods and feeds produced from plant biotechnology is a concept called 'substantial equivalence,'" explains an April 1998 promotional document for Roundup Ready soybeans addressed to farmers. "It was established in the early 1990s by the Food and Agriculture Organization of the United Nations (FAO),the World Health Organization (WHO) and the Organization for Economic Cooperation and Development (OECD)." This unanswerable argument is frequently set forth in official company documents, usually along with another designed to contribute scientific backing to it: "To establish 'substantial equivalence,' the composition of Roundup Ready soybeans was compared to conventional varieties. . . . In total, more than 1,800 independent analyses were conducted and conclusively demonstrated that the composition of RR soybeans is equivalent to other soybeans on the market. . . . In addition, feeding studies performed across the zoological spectrum (broiler chickens, dairy cattle, catfish, and rats) demonstrate the nutritional equivalence of Roundup Ready soybeans."

Thus began the final phase of the "action plan" developed, as I have noted, in October 1986. Knowing that the launch of Roundup Ready soybeans had to go off without a hitch, because it would blaze a trail for all subsequent GMOs, Monsanto decided to use the mechanism of "voluntary consultation," provided for in the FDA policy statement. Roy Fuchs, Monsanto's director of regulatory science and an assiduous attendee of UN workshops, was asked to design two studies intended to provide scientific proof that the principle of substantial equivalence had a solid basis (which confirms that, at this stage, the documents of the FAO, the WHO, and the OECD were purely theoretical and were not based on any scientific data.

The first study was designed to compare the organic composition of Roundup Ready soybeans with that of conventional soybeans, particularly by measuring levels of protein, fat, fiber, carbohydrates, and isoflavones in the two varieties—that is, all the already known components of the plant. In other words, there was no attempt to find out whether transgenic soybeans contained in their molecular structure unknown or (slightly) transformed substances due to the effects of genetic manipulation. Under the supervision of Stephen Padgette, the study was finally published in 1996 in the *Journal of Nutrition*, a reputable scientific journal, and its conclusions were unsurprising, as can be gathered from the title: "The Composition of Glyphosate-Tolerant Soybean Seeds Is Equivalent to That of Conventional Soybeans."[33]

But this study was far from universally accepted, particularly because its authors had "omitted" some data, as Marc Lappé, a noted toxicologist and the founder of CETOS (Center for Ethics and Toxics) in Gualala, California, discovered. "What did the omitted data show?" he asked in the *Los Angeles Times* in 2001. "Significantly lower levels of protein and one fatty acid in Roundup Ready soybeans. Significantly lower levels of phenylalanine, an essential amino acid that can potentially affect levels of key estrogen-boosting phytoestrogens, for which soy products are often prescribed and consumed. And higher levels of the allergen trypsin inhibitor in toasted Roundup Ready soy meal than in the control group of soy."[34]

A neophyte may find these technical details a little daunting, but I have taken the trouble to quote them here to emphasize that when it comes to food safety, one cannot be satisfied with the approximation implicit in the principle of substantial equivalence. In other words, either transgenic soybeans are exactly similar to their conventional counterparts or they are not. And if they are not, in what way are they different, and what are the possible health risks?

Precisely in order to settle the issue, Marc Lappé (who died in 2005) and his colleague Britt Bailey decided to repeat Stephen Padgette's experiment. "For our study," Bailey told me when I met her in San Francisco in October 2006, "we planted Roundup Ready soybean seeds and seeds from conventional lines, with the only difference being that the Monsanto seeds had the Roundup Ready gene. We grew the plants in strictly identical soil, with the same climatic conditions for each of the two groups. The transgenic soy plants were sprayed with Roundup following Monsanto's directions. At the end of the season we harvested the beans from the two groups and we compared their organic composition."

"What were the results?"

"We offered our study to the *Journal of Medicinal Food*, which sent it out for review. It was accepted and publication was scheduled for July 1, 1999.[35] Oddly, a week before publication, when according to normal practice the article was still under embargo, the American Soybean Association [ASA], known to be tied to Monsanto, issued a press release claiming that our study was not rigorous. We never found out the source of the leak."

I located the press release from the association (whose vice president I met soon thereafter): "ASA has confidence in the regulatory reviews of Roundup Ready soybeans conducted by U.S. and global regulatory agencies and the underlying scientific studies that found equivalence in isoflavone content between Roundup Ready soybeans and conventional soybeans."[36]

"How do you explain the fact that Monsanto found the two soybeans equivalent?" I asked Britt Bailey.

"I think the principal flaw in their study is that they did not spray the plants with Roundup, which completely invalidates the study, because Roundup Ready soybeans are made to be sprayed with the herbicide."

"How do you know?"

"Because of a blunder by Monsanto's legal department."

Britt Bailey showed me a letter from Tom Carrato, one of Monsanto's attorneys, to Vital Health Publishing, which was then about to publish a book she and Marc Lappé had written on GMOs. This letter, dated March 26, 1998, says a great deal about the company's practices. After explaining that he had learned of the imminent publication from an article in *Winter Coast Magazine*, he writes with disconcerting self-confidence: "The authors of the book assert that Roundup is 'toxic.' What do they mean by toxic? Every substance that exists, whether synthetic or found in nature, is able to produce toxicity at some dose. . . . Anyone who has consumed several cups of coffee or observed a person drinking alcohol understands the dose-response relationship and the idea of threshold. . . . These errors must be corrected prior to publication . . . because they disparage and potentially libel the product." Later in the letter, Carrato defends the study conducted by Stephen Padgette, and makes a damaging admission: "Studies of *unsprayed* [emphasis added] RR soybeans show no difference in estrogen levels. Those studies were reported in a peer-reviewed article in the *Journal of Nutrition* in January 1996."

"Anyway, the letter was effective," said Bailey, "because our publisher decided not to publish the book and we had to find another one."[37]

"Do you know whether the Roundup residues inevitably found on transgenic soybeans have been assessed from a health perspective?"

"Never. While we were writing our book, we discovered that in 1987 the authorized level of glyphosate residue on soybeans was 6 ppm. Then, strangely, in 1995, a year before Roundup Ready soybeans came on the market, the level permitted by the FDA rose to 20 ppm. I talked to Phil Errico of the toxicology branch of the EPA, and he told me: 'Monsanto provided us with studies showing that 20 ppm did not pose any health risk and the authorized level was changed.' Welcome to the United States."

To be fair, Europe hardly does any better. According to information published in *Pesticides News* in September 1999, in response to the importation of transgenic soybeans from America, the European Commission multiplied the authorized level of glyphosate residue by 200, raising it from 0.1 ppm to 20 ppm (mg/kg).

Bad Science

"Monsanto should not have to vouchsafe the safety of biotech food," Phil Angell declared in October 1998. "Our interest is in selling as much of it as possible. Assuring its safety is the FDA's job."[38] The quotation didn't even bring a smile to the face of James Maryanski, who claimed that he ate transgenic soy every day, "because in the United States, 70 percent of foods in the grocery store contain GMOs. FDA is confident that the soybean, in terms of food safety, is as safe as other varieties of soybean."

"How is the FDA confident about that?"

"It's based on all the data that the company provided to the FDA, that was reviewed by FDA scientists. And so it's not in a company's interest to try to design a study in some way that would mask results."

One would like to share Maryanski's optimism, but it is seriously open to question. At least that is the impression I had after a long conversation with Professor Ian Pryme on November 22, 2006, in his laboratory in the department of biochemistry and molecular biology at the University of Bergen, Norway. In 2003, this British scientist and a Danish colleague, Professor Rolf Lembcke (since deceased), decided to analyze the few toxicological studies that had been conducted on transgenic foods.[39] One of the studies was the second one published in 1996 by Monsanto researchers intended to assess the possible toxicity of Roundup Ready soybeans.[40]

"We were very surprised to find that there were only ten studies in the scientific literature," Pryme told me. "That's really very few, considering what's at stake."

"How do you explain it?"

"First you should know that it is very hard to get hold of samples of transgenic materials because companies control access to them. The companies require a detailed description of the research project and they are very reluctant to provide their GMOs to independent scientists for testing. When you insist, they invoke 'business confidential.' It's also very hard to get financing for studies on the long-term effects of transgenic foods. Along with colleagues from six European countries, we requested funds from the European Union, which refused on the pretext that the companies themselves had already conducted that kind of test."

"What can you say about Monsanto's study on rats, chickens, catfish, and dairy cattle?" I asked.

Pryme continued, "It's very important, because it was used as the basis for the principle of substantial equivalence and it explains in part the absence of further studies. But I have to say that it is very disappointing from a scientific point of view. If I had been asked to review it before publication, I would have rejected it, because the data provided are insufficient. I would even say that it is bad science."

"Did you try to get the raw data from the study?"

"Yes," Pryme answered, "but unfortunately, Monsanto refused to provide them on the grounds that they were business confidential. That was the first time I had heard that argument used about research data. Normally, as soon as a study is published, any researcher can ask to consult the raw data, to repeat the experiment and contribute to scientific progress. Monsanto's refusal inevitably gives the impression that the company has something to hide: either that the results were not really convincing, or they were bad, or that the methodology and protocol used were not good enough to stand up to rigorous scientific analysis. To conduct our study we had to be satisfied with the summary provided by the company to the regulatory agencies. And there are some very troubling things.

"For example, about the rat study, the authors write: 'Except for the brown color, the livers appeared normal at necropsy . . . it was not considered to be related to genetic modification.' How could they claim that without taking liver sections and examining them under the microscope to be sure that the brown color was normal? They were apparently satisfied with eyeballing the organs, which is not a scientific way of conducting a post mortem study. Likewise, the authors state: 'Livers, testes, and kidneys were weighed' and 'several differences were observed,' but 'they were not considered to be related to genetic modification.' Once again, how could they claim that? They apparently did not analyze the intestines or the stomachs, which is a very serious fault in a toxicological study. They also say that forty tissues were sampled but they don't say which ones. Besides, I only know of twenty-three tissue types that have been recorded, such as skin, bone, spleen, thyroid. What are the others?

"In addition, the rats used for the experiment were eight weeks old: too old. For a toxicological study you usually use young test animals to see whether the substance tested has an impact on the development of the growing organism. The best way of masking possible harmful effects is to use older test animals, especially because, despite the anomalies observed, the study lasted for only twenty-eight days, which is not long enough. The last paragraph of the study provides a good sense of the general impression: 'The animal feeding studies provide *some* reassurance that no major changes occurred in the genetically modified soybeans.' I don't want 'some reassurance,' but 100 percent reassurance! In fact, when you know that this study justified the introduction of GMOs into the food chain you can only be worried. But what can be done? Look at what happened recently to my colleague Manuela Malatesta," Pryme concluded.

Fear of Monsanto

I met Manuela Malatesta on November 17, 2006, at the University of Pavia. She was still traumatized by the recent events that had forced her to leave the University of Urbino, where she had worked for more than ten years. "It was all because of a study on the effects of transgenic soybeans," she told me.[41] The young researcher had done something that no one else had: she had repeated Monsanto's 1996 toxicological study. She and her research team had fed one group of mice a normal diet (control group) and another with same diet to which had been added Roundup Ready soybeans (experimental group). The test animals were followed from the time they were weaned until they died, on average two years later. "We studied the rats' organs under an electron microscope," she told me, "and we found statistically significant differences, particularly in the nuclei of the liver cells of mice fed with transgenic soybeans. Everything seemed to show that the livers had had increased physiological activity. We found similar changes in pancreas and testicle cells."

"How do you explain those differences?"

"Unfortunately, we would have liked to follow up these preliminary studies, but we couldn't because the financing stopped. So we only have hypotheses: the differences can be due to the composition of the soybeans or to the Roundup residues. Let me specify that the differences we observed were not lesions, but the question is what biological role they may play in the long run, and for that we would need another study."

"Why don't you do it?"

"Well, research on GMOs is now taboo. You can't find money for it. We tried everything to find more financing, but we were told that because there are no data in the scientific literature proving that GMOs cause problems, there was no point in working on it. People don't want to find answers to troubling questions. It's the result of widespread fear of Monsanto and of GMOs in general. Besides, when I discussed the results with some of my colleagues, they strongly advised me against publishing them, and they were right, because I lost everything—my laboratory, my research team. I had to start over from nothing in another university, with the help of a colleague who supported me."

"Do GMOs worry you?"

"Now they do. Yet at the beginning I was convinced that they didn't pose any problems, but now the secrets, the pressures, and the fear surrounding them make me doubt."

9

Monsanto Weaves Its Web, 1995–1999

> What you're seeing is not just a consolidation of seed companies, it's really a consolidation of the entire food chain.
>
> —Robert Fraley, co-president, Monsanto's Agriculture Sector, quoted in *Farm Journal*, October 1996

"As a scientist actively working in the field, I find it's very unfair to use our fellow citizens as guinea pigs." Broadcast on August 10, 1998, on ITV's documentary program *World in Action*, these few words about GMOs ruined the career of Arpad Pusztai, an internationally renowned biochemist who had worked for thirty years, from 1968 to 1998, at the Rowett Research Institute in Aberdeen, Scotland. "I think they'll never forgive me for saying that," he told me when I met him at his home on November 21, 2006. A sly grin lit up the face of the nearly eighty-year-old man.

"Who are they?" I asked, suspecting the answer.

"Monsanto, and everyone in Great Britain who blindly supports biotechnology. I would never have thought that I could be a victim of practices that recall what Communist regimes did to their dissidents."

The Accursed Potatoes

The son of a Hungarian resistance fighter opposing Nazi occupation, Arpad Pusztai was born in Budapest in 1930. When Soviet tanks invaded the Hungarian capital in 1956, he fled to Austria, where he was granted political refugee status. After obtaining a degree in chemistry, he won a fellowship from the Ford Foundation, allowing him to study in the country of his choice. He chose Great Britain, which represented for him the "country of freedom and tolerance." After earning a doctorate in biochemistry from the University of London, he was hired by the prestigious Rowett Institute, considered the best European nutrition laboratory. He specialized in lectins, proteins naturally found in certain plants that act as insecticides and protect the plants against aphid infestation. While some lectins are toxic, others are harmless for humans and mammals, such as the lectin from snowdrop plants, to which Pusztai devoted six years of his life. His expertise was so renowned that in 1995, the Rowett Institute offered to renew his contract even though he had reached retirement age, so that he could take charge of a research program financed by the Scottish Agriculture, Environment, and Fisheries Ministry.

This large contract, with funding of £1.6 million and employing thirty researchers, had the purpose of assessing the impact of GMOs on human health. "We were all very enthusiastic," Arpad Pusztai told me, "because at the time, when the first transgenic soybean crop had just been planted in the United States, no scientific studies had been published on the subject. The ministry thought our research would provide support for GMOs as they were about to arrive on the British and European markets. Because, of course, no one thought—least of all me, a strong supporter of biotechnology—that we were going to find problems." He was so enthusiastic that when Monsanto's toxicological study on Roundup Ready soybeans was published in the *Journal of Nutrition* in 1996, he thought that it was "very bad science" and that he and his team could do better. "I thought if we could show, with a scientific study worthy of the name, that GMOs were really harmless, then we would be heroes."

With the ministry's agreement, the Rowett Institute decided to work on transgenic potatoes that its researchers had already successfully developed, inserting into them the gene encoding snowdrop lectin (known as GNA). "Preliminary studies had shown that potatoes effectively resisted aphid infestation," Pusztai told me. "We also knew that in its natural state GNA was not harmful to rats, even when they absorbed a dose eight hundred times that produced by GMOs. What remained to be done was to assess the possible effects of transgenic potatoes on rats."

The protocol of the experiment provided for following four groups of rats through the period of 110 days from weaning: "In human terms, that would be the equivalent of following a child from the age of one to nine or ten, that is, during the time of rapid growth of the body." In the control group, rats were fed with conventional potatoes. In the two experimental groups, the test animals were fed with transgenic potatoes from two different lines. Finally, in the fourth group, the menu included conventional potatoes to which a quantity of natural lectin directly extracted from snowdrops had been added. "My first surprise," he recalled, "came when we analyzed the chemical composition of the transgenic potatoes. First, we found that they were not equivalent to conventional potatoes. Further, they were not equivalent among themselves, because from one line to the next, the quantity of lectin expressed could vary up to 20 percent. This was the first time I expressed doubts about whether genetic manipulation could be considered a technology, because for a classic scientist like me, the very principle of technology means that if a process produces an effect, that effect has to be strictly the same if you repeat the same process under identical conditions. In this case, apparently, the technique was very imprecise, because it did not produce the same effect."

"How do you explain it?"

"Unfortunately, I only have hypotheses that I never had the means to verify. To clearly understand the imprecision of what is inaccurately called 'biotechnology,' generally carried out with a gene gun, think of William Tell, who was blindfolded before he shot an arrow at a target. It is impossible to know where the gene that is shot lands in the target cell. I think the chance location of the gene explains the variability in the expression of the protein, in this case, lectin. Another explanation may have to do with the presence of the promoter 35S, derived from the cauliflower mosaic virus, intended to promote the expression of the protein, but no one has ever examined the side effects it might produce. The fact remains that the transgenic potatoes had unexpected effects on the rats' organisms."

"What effects did you observe?"

"First, the rats in the experimental groups had brains, livers, and testes less developed than those in the control group, as well as atrophied tissue, particularly in the pancreas and the intestine. We also found a proliferation of cells in the stomach, and that is troubling, because it can facilitate the development of tumors caused by chemical products. Finally, the immune system of the stomach was overactive, which suggests that the rats' organisms were treating the potatoes as foreign bodies. We were convinced that the process of genetic manipulation was the source of these malfunctions and not the lectin gene, whose safety in its natural state we had tested. Apparently, contrary to what the FDA claimed, the insertion technique was not a neutral technology, because by itself it produced unexplained effects."

The Arpad Pusztai Affair: Hounding the Dissident

Deeply troubled, Pusztai spoke of his worries to Professor Philip James, the director of the Rowett Institute, who was also one of the twelve members of the Advisory Committee on Novel Foods and Processes, the U.K. body charged with assessing the safety of GMOs before they were marketed. Convinced of the importance of the study's results, the director authorized Pusztai to participate in an ITV television program recorded in June 1998, seven weeks before broadcast, in the presence of the institute's public relations director. "In the interview," Pusztai explained, "I revealed no details about the study we had not yet published, but I answered frankly the questions I was asked, because I thought it was my ethical duty to alert British society to the unknown health effects of GMOs at a time when the first transgenic foods were being imported from the United States."

The European Community had adopted directive 90/220, regulating the release of GMOs in Europe, on April 23, 1990. It provided for a model procedure that was still in force eight years later (and continues in force today): to obtain authorization for the marketing of transgenic foods or plants, a company must present a technical dossier to a member state, whose national bodies assess the risks of the product for humans and the environment. After examination, the commission sends the dossier to the other member states, who have sixty days to request additional analyses if they think it necessary. Following this procedure, the commission authorized the importation of RR soybeans (as well as a Bt corn produced by Novartis) in December 1996, relying on the 1996 Monsanto study. The stakes were particularly high because, in the framework of the 1993 GATT agreements, Europe had agreed to limit the surface planted in oil-producing crops (soybeans, canola, sunflowers) to permit the sale of American stocks, forcing farmers to buy fodder from the United States.[1]

The ITV interviewer asked Pusztai: "Does the lack of tests of GMOs worry you?"

"Yes," he replied without hesitation.

"Would you eat transgenic potatoes?"

"No. And as a scientist actively working in the field, I find it's very unfair to use our fellow citizens as guinea pigs."

At first, the directors of the Rowett Institute saw nothing to criticize in the sentence repeatedly aired in promotional spots for World in Action on August 9, 1998. The next day, the Institute was flooded with interview requests, and Professor James was only too pleased to praise a study that brought such publicity. The night of the broadcast, August 10, the director could not keep himself from calling Pusztai to congratulate him for his performance on television: "He was very enthusiastic," Pusztai recalled. "Then, suddenly, everything changed."

On August 12, while a mob of reporters was waiting outside his house, Pusztai was summoned to a meeting where Philip James, accompanied by a lawyer, told him that his contract had been suspended, he would be dismissed, and the research team would be dissolved. The computers and documents connected to the study were confiscated and the telephone lines cut. Pusztai was put under a gag order under threat of prosecution. Then began an appalling disinformation campaign designed to sully his reputation and, by the same token, the validity of his warning. In several interviews, James claimed that Pusztai had made a mistake and, contrary to what he believed, had used not snowdrop lectin but another lectin called concanavalin A (con A), derived from a South American bean and known to be toxic.

In other words, the effects observed in the rats were due not to genetic manipulation but to con A, a "naturally occurring poison," as Dr. Colin Merritt, British spokesman for Monsanto, hastened to point out.[2] "Instead of rodents fed with genetically altered potatoes, Dr. Pusztai had used the results of tests carried out on rats treated with poison," according to the Scottish *Daily Record*.[3] "If you mix cyanide with vermouth in a cocktail and find that is not good for you, I don't draw sweeping conclusions that you should ban all mixed drinks," was the ironic comment of Sir Robert May, a government science advisor.[4] In France, *Le Monde* picked up this "news," which was especially strange because it involved the world's greatest specialist on lectin: "Dr. Pusztai confused data from a line of transgenic potatoes, the study of which had barely begun, and other data from experiments consisting of adding insecticidal proteins to rat food. The potatoes implicated therefore had nothing transgenic about them."[5] "It was terrible," Pusztai told me, still upset. "And I didn't even have the right to defend myself."

James attacked on a second front: he asked a committee of scientists to conduct an audit of the study. One wonders why. If the experiment was distorted by a mistake concerning the lectin used, then there was no reason to consider its results any further. And yet, on October 28, 1998, the Rowett Institute published the results of the audit: "The Audit Committee is of the opinion that the existing data do not support any suggestion that the consumption by rats of transgenic potatoes expressing GNA has an effect on growth, organ development, or the immune function. Thus the previous suggestion . . . was unfounded."[6]

But the affair had caused such a stir that the House of Commons asked "the dissident" to testify, thereby forcing James to grant him access to the data from his study. Pusztai then decided to send the data to twenty scientists around the world with whom he had worked in the course of his long career and who agreed to prepare a report comparing the data to the audit conducted for the institute. Published on the front page of *The Guardian* on February 12, 1999, the conclusions of the report were hard on the committee set up by James. After noting that the audit had deliberately ignored some results, the authors of the report specified that they "showed very clearly that the transgenic GNA potato had significant effects on immune function and this alone is sufficient to vindicate entirely Dr. Pusztai's statements."[7] They took the occasion to criticize "the harshness of his treatment by the Rowett [Institute] and even more by the impenetrable secrecy surrounding these events," and they called for a moratorium on the cultivation of transgenic crops.

The House of Commons Science and Technology Committee began its hearings a few days later. When the committee members pointed out the contradictions, James took refuge behind a new argument, one that had already been used by Monsanto spokesman Colin Merritt in an interview in *The Scotsman*: "You cannot go around releasing information of this kind unless it has been properly reviewed."[8] In other words, what the head of the Rowett Institute now criticized Pusztai for was having spoken before the study was published according to normal procedures.

The argument clearly did not persuade Dr. Alan Williams, a member of the committee. Speaking of the role of the advisory committee charged with authorizing the marketing of transgenic foods, of which James was a member, Williams addressed him with typical British irony: "There is a real problem for us here, and that is that you say that it is not right to discuss unpublished work; as I understand, all of the evidence taken by the advisory committee in that report comes from the commercial companies, all of that is unpublished. This is not democratic, is it? We cannot discuss the evidence because it is not published; there is no published evidence. So we leave it completely to the advisory committee and its good members to take all of these decisions on our behalf, where all of the evidence comes, simply, in good faith, from the commercial companies? . . . There is a hollow democratic deficit here, is there not?"[9]

Dr. Williams's remarks were at the heart of the immense controversy unleashed by the Pusztai affair, producing no fewer than seven hundred articles in the month of February 1999 alone. As the *New Statesman* observed at the time: "The GM controversy has divided society into two warring blocs. All those who see genetically modified food as a scary prospect—'Frankenstein foods'—are pitted against the defenders."[10] "Everybody over here hates us," complained Dan Verakis, Monsanto's European spokesman.[11]

Indeed, a confidential poll carried out in October 1998 at Monsanto's request, a copy of which was leaked to the press, revealed "an ongoing collapse of public support for biotechnology. . . . A third of the public is now extremely negative."[12] Seven months later, the trend was confirmed by another survey commissioned by the British government, which found that "1 percent of the public thought that GM was good for society" and that the majority of those surveyed did not trust the authorities to "provide honest and balanced information."[13]

And it had to be acknowledged that the skeptics were right. While major food distributors—including Unilever England, Nestlé, Tesco, Sainsbury, Somerfield, and the British subsidiaries of McDonald's and Burger King—publicly committed themselves to avoiding any transgenic ingredients, it was discovered that the government of Tony Blair was engaged in rather strange maneuvers to regain public confidence. According to a confidential document obtained by the *Independent on Sunday*, the government had prepared a veritable battle plan "to rubbish research by Dr. Arpad Pusztai" by "compiling a list of eminent scientists to be available for broadcast interviews and to author articles" that "will help us to tell a good story."[14] Among the scientists under consideration, the document mentioned those of the eminent Royal Society, which did indeed actively collaborate in the "rubbishing" campaign.

Monsanto, Clinton, and Blair: Effective Pressures

"The Royal Society was really ferocious," said Pusztai, and Dr. Stanley Ewen, who was sitting beside him, nodded in agreement. A renowned pathologist at the University of Aberdeen, Ewen had been involved in the study of transgenic potatoes, responsible for assessing their impact on the rats' gastrointestinal system. In a memorandum to the parliamentary committee, he had pointed out the results of his analysis: "Significant elongation of the crypt in the rats fed raw genetically modified food is the main finding. In addition I have counted the chronic inflammatory cells within the lining cells and found increased numbers of these cells in the rats fed raw genetically modified potatoes."[15]

Ewen still finds it hard to talk about the affair, which permanently destroyed his faith in the independence of science. "It felt as though the ground had given way beneath my feet," he said. "Impossible to understand: Monday, our work was wonderful, and Tuesday it was ready for the garbage heap. I myself was forced to retire, as though I had made a serious error." With a distressed air, he recounted how the Royal Society deliberately trampled on his reputation for reliability and impartiality in order to denigrate the results of the study.

On February 23, 1999, nineteen members of the Royal Society published an open letter in the *Daily Telegraph* and *The Guardian* stigmatizing the researchers who had "triggered the GM food crisis by publicizing findings that had not been subjected to peer review." This was false, because in his brief television interview Pusztai had not said a word about the results of his study, but merely called for more vigilance about GMOs in general. On March 23, the Royal Society published a critical analysis of the research, concluding that it was "flawed in many aspects of design, execution, and analysis."

Investigating this strange initiative, *The Guardian* discovered that the Royal Society had established a "rebuttal unit" whose purpose was "to mould scientific and public opinion with a pro-biotech line and to counter opposing scientists and environmental groups."[16] The Royal Society's attitude was so unusual that on May 22, 1999, *The Lancet* decided to speak out. It published an editorial declaring: "Governments should never have allowed these products into the food chain without insisting on rigorous testing for effects on health." Deliberately jumping into the controversy, it announced that it would finally publish the study by Pusztai and Ewen. Following normal procedures, it sent a copy of the article to six independent reviewers, who were not supposed to discuss the content before publication, announced for October 1999.[17]

Unfortunately, violating the established codes of conduct, one of the reviewers, John Pickett, went so far as to vehemently criticize the article in the columns of *The Independent* five days before publication.[18] Worse, he sent the proof of the article to the Royal Society, which went after Richard Horton, the editor of *The Lancet*. "There was intense pressure . . . to suppress publication," Horton told *The Guardian*, referring to a "very aggressive phone call" from Professor Peter Lachmann, former vice president and biological secretary of the Royal Society and president of the Academy of Medicine, who led him to understand that publication "would have implications for his personal position as editor" (an allegation Lachmann subsequently denied).[19]

"It's not surprising," said Ewen. "The Royal Society supported the development of GMOs from the beginning, and many of its members, like Professor Lachmann, work as consultants for biotechnology companies."*

"Monsanto among them," added Pusztai. "Besides, Monsanto was one of the private sponsors of the Rowett Institute as well as of the Scottish Agricultural Research Institute, a connection that was natural because one of its prominent members, Hugh Grant, now CEO of Monsanto, is Scottish."†

"There is no doubt in my mind that the decision to stop our work was made at the highest level," said Ewen. "I received confirmation in September 1999. I was at a dinner dance, and a Rowett Institute director was sitting at the next table. At one point, I said to him: 'Isn't it terrible, what happened to Arpad?' He answered: 'Yes, but don't you know that Downing Street called the director twice?' Then I realized there was something inter-

*According to *The Guardian*, Lachmann was a consultant for such companies as Geron Biomed, Adprotech, and SmithKline Beecham.

†In a February 16, 1999, press release, the Rowett Institute confirmed that it had signed a contract with Monsanto for a figure amounting to 1 percent of its annual budget.

national in the affair. Tony Blair's office had been pressured by the Americans, who thought our study would harm their biotechnology industry, and particularly Monsanto."

This information was indeed confirmed by a former administrator of the Rowett Institute, Professor Robert Orskov, who told the *Daily Mail* in 2003: "Phone calls went from Monsanto to Clinton. Clinton rang Blair and Blair rang James."[20]

Robert Shapiro, the Guru of Monsanto

The affair may seem incredible. And yet we've already seen how Monsanto was capable of intervening at the highest levels of government or international organizations to impose what it openly called in its activity report for 1997 "Monsanto's law."[21] When it made this odd confession, a few months before the Rowett Institute went into an uproar, the company was headed by Robert B. Shapiro, who had succeeded Richard Mahoney in April 1995, and remained CEO until January 2001.

Called "biotechnology's chief evangelist,"[22] the "image-maker,"[23] and the "guru of Monsanto,"[24] this lawyer from a well-to-do family in Manhattan was an exceptional figure in the history of the company: he was a Democrat and very close to the Clinton administration. That is presumably why the company contributed generously to the president's reelection campaign in 1996 and Clinton praised Monsanto in his State of the Union address on February 4, 1997. Soon afterward, Shapiro was appointed to the President's Advisory Committee for Trade Policy and Negotiations, which worked closely with Mickey Kantor, the trade representative and future Monsanto board member. In December 1998, Bill Clinton in person awarded the National Medal of Technology to Ernest Jaworski, Robert Fraley, Robert Horsch, and Stephen Rogers, the four inventors of Roundup Ready soybeans.

At the time, as former Secretary of Agriculture Dan Glickman has testified, the Democratic administration was enthralled by Bob Shapiro's talk about the "promises of biotechnology" that would produce a "revolution in agriculture, food, and health."[25] The Monsanto CEO painted in glowing terms the benefits of a technique that, according to him, was capable of shifting the world into the post-industrial age for the good of humanity, with a strength of conviction that even his harshest opponents acknowledge. In one of the very few interviews he granted, published in the *Harvard Business Review* on January 1, 1997, shortly after the reelection of Bill Clinton, he explained with some vigor how GMOs represented the solution for the future of the planet. After pointing out that 1.5 billion people were living in "conditions of abject poverty" and that the population would "double by sometime around 2030," he launched into an almost messianic diatribe on the consequences facing humanity. "It's a world of mass migration and environmental degradation at an unimaginable scale. At best, it means the preservation of a few islands of privilege and prosperity in a sea of misery and violence. . . . The whole system has to change. There's a huge opportunity of reinvention. . . . At Monsanto, we are trying to invent some new businesses around the concept of environmental sustainability. . . . Current agricultural practice isn't sustainable: we've lost something on the order of 15 percent of our topsoil over the last twenty years or so, irrigation is increasing the salinity of the soil, and the petrochemicals we rely on aren't renewable. Most arable land is already under cultivation. Attempts to open new farm land are causing severe ecological damage. So in the best case, we have the same amount of land to work with and twice as many people to feed. It comes down to resource productivity. . . . The conclusion is that new technology is the only alternative."[26]

Then Shapiro entered onto the philosophical portion of his presentation. Biotechnology, in his view, was an "information technology" that made it possible to replace the use of raw materials and energy, harmful to the environment, with a sophisticated use of genetic information. "Using information is one of the ways to increase productivity without abusing nature. A closed system like the Earth's can't withstand a systematic increase of material things, but it can support exponential increases of information and knowledge. If economic development means using more stuff, then those who argue that growth and environmental sustainability are incompatible are right. . . . But sustainability and development might be compatible if you could create value and satisfy people's needs by increasing the information components of what's produced and diminishing the amount of stuff."[27]

To illustrate his argument, Shapiro took the example of pesticides, 90 percent of which are dispersed into the environment at the moment of their application: "If we put the right information in the plant we use less stuff and increase productivity. . . . Information technology will be our most powerful tool."

"Can we trust the maker of Agent Orange to genetically engineer our food?" was the question posed by *Business Ethics*, "the magazine of corporate responsibility," which also interviewed Shapiro at the beginning of 1997.[28] Reading what Shapiro was saying at the time, I asked myself precisely the same question: was he sincere, and did he really believe what he said? To make up my mind, I dissected the career of the Harvard graduate, who liked to strum his guitar with Joan Baez in demonstrations against the Vietnam War. From that time he had maintained an open distaste for neckties and an unfailing attachment to the Democrats. After working in the administration of Jimmy Carter (who became an ardent supporter of biotechnology), Shapiro was hired in 1979 as legal director of the pharmaceutical company Searle, headed by none other than Donald Rumsfeld, secretary of defense for Gerald Ford from 1975 to 1977 and for George W. Bush from 2001 to 2006.

Searle was at the time in conflict with the FDA, which had decided to suspend the sale of aspartame, a highly controversial artificial sweetener, because it was suspected of causing brain tumors. Curiously, the product, sold under the name NutraSweet, was reauthorized in 1981, when Rumsfeld joined the newly elected Reagan administration. In the meantime, Shapiro, who had been in charge of handling the aspartame controversy, had been appointed head of the NutraSweet division. He negotiated with Coca-Cola the introduction of the sweetener in the new Diet Coke line of products. The story is that he carried off a major victory: he secured agreement that the name "NutraSweet"—that is, the Searle brand—would be printed on the labels with its logo (a little swirl), which prevented competitors that also made aspartame from selling it to Coca-Cola.

Monsanto bought Searle in 1985, making it the pharmaceutical division of the multinational company at the very time that Monsanto was requesting approval for the sale of bovine growth hormone. Shapiro, who often described himself as a "passionate gardener," became head of Monsanto's agricultural division in 1990 and in that position was in charge of handling Posilac, the trade name of bovine growth hormone, or rBGH.

I was troubled by this detail in his career, which cast a veil of suspicion over the ecological and Third World–friendly talk that he took up soon afterward, and I tried to contact the former Monsanto CEO. In 2006, he was head of the Belle Center of Chicago, an NGO established in St. Louis in 1984 to serve children with disabilities. In a *New Yorker* article, Michael

Specter wrote that Shapiro was "one of America's best paid executives" ($20 million in 1998); nonetheless, "he always replied to mail on the day it was sent, often within minutes."[29] He lived up to his reputation: I sent him a first e-mail on September 29, 2006, which he answered within a half hour, politely declining my request for an interview: "It's been some years since I was professionally engaged with biotechnology. . . . I no longer feel competent to speak on these subjects."

After learning that this man in his sixties, the father of two adult sons, had started a second family, on September 30 I asked him the only question that I had really set my heart on: "As the mother of three young girls, I would like to know what kind of milk you give your children: ordinary milk [sold with no distinction between the conventional and the transgenic, because they are blended and cannot be labeled] or organic milk?" The reply was almost immediate: "I have two young boys. My 10 year old is lactose intolerant, my 8 year old drinks lots of 2% milk and ice cream. We've never bought organic dairy products." When I read this, meaning that Shapiro's sons were not concerned by this affair, I could not help recalling what *Business Ethics* had written in January 1997: "It was very clear that Shapiro spoke in two voices. When discussing sustainability, he sounded hopeful. It was obvious he spoke from the heart. Yet, when responding to questions about Posilac, he reworded the queries, and provided the well rehearsed answers Wall Street investors would want to hear."[30]

The New Monsanto Will Save the World

Right after becoming CEO of Monsanto in April 1995, Shapiro launched the great "cultural revolution" that was intended to move the old chemical company into the era of "life sciences." This new concept, based on the application of molecular biology to agriculture and health, was officially presented at a Global Forum the "guru" organized in June 1995 in Chicago. Five hundred employees from all company divisions were invited to discover his new policy in a convivial atmosphere that contrasted with the company's legendary rigidity. Encouraging the participants to call him Bob, the "Renaissance man" in shirtsleeves moved the audience to tears when he spoke of the shame that some employees felt in saying what company they worked for.[31] This time was past, because the "new Monsanto" was going to "save the world." Armed with the new slogan, "Food, Health, Hope," Shapiro electrified his troops by talking of plants producing biodegradable plastics, corn supplying antibodies against cancer, canola or soybean oil protecting against cardiovascular disease. Witnesses have told of how an employee, Rebecca Tominack, excited by his speech, went up to the CEO and said, "I'm with you," took her name tag off, and put it on Shapiro in a gesture of allegiance repeated by a hundred other employees.

"I was really very impressed by Robert Shapiro's visionary speech, which made us want to work to make the world better," I was told by Kirk Azevedo, a Monsanto employee from 1996 to 1998, whom I met on October 14, 2006, in a small town on the West Coast where he was working as a chiropractor. Trained as a chemist, he had been contacted by a headhunter and resigned from Abbott Laboratories, where he had been in charge of testing new pesticides, to join what he then considered to be the "enterprise of the future." His job was to promote two varieties of transgenic cotton that Monsanto was about to launch on the market to seed dealers and California farmers: a Roundup Ready cotton and a Bt cotton, genetically manipulated to produce an insecticidal lectin (like Arpad Pusztai's transgenic potatoes) because of the insertion of a gene taken from the bacterium *Bacillus thuringiensis*.

"I was really very enthusiastic," Azevedo told me. "I did think that these two GMOs would bring about a reduction in the use of herbicides and insecticides. But the first dissonant note came three months after I was hired. I'd been invited to St. Louis to visit headquarters and participate in a training program for new hires. At one point, when I was speaking fervently in favor of biotechnology that would make it possible to reduce pollution and hunger in the world, a Monsanto vice president took me aside and said to me: 'What Robert Shapiro says is one thing, but what counts for us is making money. He talks to the public, but we don't even understand what he's talking about.'"

"Who was it?"

"I'd rather not identify him," Azevedo said hesitantly. "In any event, at the time I thought that he must be an exception. That lasted until the summer of 1997, when I experienced my second great disillusionment. I was in a field assessing an experimental plot of Roundup Ready cotton, whose cultivation was not yet authorized. With me was a Monsanto scientist, a cotton specialist. We were discussing what we would do with the cotton after it was picked. Since I was very pro-GMO, I said we should be able to sell it at the price of 'premium California,' because after all there was only one gene's difference from the original variety. That's when he told me: 'No, there are other differences; transgenic cotton plants produce not only the Roundup resistance protein but also other unknown proteins as a product of the manipulation process.'

"I was flabbergasted. There was a lot of talk at the time about mad cow disease—bovine spongiform encephalitis, and its human counterpart, variant Creutzfeldt-Jakob disease, severe pathologies caused by macroproteins known as prions. I knew that our transgenic cotton seeds would be sold as cattle fodder, and I said to myself that we hadn't even bothered to find out whether those 'unknown proteins' were prions. I told the Monsanto scientist about my concerns, and he answered that they didn't have time to worry about such things. I later tried to alert my colleagues, and bit by bit I was pushed to the sidelines. I also contacted the University of California and representatives of the state agriculture department, but I met nothing but indifference. I was so disturbed that I finally decided to resign so I wouldn't be an accomplice to such irresponsible conduct. But it wasn't an easy decision to make. When I left, I gave up a very good salary and I sacrificed tens of thousands of stock options. In fact, Monsanto buys its employees' silence."

"Now what do you think of Shapiro's speech?"

"It was hot air. When I recall the way we worked at the time, it was a constant race against the clock, and the only goal was to dominate the seed market. If you really want to save the world, you start by carefully verifying the safety of the products you're making."

The Race for Seeds

One thing had to be acknowledged about Robert Shapiro: the "visionary" was also a formidable businessman who had managed in record time to transform a chemical giant into a near monopoly operator in the international seed market. But the battle was far from over, because when Stephen Padgette's team finally had its Roundup Ready soybeans in 1993, no one at Monsanto knew what to do with them. Of course, the first instinct was to file a patent on the precious gene, but then what?

Monsanto was not a seed company, and the only solution was to sell its discovery to people in the business. Dick Mahoney, the CEO at the time, thought immediately of Pioneer Hi-Bred International, which controlled 20 percent of the American seed market (40 percent for corn and 10 percent for soybeans). Founded in Des Moines in 1926 by Henry Wallace (vice president of the United States from 1941 to 1945), the company was known primarily for having invented the hybrid corn varieties that made its fortune.

The underlying principle was that instead of allowing corn to be pollinated naturally through the air, they forced plants to inbreed to obtain pure lines with stable genetic characteristics. The results were hybrids that produced higher yields but whose seeds were practically sterile. For seed dealers, this was a godsend because farmers were forced to buy their seeds every year. This hybridization technique worked only for allogamous plants, that is, plants produced by fertilization of the ovum of one plant by pollen from *another* plant, not for autogamous plants such as wheat or soybeans, where each plant reproduces itself with its internal male and female organs. I will later describe how this detail did not escape Monsanto, which got around it by means of the patent system.

In 2002, Daniel Charles reported in detail the amazing story of Monsanto's mutation in the 1990s in *Lords of the Harvest*, which is the basis for what follows. When Robert Shapiro, who was then head of Monsanto's agricultural division, met Tom Urban, chief executive of Pioneer Hi-Bred International, to give him a sales pitch for his Roundup Ready gene, he was received coolly: "'Congratulations! You've got a gene! Guess what? We've got fifty thousand genes! . . . You don't hold the keys to the market. We do! You ought to pay *us* for the right to put your gene in our varieties!'"[32] At the time, Shapiro had no choice: after years of costly research, the company's instructions were that finally it was time to bring in some money. A first agreement was signed with Pioneer, which agreed to introduce the Roundup Ready gene into its soybean varieties. In return, remembering his success with NutraSweet for Diet Coke, Shapiro got an agreement to have "Roundup Ready" printed on the seed bags. But in the end, there was nothing to boast about. As Charles points out: "The Roundup Ready gene had become a vehicle to sell more of Monsanto's chemicals, but little more."[33]

A second set of negotiations then began over the other genetic characteristic that Monsanto had in its arsenal: the Bt gene, which was an urgent matter, because several companies were claiming authorship (leading to an interminable patent battle). In this case, the GMO was not associated with the sale of a pesticide, because the gene itself *was* the pesticide, designed expressly to kill the corn borer, a very common corn parasite. Robert Shapiro therefore secured payment for this performance from Pioneer Hi-Bred and carried away the sum of $38 million in full payment. In both cases, the amounts paid by the Des Moines seed dealer turned out to be trifling in light of the huge success that both GMOs had immediately, principally with Roundup Ready soybeans. When he became CEO of Monsanto in April 1995, Shapiro tried to renegotiate the two agreements, without success.

"It was the most rapid and enthusiastic adoption of a technical innovation in the history of agriculture," according to Charles, who reports that Roundup Ready soybeans covered 1 million acres in the United States in 1996, 9 million in 1997, and 25 million in 1998.[34] To understand the initial enthusiasm for Roundup Ready crops, you have to put yourself in the shoes of an American farmer such as John Hoffman, vice president of the American Soybean Association, considered close to Monsanto.

I met him at harvest time in 2006 on his huge Iowa farm, whose size he didn't want to reveal. "Before I used the Roundup Ready technique," he told me in the middle of a transgenic soybean field of several dozen acres, "I had to plow the earth to prepare a seedbed. Then I had to spray several selective herbicides to get rid of weeds in the course of the season. Before the harvest, I had to inspect my fields and pull up the final weeds by hand. Now, I don't plow my fields, I spray Roundup once, then I sow directly in the remains of the last harvest. This is what's called 'zero tillage,' which reduces soil erosion. Then halfway through the season, I do a second spraying of Roundup, and that's usually enough until the harvest. The Roundup Ready system allows me to save time and money."

In the summer of 1995, demonstrations were organized in the plains of the Midwest, and farmers flocked to them, drawn by these plants with a strange power. "'We'd actually let farmers run the sprayer,' says [a seed dealer]. 'And then they could drive by on the way to the coffee shop and watch the fields. It was a fantastic show. . . . They were just watching it at first. Then they couldn't believe it. And then they just wanted to buy it.'"[35] A Minnesota seed dealer says: "It was just a phenomenon, and I don't know if I'll ever see anything like it again. Farmers were just crazy to get Roundup Ready soybeans. They bought every bag."[36]

So great was the enthusiasm for RR soybeans that the major American seed dealers besieged St. Louis to get hold of the magic gene. But Shapiro had learned from his experience with Pioneer. From now on, he would control the game: to get the right to insert the gene in their varieties, seed companies had to sign a licensing agreement, which meant that Monsanto collected royalties on each transgenic seed sold. In addition, Shapiro insisted on a clause that was later attacked as improper by the antitrust authorities: companies had to sign a contract agreeing that 90 percent of the herbicide-resistant GMOs they sold would contain the Roundup Ready gene.* This was a way of cutting the ground out from under the feet of Monsanto's competitors, such as the German company AgrEvo, which was forced to give up marketing GMOs resistant to the herbicide Liberty (known as Basta in Europe) because it could not find a seed company partner.

Monsanto's CEO changed his strategy in 1996. Realizing that he had to own the seeds to earn the highest profits, he launched an ambitious program for the acquisition of seed companies, which profoundly transformed agricultural practices around the world. Shapiro didn't skimp to reach his goals: he paid $1 billion for Robert Holden's Foundation Seeds, which had a strong presence in the American corn market, with annual profits of only a few million dollars: "Overnight, Ron Holden became a very rich man."[37] Then Shapiro bought a whole string of companies: Asgrow Agronomics, the largest soybean dealer in the United States; DeKalb Genetics (for $2.3 billion), the second-largest American seed company and the ninth-largest in the world, which had many subsidiaries and joint ventures, particularly in Asia; Corn States Hybrid Services; Custom Farm Seed; Firm Line Seeds (Canada); the British companies Plant Breeding International and Unilever; Sementes Agroceres, a leading force in the Brazilian corn market; Ciagro (Argentina); Mahyco, principal supplier of cotton seeds in India, along with Maharashtra Hybrid Seed Company, Eid Parry, and Rallis, three other Indian companies; the South African Sensako (wheat, corn, cotton); National Seed Company (Malawi); Agro Seed Corp (Philippines); not to mention the international division of Cargill, the largest seed dealer in the world, with branches in Asia, Africa, Europe, and South and Central America, that Monsanto bought for $1.4 billion.

In two years, Shapiro had spent more than $8 billion and made Monsanto the second largest seed company in the world after Pioneer.† To finance this costly program of acquisitions, it had sold its chemical division to Solutia in

*The percentage was later reduced to 70 percent after the intervention of regulatory authorities.
†Monsanto continued its acquisitions in the early 2000s. With the purchase of Seminis (vegetable seeds) in 2005, the company became the largest seed company in the world.

1997. But that was not enough: it had had to incur record indebtedness, backed by the stock market, which still believed at the time in the promise of biotechnology. Monsanto's stock price climbed 74 percent in 1995 and 71 percent in 1996. Investors blindly followed the "guru of St. Louis" until the false move in 1998 that initiated his fall from grace.

The Terminator Patent: One Step Too Far for Monsanto

On March 3, 1998, a brief article in the *Wall Street Journal* announced that the USDA (then headed by Dan Glickman) and the Delta and Pine Land Company of Mississippi, the largest American cotton seed company, had jointly obtained a patent entitled "Control of Plant Gene Expression." Behind this mysterious title lay a technique making it possible to genetically modify plants so that they produced sterile seeds. Developed by Melvin Oliver, an Australian scientist working in the USDA research laboratory in Lubbock, Texas, the technique was also called the "Technology Protection System" (understood to be transgenic), because it was designed to prevent farmers from resowing part of their crop, forcing them to buy seeds every year and pay royalties to GMO manufacturers. Concretely, the plant had been manipulated to produce a toxic protein when its growth was complete that made its seeds sterile.

Hope Shand, research director of the Rural Advancement Foundation International (RAFI), a Canadian NGO since renamed the ETC Group (Erosion, Technology, Concentration), which fights for the protection of biodiversity and against the perverse effects of industrial agriculture, came across the article in the *Wall Street Journal* by chance. She immediately informed her boss, Pat Mooney, who said, "It's Terminator!" referring to the legendary robot played by Arnold Schwarzenegger. The expression stuck permanently to designate the sterilization technique and, beyond that, the overall aim of GMO producers. "You understand," Mooney told me when I met him in Ottawa in September 2004, "this technique was a direct threat to food security, especially in developing countries where more than 1.5 billion people survive by saving seeds. Imagine that Terminator plants crossbreed with neighboring crops and make the seeds gathered by peasants sterile. It would be a catastrophe for them, but also for the biodiversity they maintain precisely because they continue to replant every year local varieties adapted to their climate and their soil."

On March 11, 1998, RAFI published a communiqué titled "Terminator Technology: A Global Threat to Farmers, Biodiversity, and Food Security." But it went practically unnoticed. "In fact," Mooney said with a smile, "it was thanks to Monsanto that our campaign had worldwide success." Two months later, Shapiro announced that he was in negotiations to acquire Delta and Pine for $1.9 billion. The news caused an international uproar, because Monsanto would be taking over the Terminator patent. NGOs concerned with ecology or development were not the only ones to react; disapproval was also expressed by the Rockefeller Foundation (which had sponsored the green revolution in the 1960s and generally supported biotechnology) and the Consultative Group on International Agricultural Research (CGIAR), which publicly promised never to use Terminator in its seed programs. Feelings ran so high that the UN Convention on Biological Diversity voted for a moratorium—still in force ten years later—on field tests and the commercial use of Terminator. The crowning touch was that the antitrust division of the U.S. Justice Department challenged the acquisition.*

For Monsanto, the timing could not have been worse. Since the fall of 1997, all indicators in Europe had turned red. The first shipments of transgenic soybeans had been blocked in European ports on the initiative of Greenpeace, which was conducting a very effective campaign against "Frankenfood." Fresh from its success in North America, where it had been able to avoid the labeling and segregation of GMOs, the company did not expect that Greenpeace would bring the machine to a halt. On May 26, 1998, the EU adopted Regulation 1139/98, ratifying the establishment of a labeling procedure for transgenic products. Even earlier in the year, Monsanto had convened emergency committees in St. Louis, Chicago, London, and Brussels. The decision was made to launch a massive advertising campaign in early June 1998 in Germany, France (costing 25 million francs), and Great Britain (at a cost of £1 million).

Designed by the English advertising agency Bartle Bogle Hegarty, the campaign used the same basic slogan in all three countries: "Food biotech-

*Monsanto did not actually acquire Delta and Pine, and the patent, until 2006.

nology is a matter of opinions. Monsanto believes you should hear all of them." Then came the addresses and phone numbers of the company's principal opponents, such as Friends of the Earth and Greenpeace. In France, the first ad adopted a condescending tone: "69 percent of the French are suspicious of biotechnology, 63 percent say they don't know what it is. Fortunately, 91 percent know how to read." Other messages adopted the messianic vision of Robert Shapiro, with his trademark moralizing tone: "As we stand on the edge of a new millennium, we dream of a tomorrow without hunger. To achieve that dream, we must welcome the science that promises hope. Biotechnology is one of tomorrow's tools today. Slowing its acceptance is a luxury our hungry world cannot afford." In an interview with the magazine *Chemistry and Industry*, Jonathan Ramsay, a Monsanto executive, summed up well the spirit of the campaign, which many considered very arrogant: "We will have succeeded if biotechnology becomes less the subject of Luddite superstition and more the subject of serious and informed public debate."[38]

The campaign in Great Britain flopped immediately thanks to the intervention of the Prince of Wales, a champion of organic farming. As soon as the campaign was launched, he published an article in the *Daily Telegraph* titled "The Seeds of Disaster": "I have always believed that agriculture should proceed in harmony with nature, recognising that there are natural limits to our ambitions. . . . We simply do not know the long-term consequences for human health and the wider environment of releasing plants bred in this way. We are assured that these new plants are vigorously tested and regulated, but the evaluation procedure seems to presume that unless a GM crop can be shown to be unsafe, there is no reason to stop its use. . . . I personally have no wish to eat anything produced by genetic modification, nor do I knowingly offer this sort of produce to my family or guests."[39] The prince's words were reported in all British newspapers, forcing Monsanto to acknowledge its mistakes, proof that the matter was serious. "We barged in," Toby Moffett, vice president for international government affairs, admitted, "like someone barging in on someone's private party. We weren't European enough."[40]

It was in this context that the Arpad Pusztai affair exploded. To crown Monsanto's bad luck, the day after the broadcast of the Pusztai interview in August, the British Advertising Standards Authority received four complaints against Monsanto for deceptive advertising: in one of the campaign ads, the company claimed that its GMOs had received regulatory approval in twenty countries, including the United Kingdom.[41] Piling on the mishaps, in September, the British magazine *The Ecologist* published a special sixty-five-page feature recounting the entire history of the company from its founding in 1905.[42] The fourteen thousand copies of the first printing were pulped by Penwells, the printer who had worked for the magazine for twenty-five years, because of "pressures" whose source they never publicly identified. Zac Goldsmith, editor of *The Ecologist*, had to find another printer, but two major British newsagents refused to distribute the new copies.[43]

CEO Musical Chairs

In any event, Robert Shapiro's days of glory were at an end. Starting in the fall of 1998, Monsanto went into decline on Wall Street: "Monsanto stock has lost more than a third of its value in the last 14 months, and analysts believe that company executives could be forced into radical changes, possibly including breaking Monsanto into pieces."[44] Around the same time, *Le Monde* wrote: "Monsanto is now nothing but a kind of giant start-up in plant biotechnology, with revenues of $8.6 billion and losses of $250 million in 1998. Its recent numerous acquisitions in seed companies, sometimes paid for at premium prices, have cut into its profits. Investors are beginning to shun the company . . . and yesterday's friends are turning away for fear of being discredited in turn."[45]

The rout was so complete that Shapiro was forced to declare a cease-fire with his worst enemies: on October 6, 1999, he agreed to participate in a business conference organized by Greenpeace in London. Unable (or not daring) to appear in person, his presentation was recorded in St. Louis and transmitted by satellite onto a giant screen, where his face appeared "drawn and ashen," according to the *Washington Post*.[46] Making amends in front of a stunned audience, the CEO, who would resign some months later, said: "We have probably irritated and antagonized more people than we have persuaded. Our confidence in this technology and our enthusiasm for it has, I think, been widely seen—and understandably so—as condescension or indeed arrogance." Then, addressing Peter Melchett, executive director of Greenpeace UK and a former agriculture minister, he promised "not to commercialize the technologies popularly known as terminator or sterile seed technologies," and continued: "As we work to help develop constructive answers to all the questions that people around the world have at the dawning of this new technology, we are committed to engage openly, honestly and non-defensively in the kind of discussion that can produce good answers for all of us. . . . To me, that means, among other things, listening carefully and respectfully to all points of view."

As he spoke those words, Shapiro was desperately seeking a partner to save the company. First, he held discussions with American Home Products, then with DuPont, but the deals fell through. Finally, on December 19, 1999, Monsanto announced its merger with Pharmacia and Upjohn, originally a Swedish pharmaceutical company based in New Jersey. "The terms of the merger signaled the failure of Monsanto's guiding vision and of its creator, Robert Shapiro," according to Michael Watkins, a researcher at Harvard Business School.[47] Renamed Pharmacia, the new corporation was interested primarily in Searle, Monsanto's pharmaceutical division, whose value was then estimated at $23 billion (it manufactured Celebrex, a leading medicine for arthritis). But it soon sought to separate from the agrichemical division of Monsanto, known as "the new Monsanto," which it finally did in the summer of 2002 (at the same time that Pharmacia was absorbed by Pfizer).

The messianic vision of Robert Shapiro, who had dreamed of a company dedicated to the life sciences, was well and truly buried. When he left the company after the merger with Pharmacia in late 1999, the firm displayed its true face: it was indeed the largest supplier in the world of transgenic seeds, but it got 45 percent of its revenues from Roundup, which was threatened by the arrival of generics. Shapiro was replaced by the Belgian Hendrik Verfaillie, who was in turn forced to resign in December 2002 because of "poor financial performance."[48] He was succeeded by the Scotsman Hugh Grant (still CEO in early 2008), who had the delicate task of getting things back on an even keel, while GMOs enjoyed anything but universal support in North America, not even in farmers' fields.

10

The Iron Law of the Patenting of Life

> Monsanto Company activities and the use of its products positively affect agricultural sustainability.
>
> —Monsanto, *Pledge Report*, 2005

"One of my biggest concerns is what biotechnology has in store for family farmers," Dan Glickman declared in the July 13, 1999, speech that so irritated his government colleagues involved with foreign trade. "We're already seeing a heated argument over who owns what. Companies are suing companies over patent rights even as they merge. Farmers have been pitted against their neighbors in efforts to protect corporate intellectual property rights. . . . Contracts with farmers need to be fair and not result in a system that reduces farmers to mere serfs on the land or create an atmosphere of mistrust among farmers or between farmers and companies."

The Weapon of Patents

When he spoke these iconoclastic words, Bill Clinton's secretary of agriculture was touching on one of the subjects at the heart of opposition to GMOs: the subject of patents. "We have always criticized the doubletalk of biotechnology companies," Michael Hansen of Consumers Union told me. "On one hand, they say there is no need to test transgenic plants because they are exactly the same as their conventional counterparts; on the other, they file for patents, on the grounds that GMOs are unique creations. You have to make up your mind: either Roundup Ready soybeans are identical to conventional soybeans, or else they're not. They can't be both depending on Monsanto's interests."

Before the late 1970s it would have been inconceivable to file a patent application for a plant variety, even in the United States, where the 1951 patent law clearly provided that patents applied exclusively to machines and industrial processes, but in no case to living organisms, hence not to plants. The patent system was at its origin a tool of public policy intended to stimulate technical innovations by granting the inventor a monopoly on the manufacture and sale of a product for a period of twenty years. "The criteria for granting patents are usually very strict," according to Paul Gepts, a researcher in the Department of Molecular Biology at the University of California, Davis, whom I interviewed in July 2004. "They are three in number: the novelty of the product, that is, the fact that the product did not exist before the inventor created it; the fact that it is not obvious; and its usefulness for industry. Before 1980, the legislature had excluded living organisms from the field of patents, because it thought they could under no circumstances satisfy the first criterion: even if humans intervened in their development, living organisms exist *before* human action and, moreover, they can reproduce on their own."

With the advent of genetic manipulation, the question of plant varieties "improved" by the technique of genetic selection described in Chapter Seven arose. Concerned with recovering their investments, seed companies won legislation which granted to their varieties what was called "plant variety protection," enabling them to sell user licenses to dealers or to include a kind of "tax" in the price of their seeds.* But a certificate of plant variety protection was only a distant cousin of a patent, because it did not prohibit farmers from keeping part of their harvest to sow their fields the next year, nor researchers such as Paul Gepts or breeders from using the variety concerned to create new ones. This was known as the breeder's and research exemption.

Everything changed in 1980, when the U.S. Supreme Court issued a decision with serious consequences declaring a transgenic microorganism

*The system is guaranteed by the UPOV agreements (Union for the Protection of New Varieties of Plants), signed by thirty-seven countries in 1973.

patentable. The case had begun eight years earlier when Ananda Mohan Chakrabarty, a geneticist working for General Electric, had filed a patent application for a bacterium that he had been altered to enable it to consume hydrocarbons. The U.S. Patent and Trademark Office had logically rejected the application according to the terms of the 1951 law. Chakrabarty appealed and won in the Supreme Court, which stated: "Anything under the sun that is made by man can be patented."

This startling decision had opened the way to what has been called the "patenting of life": based on U.S. precedents, the European Patent Office in Munich granted patents on microorganisms in 1982, on plants in 1985, on animals in 1988, and on human embryos in 2000. Theoretically, these patents were granted only if the living organism had been altered by genetic engineering, but in reality the process has gone beyond GMOs alone. Patents have now been granted for non-transgenic plants, particularly if they have medicinal properties, in total violation of existing laws. "Ever since biotechnology came on the scene, the common-law system of patents has been abused," Christoph Then, the Greenpeace representative in Munich, told me in February 2005. "To get a patent, it is no longer necessary to present a real invention; often all you need is a simple discovery. Someone discovers a therapeutic use for a plant, the Indian neem tree, for instance, describes it, isolates it from its natural context, and files a patent application for it. The deciding factor is that the description be done in a laboratory, and no attention is paid to the fact that the plant and its virtues have been known by others for thousands of years."[1]

The U.S. Patent and Trademark Office grants more than seventy thousand patents a year, about 20 percent of which involve living organisms. It took a long struggle for me to get an interview with a representative of this huge institution, which is under the authority of the Commerce Department and employs seven thousand agents. A citadel in the Washington suburbs, the Patent Office is a strategic location for a company like Monsanto, which secured 647 patents associated with plants between 1983 and 2005.

"The Chakrabarty case opened the door to a very exciting period," said John Doll of the biotechnology department when I met him in September 2004. "We now grant patents on genes and transgenic plants and animals, any product of genetic engineering."

"But a gene is not a product," I said, a little taken aback by his triumphant tone.

"Sure," he agreed, "but once a company has been able to isolate the gene and describe its function, it can get a patent."

The New Agricultural Order

I have already described how, as soon as Monsanto researchers had managed to cobble together the genetic cassette allowing the creation of Roundup-resistant soybeans, the company filed a patent application and received the patent without difficulty. The patent ran until 2004 in the United States. In June 1996, the European Patent Office in turn granted a patent to RR soybeans, which applies by extension to any plant variety into which the cassette can be inserted: "maize, wheat, rice, soybean, cotton, sugar beet rapeseed, canola, flax, sunflower, potato, tobacco, tomato, lucerne, poplar, pine, apple, and grape," which tells a lot about the company's plans.*

Monsanto then had to find the means to enforce its intellectual property rights. One might think that the strategy of first selling user licenses to seed dealers and then acquiring the principal seed companies would amply secure its return on investment, but this was not the case. Monsanto's real problem was farmers themselves, who around the world still had the annoying habit of saving part of their crop to replant it (except for hybrids, which do not include autogamous plants such as soybeans and wheat). "In some countries, farmers commonly save seed for planting the following year," cautiously noted Monsanto's 2005 *Pledge Report*, which the company has published periodically since the creation of the "new Monsanto." "When the seed contains a patented trait, such as the Roundup Ready trait, this traditional practice creates a *dilemma* for the seed company that developed the variety."[2] In the 10-K form that has to be sent to shareholders and filed with the SEC every year, the language was more direct. Under the heading "Competition," the company stated in 2005: "The global markets for our products are highly competitive. . . . In certain countries, we also compete with government-owned seed companies. Farmers who save seed from one year to the next also affect competitive conditions."

The company's language seems to suggest that the practice of saving seeds exists only in distant and backward countries. This was so far from being the

*Patent EP 546090, titled "Glyphosate tolerant 5-Enolpyruvylshikimate-3-Phosphate Synthases."

case that when Robert Shapiro came up with the brilliant idea of having all farmers who bought RR soybean seeds sign a "technology use agreement," he encountered a good deal of resistance. The agreement, which dealers were required to present, provided for payment of a technology fee, set first at $5 and then at $6.50 per acre of soybeans, and, most important, a commitment not to replant any harvested seeds the following year. Another clause required growers to use only Monsanto's Roundup, not any of the many generics on the market, after the expiration of the patent in 2000.

The terms of the contract that must be signed are still draconian: farmers who violate it risk having to pay a heavy penalty or being sued in state or federal court in St. Louis (which has certain advantages for the company). Monsanto also assumes the right to review customers' accounts going back three years and to inspect their fields at the slightest suspicion: "If Monsanto reasonably believes that a grower has planted saved seed containing a Monsanto genetic trait, Monsanto will request invoices or otherwise confirm that the fields in question have been planted with newly purchased seed. If this information is not provided within 30 days, Monsanto may inspect and test all of the grower's fields to determine if saved seed has been planted."[3]

The provision also covered seed dealers, one of whose activities used to be to clean the seeds farmers had harvested before they could replant them, by removing the chaff. In *Lords of the Harvest*, Daniel Charles tells of an Ohio seed dealer who was forced against his will to post a notice in his barn that was supposed to protect him from growers Monsanto called "pirates": "IMPORTANT INFORMATION FOR INDIVIDUALS SAVING SEED AND REPLANTING . . . Seed from Roundup Ready soybeans cannot be replanted. It is protected under U.S. patents 4,535,060; 4,940,835; 5,633,435 and 5,530,196. A grower who asks to have Roundup Ready seed cleaned is putting the seed cleaner and himself at risk."[4] "In the end," Charles remarks, "most farmers went along. They signed, grumbled, and joined the new agricultural order."[5] According to Peter Carstensen, a professor at the University of Wisconsin Law School, the practice instituted by Monsanto effected a "dual revolution." "First," he told me when I met him in October 2006, "it had the right to patent seeds, which was absolutely prohibited before the advent of biotechnology; second, it extended the rights of the manufacturer granted by patents. For that I would adopt the image that Monsanto likes to use. It compares a transgenic seed to a rental car: when you've finished using it, you return it to the owner. In other words, the company doesn't sell seeds, it just *rents* them, for one season, and it remains the permanent owner of the genetic information contained in the seed, which is divested of its status as a living organism and becomes a mere commodity. Finally, farmers became users of Monsanto's intellectual property. When you realize that seeds are the basis for feeding the world, I think there are reasons to be worried."

"But what means does Monsanto have to enforce its contract?"

"They're huge. I was stunned when I found out that they'd hired the Pinkerton Detective Agency.* Monsanto pays its agents to comb the countryside looking for cheaters, and if necessary it seeks out informants. The company set up a toll-free number where anyone can denounce his neighbor. It spends a lot of money to enforce its rule in the fields."

Of course, all this could have been avoided if Robert Shapiro had been able to use the Terminator technique, which would have allowed him to resolve the company's "dilemma" without spending a penny and above all without having to set up a very unpopular war machine.

The Gene Police

"Biotech crops are protected by U.S. patent law," John Hoffman, vice president of the American Soybean Association, told me with his perpetual smile. "And so I may not in any way save seed to replant the following year. It's something that is a protection for Monsanto, for biotech companies. Because they literally invest millions and millions of dollars to produce this new technology we are very happy to use." Listening to this Iowa farmer brought to mind Hugh Grant, the CEO of Monsanto, who said the same thing in an interview with Daniel Charles: "We are interested in protecting our intellectual property, and we make no apologies for that. . . . It's as hard as that. There's a gene in there that's the property of Monsanto, and it's illegal for a farmer to take that gene and create it in a second crop. It's necessary from the point of view of return on investment, and it's against the law."[6]

*Notorious for its violent methods, like those of a private militia, particularly when it was hired to break strikes in the late nineteenth century. The Pinkerton National Detective Agency was founded in 1850 by Alan Pinkerton, who had his moment of glory when he foiled an assassination attempt against President Abraham Lincoln, who hired his agents to ensure his security during the Civil War. Helped by its logo—an eye with the slogan "We never sleep"—the agency was hired by companies to infiltrate unions and factories with methods summed up in the expression "bloody Pinkerton," designating a strike-breaking cop.

"And how can Monsanto know that someone, for instance, replanted harvested seeds?" I asked Hoffman.

"I'm not sure how to answer that, no. That's a good question for Monsanto."

Unfortunately, as I said earlier, Monsanto executives refused to see me, as I was told by the company's public relations director, Christopher Horner. I would have been interested in interviewing Horner because, according to an article in the *Chicago Tribune*, he was the one who had to come to the defense of his employer when the Center for Food Safety in Washington published a very disturbing report in November 2004. Titled *Monsanto vs. U.S. Farmers*, this very detailed eighty-four-page document confirmed the existence of what is known in North America as the "gene police," operated by Pinkerton in the United States and Robinson in Canada.[7] It also reported that the company had been conducting a veritable witch hunt in the American prairies since 1998, leading to "thousands of investigations, nearly 100 lawsuits, and numerous bankruptcies."[8]

"The number of farmers sued represents a minuscule number of the 300,000 or so who use the company's technology," Horner retorted. "Lawsuits are the company's last resort."[9] But Joseph Mendelson, legal director of the Center for Food Safety, criticized the company's "dictatorial methods." He claimed it was capable of anything to "impose its control over all phases of agriculture." The report he supervised does chill the blood. After noting that 85 percent of the soybeans grown in the United States in 2005 were transgenic, along with 84 percent of canola, 76 percent of cotton, and 45 percent of corn, it goes on to say: "No farmer is safe from Monsanto's heavy-handed investigations and ruthless prosecutions. Farmers have been sued after their field was contaminated by pollen or seed from someone else's genetically engineered crop; when genetically engineered seed from a previous year's crop has sprouted, or 'volunteered,' in fields planted with non–genetically engineered varieties the following year; and when they never signed Monsanto's technology agreement but still planted the patented crop seed. In all of these cases, because of the way patent law has been applied, farmers are technically liable."

To conduct its study, the CFS consulted data supplied by the company itself, which frequently publicizes the cases of "seed piracy" it has detected in the country—an unusual degree of transparency designed to dissuade anyone tempted to violate its iron law. In 1998, for example, the company investigated 475 cases of "piracy," and up to 2004, the annual average was more than 500. The CFS compared these data with a list of lawsuits filed against American farmers by Monsanto, compiled by the Administrative Office of the U.S. Courts, which by 2005 had recorded ninety suits.[10] The average damage amount won by the company was $412,259, with a high of $3,052,800, for a total of $15,253,602 (in a few exceptional cases growers were exonerated). The suits led to the bankruptcy of eight farmers. Mendelson told me, "These numbers are only the tip of the iceberg, because they cover only the rare cases that went to court. The vast majority of farmers who were sued, very often unjustly, decided to negotiate a settlement because they were afraid of the costs of a trial against Monsanto. And none of these settlements show up because they all contained a confidentiality clause. That's why we were able to analyze only the cases that ended with a verdict."

The CFS report discloses that Monsanto has an annual budget of $10 million and a staff of seventy-five to conduct its "investigations." Its primary source of information is the toll-free number 1-800-ROUNDUP, which the company officially inaugurated on September 29, 1998, in a formal press release: "Dial 1-800-ROUNDUP; tell the rep that you want to report some potential seed law violations or other information. It is important to use 'land lines' rather than cellular phones due to the number of people who scan cellular calls. You may call the information in anonymously but please leave your name and number if possible for any needed follow up."[11] According to Daniel Charles, the tip line received fifteen hundred calls in 1999, five hundred of which triggered an investigation.[12] Questioned about the line, criticized for "fraying the social fabric that holds farming communities together," in the measured words of the *Washington Post*, Karen Marshall, a spokesperson for Monsanto, replied simply: "This is part of the agricultural revolution, and any revolution is painful. But the technology is good technology."[13]

"We Own Anybody That Buys Our Products"

Most farmers who had lost cases contacted by the CFS told the same story: one day an agent, usually a Pinkerton man, knocked on their door, sometimes accompanied by the police. He asked to see their invoices for seeds and herbicides and demanded that he be allowed to go into their fields, where he took plant samples and photographs. The tone was often threatening, even brutal. Sometimes no agent ever appeared, but the grower was sent a summons on the basis of a "dossier" made up of aerial views and analyses of plants taken from the farmer's property without his or her knowledge. Not infrequently, farmers who were sued denied that they could be bound by a technology agreement (twenty-five out of ninety), because the dealer who sold them the seed had never talked about it, because they signed without really reading it, or because the practice was so out of the ordinary. This was the case for Homan McFarling, a Missouri farmer sued in 2000 for having "saved RR soybean seed," something he never denied. The trial verdict required him to pay 120 times the cost of the saved seed, or $780,000, according to the terms of the agreement, which he didn't even remember signing and of which he didn't have a copy. He appealed and, unusually, won a reduction in damages: the court questioned "the constitutionality of a contract asking for enormous damages for what was a very small actual loss."[14] The amount he finally paid is unknown.

Others were penalized even though they didn't know they were growing GM crops. For example, Hendrik Hartkamp, a native of the Netherlands, bought a ranch in Oklahoma in 1998. On the property he found a store of soybean seeds, which he planted. On April 3, 2000, he was sued by Monsanto for "patent law violation," because some of his seeds were transgenic. After ruining himself in conducting his defense, he sold his ranch at a loss and left the United States for good. "The terrible thing," Joseph Mendelson told me, "is that courts don't distinguish between those who knowingly reuse their seeds and those who did not plant GMOs intentionally. The only thing that counts is that the gene was found in a field: whatever the reason, the owner of the field is held liable." When a farmer claimed that he had never signed a contract but settled for $100,000 (hence remaining anonymous), a Monsanto representative retorted with remarkable frankness: "We own you—we own anybody that buys our Roundup Ready products."[15]

The CFS report also reveals that for at least six of the ninety suits filed by Monsanto the agreement presented by the company had a forged signature, "a practice documented as common among seed dealers." This happened, for example, to Eugene Stratemeyer, an Illinois farmer who fell into a trap set by an "inspector": in July 1998, a man appeared at his farm and asked to buy a small quantity of seeds. Since the planting season was over, he explained that he wanted to do an erosion test. Stratemeyer agreed to help him out. Ordered to pay damages of $16,874.28 for patent infringement, Stratemeyer countersued Monsanto for use of forgery.

When farmers decide to defend themselves by publicly challenging the prohibition of replanting part of their crop, they leave themselves open to harassment or even a carefully orchestrated campaign of slander in the media and in the eyes of all agricultural intermediaries. This is what happened to Mitchell Scruggs, a Mississippi farmer who had always admitted saving RR soybean and Bt cotton seeds. He saw this as an inalienable right that he defended on principle, but also because of the financial implications of Monsanto's requirement. His calculation was simple: In 2000 he grew soybeans on 13,000 acres, 75 percent of them transgenic. To sow one acre with RR soybean seed, he had to pay $24.50 for a fifty-pound bag, compared to $7.50 for conventional soybean seed. To illustrate the "huge profits earned by Monsanto," he pointed out that if he decided to sell legally the surplus of his conventional crop as seeds, he would get $4 a bag.[16] For Bt cotton, he said, the ratio was one to four between conventional and transgenic seeds.

Ordered to pay damages of $65,000 in 2003, Scruggs initiated a class action suit accusing Monsanto of antitrust violations and asking that GMOs be subject to the usual plant variety protection system. Because he had openly resisted "Monsanto's law," his life became infernal: company agents had gone so far as to buy an empty lot across the street from his farm supply store where they set up a surveillance camera, and helicopters frequently flew over his property.[17]

Matters sometimes turned tragic, ending in prison terms. In January 2000, for example, Ken Ralph, a Tennessee farmer, was sued for saving forty-one tons of transgenic soybean and cotton seed. Judge Rodney Sippel of the U.S. District Court in St. Louis ordered Ralph to pay preliminary damages of $100,000 and required that he keep the seed in question so that the exact harm suffered by Monsanto could be assessed. At the end of his rope, even though he maintained that the signature on the agreement presented by the company was a forgery, Ralph decided to burn the stock. "We're tired of being pushed around by Monsanto. We are being . . . drug down a road like a bunch of dogs," he told the Associated Press.[18] Sippel finally ordered him to pay $1.7 million in civil damages, and, following a guilty plea, another judge sentenced him to eight months in prison and further damages of $165,469 for "obstruction of justice and destruction of evidence."

The case caused a stir, because it brought to light another of the company's abusive practices: the technology agreements contained a clause providing that in case of a dispute, proceedings were *exclusively* to be brought before state or federal court in St. Louis. For victims around the country this meant extra expenses in the conduct of their defense. Most important, it gave Monsanto what the *Chicago Tribune* called in 2005 a considerable "hometown advantage."[19] Established in its domain for more than a century, the company was used to working with the same law firms, including Husch and Eppenberger.[20] It turns out that Judge Sippel, known for his hard line against "pirates," had begun his legal career at Husch and Eppenberger.[21]

It should also be pointed out that in 2001, when discontent was spreading in American prairie farms against the patenting of seeds, John Ashcroft, then George W. Bush's attorney general, who had also been governor of Missouri from 1983 to 1994, asked the Supreme Court for a ruling on the question. On December 10, in an opinion written by Clarence Thomas (formerly, it will be recalled, an attorney for Monsanto) the court decided 6–2 in favor of the patenting of seeds.[22]

Everyone Is Afraid

"Patents have changed everything," said Troy Roush, an Indiana farmer who was a victim of the gene police, when I met him on his Van Buren farm in October 2006. "I really advise European farmers to think very hard before they get into transgenic crops. Afterward, nothing will be the same." Hearing this six-foot-tall rugged man say these words while holding back both tears and anger was deeply moving.

His nightmare began in the fall of 1999 with a visit from a "private detective from Monsanto," who told him he was "doing an investigation of farmers who save their seed." That year, Roush, who ran a family farm with his brother and his father, had planted five hundred acres of RR soybeans for a seed company with which he had signed a contract.* He had also planted twelve hundred acres of conventional soybeans with seeds that he had saved from his preceding harvest.

*The company, which had inserted the gene into one of its varieties, paid him for multiplying the seeds the company would sell to other farmers.

"It was very easy to tell which fields were under contract, as the contract clearly stipulated," he told me. "I offered to let the detective consult the documents and my herbicide invoices, but he refused." In May 2000, he was sued; supporting Monsanto's claim was a topographical map and analyses of samples taken from his property without his permission. "There were several glaring mistakes. For example, one of the suspected fields was in reality planted with conventional corn for the Weaver Popcorn Company, which I was easily able to prove."

"Why did you negotiate a settlement with Monsanto?" I asked.

"We had already spent $400,000 to establish our innocence," he answered. "And after two and a half years, the family was totally wiped out. I no longer had the strength to face a trial with an uncertain outcome, because precedent unfortunately favors Monsanto, which has unlimited resources for this kind of case and has everything under control. If the company had won, we would have lost everything, because it would have taken everything. Everything. Also, when I asked my lawyer what I would gain from going to trial, he told me: 'Just the glory of being found innocent.'"

In the middle of this conversation David Runyon, another Indiana farmer who had been visited by "detectives" in 2003, came into the room. The detectives had left a business card with the name "McDowell and Associates" and a startling logo: a large M superimposed on a row of men wearing capes and black hats. According to him, these were Monsanto agents claiming to have an agreement with the Indiana Department of Agriculture authorizing them to inspect the fields of farmers suspected of "piracy." David Runyon wrote immediately to Senator Evan Bayh, who checked the claim and confirmed that it was a lie, in a letter that I have a copy of.

"Patents ruined the life of rural communities," David Runyon told me, obviously very upset. "They destroyed trust between neighbors. Personally, I talk to only two farmers these days. And before I agreed to meet with you or even talk to you on the phone, I checked on Google [to see] who you were."

"Farmers are really afraid?"

"Of course they're afraid," Roush answered. "It's impossible to defend yourself against that company. You know, in the Midwest, the only way to survive with the profit margins of farming constantly going down is to increase the size of your land. For that to happen, a neighbor has to leave. So, a phone call to the snitch line, and you never know."

"You don't feel safe from another charge?"

"Certainly not," Runyon answered. "First of all, because in Indiana we're like the last of the Mohicans, since we still grow conventional soybeans in the middle of a transgenic empire. And also because our fields may be contaminated by nearby GMOs. That's what happened to my neighbor."

He took out some photographs showing a field of yellowed and stunted soybean plants, dotted with green plants. "This plot of conventional soybeans was mistakenly sprayed with Roundup by my neighbor's son, who mixed up different plots. All the green plants are Monsanto soybeans. I calculated that the contamination amounted to 15 percent."

"How is that possible?"

"In the United States, the distribution channels for the two kinds of soybeans are not separate," said Runyon. "My neighbor's conventional seeds could have been contaminated by transgenic grains left in the combine that had previously worked in a Roundup Ready field, or at the dealer during seed cleaning. It's also possible that GM pollen was spread by insects or by the wind. My neighbor has just realized that Monsanto can sue him for patent infringement."

"That's right," Roush agreed. "That's what happened to our Canadian colleague Percy Schmeiser."

Percy Schmeiser: A Rebel in Big Sky Country

Born in 1932 in Bruno, a little town of seven hundred in the heart of Saskatchewan, Canada, Percy Schmeiser is "Monsanto's nightmare, the pebble in its shoe," according to a reporter for *Le Monde*, Hervé Kempf.[23] A descendant of European pioneers who had settled in the North American prairies in the late nineteenth century, the man is a fighter—a "survivor," as he likes to say—who more than once has come close to having his energy sapped by his experience. He survived, for example, a severe work accident that disabled him for years, as well as virulent hepatitis contracted in Africa. For, along with his activities as a farmer, the prairie rebel is a man of action and a practicing Catholic: he was mayor of his town for a quarter century, then a representative in the provincial assembly, and he went on numerous humanitarian missions; he and his wife did not hesitate to entrust their five children to their grandparents, so that they could spend time helping people in Africa and Asia. Schmeiser is also a sportsman who, during the long winter cold, has climbed Kilimanjaro, and attempted Everest three times without success.

Unfortunately, I was unable to meet him, because when I went to Saskatchewan in September 2004, he was, I believe, in Bangkok, in response to one of the many invitations from around the world he has been receiving since he became the "man who rebelled against Monsanto."[24]

The case of this farmer, who had been working a fifteen-hundred-acre family farm for fifty years, began in the summer of 1997. He had just sprayed the ditches bordering his canola fields with Roundup, and he realized that his work had done practically no good: many plants that had germinated outside his area of cultivation resisted the spraying. Intrigued, he contacted a Monsanto representative, who told him that this was Roundup Ready canola, put on the market two years earlier. The months went by, and in the spring of 1998, Schmeiser, who was known throughout the region as an expert breeder of canola seeds, replanted seeds from his previous crop. When he was preparing to harvest the crop in August, he was contacted by a representative of Monsanto Canada who informed him that inspectors had detected transgenic canola in his fields and proposed that he enter into a settlement to avoid being sued.

But Schmeiser refused to give in. He turned over documents to his lawyer proving that he had bought a field in 1997 that had been planted with Roundup Ready canola. He also explained that the plant had the strength of a weed, the very light seed was able to invade the surrounding prairies at the speed of the wind and be carried for miles by birds, and seeds could lie dormant in the soil for more than five years. Observing that the transgenic canola was mostly found on the edges of his fields, he concluded that they must have been contaminated by his neighbors' GM plantings or by grain trucks passing by on the road. Schmeiser's resistance was, of course, stimulated by the revelation of Monsanto's harsh practices, including the spraying of Roundup by helicopter of fields of farmers suspected of "piracy," according to what Ed and Elizabeth Kram, a farming couple in the province, said in August 1998. This was an action that was at least "strange," and one that Monsanto has never denied, as Hervé Kempf reports, "also acknowledging in a statement to the police that its agents had taken samples of canola from Ed Kram for laboratory analysis."[25]

Monsanto Canada, in any case, was adamant. Displaying to the press the analyses of the samples it claimed to have taken (without his knowledge) from Schmeiser's farm, which contained a level of contamination greater than 90 percent, the company decided to file suit while continuing to pressure Schmeiser to settle.[26] "During 1999, Schmeiser told Kempf "we were often watched by men in a car, who said nothing, did nothing, but were just there, looking. Once they stayed three days in a row. When you walked toward them, they sped away. We also got anonymous phone calls, people who said: 'We're going to get you.' We were so afraid I bought a rifle that I kept in the tractor when I was working in the fields."[27]

The case finally came to trial in the provincial capital, Saskatoon, in June 2000. Judge Andrew McKay issued his decision on March 29, 2001, provoking stupefaction among all Schmeiser's supporters. The judge determined that in sowing his fields with seeds harvested in 1997, which he "knew or ought to have known are Roundup tolerant," Percy Schmeiser had infringed Monsanto's patent. He stated that "the source of the Roundup-resistant canola in the defendant's 1997 crop is really not significant for the resolution of the issue," and that "a farmer whose field contains seed or plants originating from seed spilled into them, or blown as seed, in swaths from a neighbor's land, or even growing from germination by pollen carried into his field from elsewhere by insects, birds, or by the wind, may own the seed or plants on his land even if he did not set about to plant them. He does not, however, own the right to the use of the patented gene, or of the seed or plant containing the patented gene or cell." This is so because "growth of the seed, reproducing the patented gene and cell, and sale of the harvested crop constitutes taking the essence of the plaintiff's invention and using it without permission."[28]

The judge thereby rejected out of hand the defense argument that Monsanto's interest in using the "essence" of GMOs was to be able to apply Roundup to crops, which Schmeiser had not done, as his herbicide invoices showed. He did not consider the fact that to take its samples, Monsanto had had to enter the farmer's property illegally, nor that the tests conducted by experts that Schmeiser had consulted showed a significantly lower level of contamination. As Kempf rightly pointed out, "the decision is extraordinary: it means that a farmer infringes the patent of any company producing GM seeds whenever his land is contaminated by transgenic plants." The decision obviously pleased Monsanto: "This is very good news for us," said Trish Jordan, a representative of Monsanto Canada. "What the judge found was that Mr. Schmeiser had infringed on our patent, and awarded us damages."[29]

They amounted to $15,450 Canadian, or $15 per acre harvested in 1998, though only part of the harvest was contaminated. Monsanto was also awarded legal costs.

Schmeiser appealed, but Judge McKay's decision was upheld on September 4, 2002. But Schmeiser, who had already sacrificed his pension and some of his land to carry on his defense (which cost $200,000 Canadian), did not give up. "This is no longer the Schmeiser case," he said, "it's the case of all the farmers in the world."[30] He appealed to the Supreme Court of Canada, which on May 21, 2004, issued a decision eagerly awaited by everyone worried by the progression of GMOs: by a 5–4 ruling, the court upheld the two previous decisions but, oddly, found that Schmeiser had to pay neither damages nor Monsanto's legal costs. The substantive finding was dramatic, because it confirmed that farmers were responsible for transgenic contamination of their fields, but the decision also suggested that the justices were troubled at the outcome. "With one hand they give and with the other hand they take away," said Richard Gold, an intellectual property specialist at McGill University in Montreal.[31] But Monsanto saw it as a victory that it would not fail to exploit in the future. "The ruling affirms the way that we do business," said Jordan.[32]

When GMO Contamination Produces Superweeds

I have been constantly impressed by Monsanto's capacity to say one thing and do the exact opposite. At the very time it was harassing Percy Schmeiser, its public relations department wrote in its *Pledge Report*: "In cases of unintended appearance of our proprietary varieties in a farmer's fields, we will surely work with the farmer to resolve the matter to the satisfaction of both the farmer and Monsanto."[33] So much for the window dressing designed to reassure shareholders and possible customers. On the ground, the reality was entirely different, for GMO contamination had become a major problem on the North American prairies.

"GM canola has, in fact, spread much more rapidly than we thought it would. It's absolutely impossible to control," said Professor Martin Entz of the University of Manitoba in 2001. "It's been a great wake-up call about the side effects of these GM technologies."[34] The same year, Professor Martin Phillipson observed: "Farmers in this province are spending tens of thousands of dollars trying to get rid of this canola that they didn't plant. They have to use more and more powerful pesticides to get rid of this technology."[35] These two statements were quoted in *Seeds of Doubt*, a report published in September 2002 by the Soil Association (a British association for the promotion of organic farming founded in 1946), which presented a very detailed description of transgenic crops in North America: "Widespread GM contamination has severely disrupted GM-free production including organic farming, destroyed trade, and undermined the competitiveness of North American agriculture overall. GM crops have also increased the reliance of farmers on herbicides and led to many legal problems."[36]

A study commissioned by the Saskatchewan Agriculture Department, for example, found in 2001 that pollen from Roundup Ready canola could travel a distance of at least eight hundred yards, eight times the distance recommended by authorities between GM and conventional crops.[37] The result was that the U.S. body certifying organic food acknowledged in the *Western Producer* in 2001 that it was practically impossible to find canola, corn, or soybean seed that had not been contaminated by GMOs. In the same article, the Canadian Seed Trade Association admitted that all conventional varieties had been contaminated to a level of at least 1 percent by GMOs.[38] One wonders what the situation is eight years later.

In any event, anticipating the uncontrollable effects of transgenic contamination, the principal agricultural insurance companies in the United Kingdom announced in 2003 that they would refuse to insure producers of GM crops against this risk, which they compared to the problems of asbestos and terrorist acts, because of the unforeseeable costs it might bring about. In a survey published in *The Guardian*, insurance companies such as National Farm Union Mutual, Rural Insurance Group (Lloyd's), and BIB Underwriters Ltd (Axa) said they "felt that too little was known about the long-term effects of these crops on human health and the environment to be able to offer any form of cover."[39]

But one thing was certain: in North America, GMO contamination had caused "a morass of litigation," in the words of the Soil Association, "embracing all levels of the industry: farmers, processors, retailers, consumers, and the biotechnology companies," with disputes among them all arising whenever an unwanted GMO appeared anywhere.[40] To illustrate the insoluble absurdity of the situation, *Seeds of Doubt* gave the example of the contamination of a shipment of conventional Canadian canola, inspected in

Europe in May 2000 because a Monsanto transgene had been detected in it. The Advanta seed company in Canada had to destroy thousands of acres, indemnify its growers, and then shift its seed production from west to east in Canada, where it judged it could better protect itself from cross-pollination, and all of this was followed by a wave of lawsuits.[41]

The problems posed by transgenic contamination are not only legal but also environmental. When a transgenic canola seed is blown by the wind, for example, into a wheat field, the farmer considers it a weed that he finds it very hard to get rid of: "as this canola is resistant to Roundup, a total herbicide, the only way to get rid of it is to pull it up by hand or use 2-4D, an extremely toxic herbicide."[42] Likewise, a GMO producer who wants to rotate his crops by alternating, for example, Roundup Ready canola with Roundup Ready corn, can also confront this problem, intensified by the specificity of canola: because its pods ripen at uneven rates, producers have adopted the habit of cutting the plants and drying them in the fields before harvesting the seeds. Unfailingly, thousands of seeds stay in the ground and germinate the following year, or even as much as five years later. This has been dubbed "volunteer" or "rebel" canola, which is in fact a "superweed."

GMOs Mean Ever More Herbicides

The irony of the story is that Monsanto understood very early on the financial interest these "rebel" plants might represent. On May 29, 2001, the company was awarded patent 6,239,072 covering a "tank mixture" that would "allow control of glyphosate-susceptible weeds and glyphosate-tolerant volunteer individuals."[43] As the Soil Association report points out, "the patent will enable the company to profit from a problem that its products had created in the first place."[44]

Considering developments in the North American prairies, one might expect that this "tank mixture" will become the company's next cash cow. The development of superweeds has in fact become one of the major headaches of North American agronomists, who have observed that they may emerge in one of three ways. In the first case, which has just been described, they are Roundup-resistant "volunteers" whose destruction requires the use of more potent herbicides. In the second case, GMOs cross with "adventitious" plants (the technical term for weeds) that are genetically close, transferring to the weeds the gene for Roundup resistance. This happens particularly with canola, a natural hybrid of turnip and cabbage, able to exchange genes with related wild species, such as wild radish, mustard, and arugula, that farmers consider weeds. A study conducted by Mike Wilkinson of the University of Reading confirmed in 2003 that the flow of genes between canola and wild turnip (*Brassica rapa*), one of the most widespread adventitious plants, was very common, which indicated that "cross-pollination between GM plants and their wild relatives is inevitable and could create hybrid superweeds resistant to the most powerful weedkillers," as the *Independent* pointed out.[45]

The third case in which superweeds appear is simply because, having been sprayed exclusively by Roundup several times a year, year after year, weeds develop resistance to the herbicide. Oddly, even though the company has had long experience with herbicides, it has always denied this phenomenon: "After 20 years of use, there are no reports of any weedy species developing resistance to Roundup herbicide," claims an advertisement extolling the virtues of RR soybeans.[46] Similarly, in its 2005 *Pledge Report*, the company continues to assert that transgenic crops "allow growers to use less herbicide."[47]

"Untrue," says the American agronomist Charles Benbrook in a study published in 2004 titled "Genetically Engineered Crops and Pesticide Use in the United States: The First Nine Years."[48] According to him, the claim of reduction in the use of herbicides was valid for the first three years following the introduction of GM crops in 1995, but not after 1999. "The increased herbicide use . . . should come as no surprise," he explains. "Weed scientists have warned for about a decade that heavy reliance on HT [herbicide-tolerant] crops would trigger changes in weed communities and resistance, in turn forcing farmers to apply additional herbicides and/or increase herbicide rates of application. . . . Farmers across the American Midwest look back fondly on the initial efficacy and simplicity of the Roundup Ready system and many miss the 'good old days.'"

Charles Benbrook knows his subject: after working as an agriculture expert in the Carter White House and then on Capitol Hill, he was head of the agriculture division of the National Academy of Sciences for seven years before setting up his own independent consulting firm in Sandpoint, Idaho. Since 1996 he has been carefully studying the data on herbicide use recorded by the National Agricultural Statistics Service (NASS), a division of USDA, comparing them with the data supplied by Monsanto, which he considers "misleading and dishonest."[49] In a 2001 article, he had already noted that "total herbicide use on RR soybeans in 1998 was 30 percent or more greater on average than on conventional varieties in six states, including Iowa where about one-sixth of the nation's soybeans are grown."[50]

In his 2004 study, he observed that the quantity of herbicides sprayed on the three principal crops in the United States (soybeans, corn, and cotton) had grown by 5 percent between 1996 and 2004, amounting to 138 million additional pounds. Whereas the quantity of herbicides used for conventional crops had continually decreased, the quantity of Roundup had gone in the opposite direction, as Monsanto in fact congratulated itself for on its 2006 10-K form: after noting that glyphosate sales accounted for $2.20 billion in revenues in 2006, compared to $2.05 billion in 2005, the company stated that "any further expansion of crops with our Roundup Ready traits should also incrementally increase sales of our Roundup products."

These results were the fruit of a strategy that had long been planned. The company's annual report for 1998 stated: "A key factor in volume growth for Roundup is a strategy based on price elasticity, with selective price reductions followed by larger percentage volume increases." When it was pointed out that this development was proof that GMOs do not reduce herbicide use, the company replied that it was to be expected that Roundup sales would increase because the surface planted in Roundup Ready crops was continually growing. Nine years after first being marketed, transgenic crops did cover nearly 125 million acres in the United States, 73 percent of which were Roundup Ready (another 23 percent was Bt), but these areas had already been cultivated before the advent of GMOs, and hence sprayed with pesticides.[51]

In addition, according to Charles Benbrook, the end of Monsanto's monopoly on glyphosate in 2000 produced a price war that brought the price of Roundup down by at least 40 percent, although the company's revenues were not adversely affected. Finally, he writes, "reliance on a single herbicide, glyphosate, as the primary method for managing weeds on millions of acres planted to HT varieties remains the primary factor that has led to the need to apply more herbicides per acre to achieve the same level of weed control."[52] He noted that before the introduction of GMOs, scientists had identified only two glyphosate-resistant weeds—rigid ryegrass in Australia, South Africa, and the United States, and goosegrass in Malaysia—but that there were now six on American territory alone, led by horsetail, which had become a veritable plague on the prairie, and Palmer pigweed varieties such as waterhemp and ragweed. For example, a University of Delaware study showed that horsetail plants taken from RR soybean fields survived ten times the recommended dose of Roundup.[53] In addition to those weeds already identified as Roundup-resistant, there is a whole list of glyphosate-tolerant weeds, that is, not yet resistant but for which doses have to be multiplied by three or four to get rid of them.

The Dark Side of Biotechnology

"Specific weed resistance can reduce a farm's rentable value by 17 percent." This was one of the conclusions of a 2002 report from Syngenta, a Swiss company that was one of Monsanto's principal competitors, sent to all its agricultural customers.[54] Relying on a survey of American farmers, the chemical and biotech giant reported that 47 percent of them favored a return to "crop and chemical rotation." As Charles Benbrook noted in early 2002, the decline in profitability was not the only "bad news" about what he called the "dark side" of biotechnology, which "scientists are now unraveling and farmers are just learning about."[55]

First, contrary to what Monsanto has always claimed in its advertising, it is not true that "under comparable growing conditions, the yields for these new lines are expected to be equivalent to other top-yielding varieties."[56] "Unfortunately, we proved the opposite," Roger Elmore, an agronomist, told me. In 2001, he and colleagues at the University of Nebraska published a study on the subject.[57] Now at the University of Iowa, near where I met him at his home in October 2006, he told me: "We conducted this study, for two years and in four different locations, because we had received information from various states indicating that transgenic soybeans had lower yields than related conventional varieties. Our results prove that yields decline by at least 5 percent."

"How do you explain it?" I asked, scrutinizing his chart.

"It's what we call 'yield drag.' We had two hypotheses that might explain the drag affecting the yield of transgenic plants: either it was due to the effect of Roundup on plant metabolism, or it was the result of genetic manipulation. To test the first hypothesis, we grew three groups of RR soybeans from the same strain, one of which was sprayed with Roundup, a second with ammonium sulfate, a product that stimulates the action of herbicides, and the third with water. The yield in all three cases was exactly the same, fifty-five bushels an acre. So it's genetic manipulation that explains yield drag. Apparently, the violent insertion of the gene disturbs the productive capacity of the plant."

"So transgenic soybeans are not the same as the conventional variety?"

"That's what our study shows."

"How did Monsanto react?"

"Let's say the company wasn't really eager to have us publish it," he answered with the necessary caution.

"But hadn't they done a study of the yield of their own soybeans?"

"The data they supplied were very weak from the scientific point of view and answered more to needs that were, let's say, commercial."

The results of Elmore's study thus confirmed the meta-analysis carried out by Charles Benbrook in which he had gone through 8,200 yield measurements made by U.S. university agriculture departments in 1998. They showed that yield drag on average was 6.7 percent, with peaks of 10 percent, particularly in the Midwest, which amounted to a loss of 80 million to 100 million bushels of soybeans for the year 1999 alone.[58]

As Benbrook pointed out, yield drag turned into a genuine catastrophe because of another phenomenon brought to light by researchers from the University of Arkansas in 2001.[59] They found that Roundup affects the rhizobium bacteria present in the soybean roots, which assist in growth by the fixation of atmospheric nitrogen. The sensitivity of the bacteria to the herbicide would explain the decline in yield of RR soybeans, which might reach 25 percent in a dry spell. "Unfortunately," according to Benbrook, "it now appears that RR crops are more vulnerable to certain diseases, especially when plants are battling other sources of stress caused by, for example, excessive cold or high pest pressure, or a mineral or microbial imbalance in the soil. These plant health problems arise because the genetic material moved into RR crops to make them tolerant of Roundup modifies the normal functioning of a key biochemical pathway that also happens to trigger and regulate a plant's immune response." He went on to say: "Unfortunately this information was only available after 100 million acres of RR soybeans had already been planted in America."[60]

A careful review of scientific and agricultural journals reveals that problems with Roundup Ready crops have been common around the country (similar problems with Bt plants will be discussed later). In 1999, for example, scientists in Georgia were contacted by soybean producers complaining that the stems of their plants were splitting for unknown reasons, leading to extremely low yields. Their study revealed that transgenic soybeans produce 20 percent more lignin than conventional soybeans, which, at higher than normal temperatures, made the stems exceptionally fragile.[61]

An Economic Disaster

"There's profit in your fields. Unleash it with Asgrow Roundup Ready soybeans." This ad published by a Monsanto subsidiary in a farm magazine in January 2002 did not convince the Soil Association, which wrote in *Seeds of Doubt*: "The evidence we have gathered demonstrates that GM food crops are far from a success story. In complete contrast to the impression given by the biotechnology industry, it is clear that they have not realised most of the claimed benefits and have been a practical and economic disaster."

Monsanto was quick to reply to this stinging indictment that one could expect nothing less from one of the principal European organizations for the promotion of organic farming. But this assessment was also that of researchers who had taken the trouble to consider *all* aspects of transgenic agriculture to determine whether, from a strictly economic point of view, the effort paid off. Michael Duffy, a University of Iowa economist, for example, conducted a study in cooperation with the National Agricultural Statistical Service of USDA. He went through the accounts of the state's farmers item by item, comparing production costs and revenues for RR soybeans (108 fields) and conventional soybeans (64 fields) in the 2000 harvest. The result was beyond question: if all factors of production were taken into account (cost of seeds, herbicide use, yield, fuel costs, fertilizer, and so on), producers of transgenic soybeans *lost* $8.87 per acre compared to $0.02 for producers of conventional soybeans.[62] It should be noted that this study was conducted in the midst of a price war on herbicides that had lowered costs and at a time when weeds were not Roundup resistant. Michael Duffy also compared earnings from Bt corn and conventional corn and came to a similar conclusion: $28.28 loss per acre for the former and $25.02 loss for the latter.

One might be surprised that farmers *lost* money in producing in all cases. This was precisely another drawback of GMOs, which had produced a collapse of American exports to Europe and a resulting price decline. Under consumer pressure, the European Commission, which had at first unhesitatingly authorized the importation of transgenic soybeans, corn, and canola from the United States and Canada, had had to backtrack and declare a five-year moratorium on GM crops on June 25, 1999, followed by required labeling of GM products on October 21.[63] These two decisions, which were vigorously challenged on the other side of the Atlantic, created confusion in the American prairies, where grain dealers asked farmers to deliver their transgenic and conventional crops separately, with a bonus for the conventional crops.

According to the *Washington Post*, there was growing anger, especially in exporting states such as Iowa and Illinois, where farmers had a persistent sense of having been bamboozled: "American farmers planted [gene-altered crops] in good faith, with the belief that the product is safe and that they would be rewarded for their efforts. Instead they find themselves misled by multinational seed and chemical companies and other commodities associations who only encouraged them to plant increased acres of [these crops] without any warning to farmers of the dangers associated with a crop that didn't have consumer acceptance."[64]

In the meantime, the harm had already been done: according to the Department of Agriculture, corn exports to Europe fell by 99.4 percent between 1996 and 2001, amounting to an annual loss of $300 million. Likewise, while Europe had absorbed 27 percent of soybean exports in 1998, the figure fell to 7 percent in 1999. And Canada, the world's largest exporter of canola, lost its entire European market, not only for canola, but also for honey.[65]

As a consequence, to save its farmers' earnings, the American government had to provide special subsidies, estimated at $12 billion between 1999 and 2002.[66] In May 2002, the Senate passed a new farm bill providing $180 billion in subsidies for the following ten years, a way "to mask the economic failure of GM crops from farmers," in the killing words of the Soil Association.

This context lay behind the conflict early in the new century between Canadian and U.S. farmers and Monsanto, which for once suffered a serious setback in its strategy to spread GMOs when it had to give up its transgenic wheat.

11

Transgenic Wheat: Monsanto's Lost Battle in North America

> We will listen carefully to diverse points of view and engage in thoughtful dialogue to broaden our understanding of issues in order to better address the needs and concerns of society and each other.
>
> —Monsanto, *Pledge Report*, 2001–2002

The story goes that the champagne corks popped in the Greenpeace office in Ottawa on May 10, 2004, and elsewhere among its allies in North America. That day, Monsanto announced in a laconic press release: "The company is deferring all further efforts to introduce Roundup Ready wheat" after "extensive consultation with customers and leaders in the wheat industry."[1] "Dialogue leads to wheat decision," it claimed in the 2004 *Pledge Report*.[2]

Monsanto Flops with Wheat

This evasive language masked an extraordinary struggle that had led to the greatest defeat ever suffered by Monsanto. For the first time in its history, the company had been forced to give up the marketing of a product for which it had invested several hundred million dollars in research and development. When I met him in October 2004, Dennis Olson, an economist with the Institute for Agriculture and Trade Policy in Minneapolis, who had been a very active participant in the American campaign against Roundup Ready wheat, told me: "For us, it was an unexpected victory that confirmed the economic failure of transgenic crops. It was especially symbolic because it had been won in North America, where GMOs were born, and thanks to the decisive support of the people who grew them."

And yet, when the company announced right before Christmas in 2002 that it had filed simultaneous requests in Ottawa and Washington for authorization to market a Roundup-resistant spring wheat variety, it seemed like a done deal, because it was operating in conquered territory. When Monsanto filed the requests it forgot a detail that would be fatal: until then, all its GMOs involved crops used primarily as fodder or for the manufacture of oils and clothing (soybeans, canola, cotton), less frequently for direct human consumption (corn). But with wheat, a mythic plant if there ever was one, it was another story: in altering the golden grain that covers nearly 20 percent of the cultivated land on the planet and is the basic nourishment for one person in three, it was touching on a symbol—cultural, religious, and economic—that was born with agriculture ten thousand years ago somewhere in Mesopotamia.[3]

And this symbol was also the daily bread—literally and figuratively—of the powerful grain farmers of North America, who cultivated the red spring wheat into which Monsanto had inserted its Roundup Ready gene. Known as the "king of wheats" because of its exceptional protein and gluten content, it is grown in four northern U.S. states—North and South Dakota, Montana, and Minnesota—and across the border in the plains of Saskatchewan in western Canada, where 15 million of Canada's 25 million acres of wheat are grown, and which is also the home of Percy Schmeiser, the herald of resistance to GMOs. Obviously, these great wheat growers also produced transgenic soybeans, corn, and canola, but when they opposed the latest manifestation from the tinkerers in Missouri, they did so primarily for economic reasons. "Canada exports 75 percent of its annual wheat production, which on average amounts to 20 million tons," I was told by Ian McCreary, vice president of the Canadian Wheat Board (CWB), run by producers, which controls the marketing of all grain produced in the prairies, by authority of a 1935 federal law. "That represents around €2 billion in revenues every year. And all our foreign customers, led by Japan and Europe, have clearly stated they did not want transgenic wheat. If Monsanto's wheat had been marketed, the 85,000 grain framers in western Canada could have gone out of business."

Ian McCreary, who is forty-two, runs a seventeen-hundred-acre farm near Bladworth, in the heart of the vast, flat, dreary province known as the breadbasket. When I met him in September 2004, he and his wife, Mary, were making final adjustments to their combine. It looked like the end of the world, with thousands of acres of wheat stretching to the horizon glittering under a steel-blue sky toward which were raised huge grain elevators dotting the prairie like Lego pieces.

"We're far away from everything here," McCreary said with a smile, after saying grace before the family lunch. "Transport costs are astronomical, and to make a living we have to concentrate on the quality of our wheat, which is highly valued by millers around the world; they blend it with varieties of lower baking quality. As they did with canola and corn, GMOs would have created price declines and we can't let ourselves sell wheat for fodder."

"But Monsanto says its wheat would have taken care of the weed problem," I said.

"Unlike soybeans, weeds are not really a problem for wheat. I think it was Monsanto that had a problem: its Roundup patent had just expired and the company wanted to make up for it by selling herbicide and seeds for one of the largest food crops in the world. As for wheat growers, they were afraid that Roundup Ready wheat would increase herbicide costs because 'volunteers' would show up, not to mention the exorbitant cost of patented seeds: in the plains we usually keep our wheat seeds for at least ten years before buying new ones."

And so the powerful CWB ended up campaigning alongside Greenpeace and the Council of Canadians (the country's largest citizens' organization), "two organizations it has clashed with in the past," as the *Toronto Star* remarked, "to present a united front opposing GM wheat."[4] The article quotes a letter from Rank Hovis, the leading British flour miller, to the CWB: "If you do grow genetically modified wheat, we will not be able to buy any of your wheat, neither the GM nor the conventional. . . . We just cannot sell it." At the same time, Grandi Molini Italiani, the leading Italian miller, sent a similar message to North American wheat growers.[5] They were soon joined by the powerful association of Japanese millers, whose executive director, Tsutomu Shigeta, predicted a "collapse of the market" if Monsanto's wheat were to invade the plains, because the majority of consumers didn't want it.[6] (In May 2003, a survey conducted by the Western Organization of Resource

Councils had found that 100 percent of Japanese, Chinese, and Korean importers contacted would refuse to buy transgenic wheat.)

In the United States, half of whose wheat is exported, for an annual revenue of approximately $5 billion, the message was heard loud and clear by all grain growers, including those who did not grow spring wheat. "The impact on the market concerns all producers," explained Alan Tracy, president of U.S. Wheat Associates, who had been shaken by a study published in October 2003 by Robert Wisner, a University of Iowa economist.[7] Wisner had examined the impact the marketing of the new GMO would have on the wheat economy, and his conclusions were very dark: a decline of 30 to 50 percent in red spring wheat exportations and even more for other hard wheat varieties, a two-thirds fall in prices, loss of jobs throughout the sector, and a wave of repercussions throughout rural life. "A large majority of foreign consumers and buyers do not want transgenic wheat," Wisner said. "Whether they are right or wrong, consumers are the driving force in counties where labeling permits choice."[8]

Hundreds of farmers, who had applauded the arrival of GMOs less than ten years earlier, were seen traveling around the northern Great Plains to "fight against biotechnology." In North Dakota and Montana, the resistance had "solidified into a political movement," which demanded a moratorium on Monsanto wheat.[9] The company moved heaven and earth to block these initiatives. To bring the wandering sheep back into the fold, it went so far as to charter a plane to bring a delegation of North Dakota rebels to its Missouri headquarters, where they were received by Robert Fraley, one of the inventors of RR soybeans, who had been promoted to a position as vice president. He "seemed to imply that farmers opposing Monsanto might be advancing the agenda of radical environmental groups." "At that point," said Louis Kuster, one of the farmers who had been at the meeting, "I . . . was a little bit angry and I looked right straight at him . . . and I said, 'You're not talking to the Greens here today. . . . We need to make money, too.'"[10]

The Attack on Bt Plants: The Misfortunes of the Monarch Butterfly

To fully understand the 2003 revolt of North American farmers, it has to be set in the context of the time, which was not very favorable for Monsanto.

As the French sociologists of science Pierre-Benoît Joly and Claire Marris pointed out in that year, resistance to GMOs was built up around "trials" and "themes" that had specific characteristics on either side of the Atlantic and that converged at the beginning of the new century, leading to a shared rejection of Roundup Ready wheat.[11]

In Europe, the first issue that spawned the anti-GMO movement was the mad cow crisis, which broke out in 1996, at the time when the first shipment of RR soybeans were arriving from the United States. The campaign Greenpeace organized against GMOs won support particularly because it was rooted in the cataclysm of the fatal prion, which had revealed the inability of government institutions to measure the risks of intensive agriculture and the system of industrial production of food. As Joly and Marris note, "On November 1, 1996, *Libération* printed the headline 'Warning: Mad Soybeans,' which clearly points to the importance of the mad cow crisis as a precedent strongly influencing the way in which GMOs were represented."[12]

Combined with the rising power of the anti-globalization movement that denounced the control of multinationals such as Monsanto over world agriculture (consider, for example, the events surrounding the WTO summit in Seattle in December 1999), the theme of junk food underlay the sympathy felt by the French for the people who, alongside the peasant leader José Bové, tore down the McDonald's in Millau in August 1999 and tore up transgenic test plots.

In the United States, where junk food was a way of life, what was on the consumer's plate was not a mobilizing theme during the entire "calm period" that accompanied the "large-scale spread of GMOs." But when Terminator, and more broadly the patent issue, caused the first stirrings in the countryside, two other sets of events shifted public opinion, which suddenly began to question the reliability and impartiality of regulatory agencies in their management of the risks associated with products derived from biotechnology. The first of these involved the monarch butterfly, a migratory insect with orange wings that became the most effective symbol for the anti-GMO cause in the United States.

On May 20, 1999, *Nature* published a study conducted by John Losey, a Cornell University entomologist.[13] Along with two colleagues, he had studied the effects on butterfly larvae of a Bt corn variety produced by Novartis (now Syngenta) that was supposed to fight the corn borer, a plant parasite.

Recall that Bt plants—of which Monsanto was the largest producer—took their name from a bacterium found naturally in the soil, *Bacillus thuringiensis*, which produces a substance that works like an insecticide. Isolated in 1901 by a Japanese bacteriologist who had observed that it infected and killed silkworms, this bacillus is used in spray form by organic farmers, because the toxin the bacteria produce has the property of rapidly decaying in sunlight, allowing for selective use with no consequences for the environment or for untargeted insect populations. But biotechnology had completely changed things. Insertion of the gene that coded for the toxin meant that the toxin was expressed *permanently throughout the plant*, creating the risk of affecting *all* insect populations, the useful as well as the harmful, as for example the chrysopa, a predator of the corn borer that Bt was supposed to combat. When Losey conducted his research on the monarch butterfly, various studies had already showed that Bt crops could be fatal for beneficial insects such as ladybugs, as well as microorganisms in the soil and insect-eating plants.[14]

In its lab the Cornell team had fed monarch butterfly larvae with milkweed leaves, their favorite diet, dusted with Bt corn pollen. "Four days later, 44 percent of the larvae had died, and the survivors had lost their appetite. But none of the larvae exposed to leaves dusted with natural pollen had died."[15] The study caused a stir in North America, and the very day on which it was published, the European Commission announced the suspension of requests for authorization for the marketing of several Bt varieties, including Monsanto's. Christian Morin, the Novartis spokesman, defended the company: "These were laboratory observations, in conditions that placed the monarch in extreme circumstances," and he asked that the experiment be repeated in the field.[16] But nothing was to be done; the misfortunes of Americans' beloved butterfly delivered the first blow against corn exports to Europe, which collapsed. Margaret Mellon of the Union of Concerned Scientists was indignant: "Why is it that this study was not done before the approval of Bt corn? This is 20 million acres of Bt corn too late. This should serve as a warning that there are more unpleasant surprises ahead."[17]

GMO producers led by Monsanto organized a response by conducting a campaign "downplaying and, in some instances, ridiculing the study," and if necessary making statements that "were misleading, fanciful, and betrayed an ignorance of the monarch's natural history," as Lincoln Brower, who had been working on the butterfly since 1954, wrote in a 2001 article.[18] This very well-informed article shows how a scientific debate can be completely perverted by private interests with the complicity of government institutions and elements of the scientific community: "In the ongoing debate over the Cornell findings, the scientific process has been spun, massaged, and manipulated by the agricultural industry . . . losing sight of a larger, more serious issue: the real danger that genetically engineered crops will accelerate . . . the impoverishment of biological diversity." Along the way, he notes that the intensive use of Roundup has caused the disappearance of wildflowers such as milkweed, on which the monarch depends for survival.

He then recounts the process of manipulation that he witnessed. In the days following the publication of the Cornell study, the leaders of the biotechnology industry decided to create a consortium, which they named the Agricultural Biotechnology Stewardship Working Group (ABSWG), whose mission was to sponsor university research similar to that conducted by John Losey. On November 2, 1999, when these studies were still in their preliminary stages, the ABSWG organized a conference in Chicago that was supposed to present an open debate on the delicate question. Participants included a number of researchers financed by the consortium, but also independent figures such as Lincoln Brower and Carol Yoon, a science reporter for the *New York Times*. Although the discussions had barely begun, Yoon was informed that a press release from the Biotechnology Industry Organization had been received by the *Times* that morning, with an unequivocal title: "Scientific Symposium to Show No Harm to Monarch Butterfly."[19] Flabbergasted, Yoon asked the participants if they had received word of this press release, and they uniformly replied no. She reported the rather revealing anecdote,[20] but all other newspapers blindly reproduced the false claims of the press release.[21]

However, the Cornell team's results were confirmed by a University of Iowa study published on August 19, 2000, in the journal *Oecologia*.[22] John Obrycki, who directed the research, conducted in the field with milkweed leaves gathered in proximity to transgenic crops, commented: "We found that after five days exposure to Bt pollen, 70 percent of monarch butterfly larvae died."[23] The debate was relaunched at the time, but it was soon overwhelmed by the greatest health and environmental scandal that GMOs had provoked so far.

The StarLink Debacle

On September 18, 2000, Friends of the Earth issued a press release that triggered a veritable cataclysm. The American ecological association announced that it had analyzed samples of products containing corn (chips, tacos, cereals, cornmeal, soups, pancakes) bought in supermarkets and that the tests had detected traces of StarLink, a Bt corn variety produced by Aventis that was banned for human consumption.* In order to increase the insecticidal function of its GMO, the company had introduced a Bt protein (Cry9C) that was particularly heavy and stable, but which was "suspected of causing allergies because it has a heightened ability to resist heat and gastric juices, giving more time for the body to overreact," as the *Washington Post* reported.[24] The EPA had limited the sale of this Bt corn variety to animal feed and ethanol production. But grain dealers who were not aware of the regulatory subtlety had mixed StarLink with other yellow varieties of the grain.

Before discussing the consequences of this appalling affair, I would like to point out how revealing it is of what Joly and Marris call the "inadequacy of the American regulatory framework."[25] It will be recalled that after publishing its "coordinated framework for the regulation of biotechnology," the Republican administration had distributed responsibilities among the three principal regulatory agencies: the FDA was responsible for transgenic food, the EPA for GMOs that functioned as pesticides, and the USDA for transgenic crops. The result of this arbitrary division was that Bt plants, some of which ended up on consumers' plates, were under the jurisdiction not of the FDA but of the EPA, because they were considered pesticides.

This paradox, which explains the StarLink catastrophe, was brilliantly demonstrated by Michael Pollan in the *New York Times* in 1998.[26] He tells how he "planted something new in [his] vegetable garden," a Bt potato recently put on the market by Monsanto, called "New Leaf," which was supposed to "produce its own insecticide." In the instructions for use he found that the potato had been registered as a pesticide by the EPA, and was sur-

*At the time, Aventis was a European pharmaceutical group created in 1999 by the merger of the German company Hoechst, the French companies Rhone-Poulenc and Roussel-Uclaf, the American companies Rorer and Mario, and the British company Fisons. In 2004, it was acquired by Sanofi-Synthélabo, which became Sanofi-Aventis.

prised that the label provided information on its organic composition, the nutrients, and "even the trace amounts of copper" it contained, but did not say a word about the fact that it was a product of genetic engineering nor even that it "contain[ed] an insecticide." He then decided to call James Maryanski, the biotechnology coordinator at the FDA, who told him: "Bt is a pesticide, so it's exempt" from FDA regulation and therefore falls under the jurisdiction of the EPA. Since Pollan was going to eat his potatoes, he wanted to know if the EPA had tested their food safety. "Not exactly," Maryanski replied. As the name indicates, "pesticides are toxic to something," so the EPA instead establishes human "tolerances." Pollan then called the EPA, where he was told that since "a New Leaf is nothing more than the sum of a safe potato and a safe pesticide," the agency thought it posed no human health risks. Pollan goes on: "Let us assume that my potatoes are a pesticide—a very safe pesticide. Every pesticide in my garden shed—including the Bt sprays—carries a lengthy warning label. The label on my bottle of Bt says, among other things, that I should avoid inhaling the spray or getting it in an open wound. So if my New Leaf potatoes contain an EPA-registered pesticide, why don't they carry some such label?"

It would be hard to find a better illustration of the aberrant nature of the American regulatory system: the EPA, after being alerted to the possible allergenic effects of StarLink corn, decided to restrict its use to animal feed instead of simply banning it. It is worth noting the complete indifference of the FDA to this question; a letter sent by Alan Rulis on May 29, 1998, to AgrEvo, the Aventis subsidiary that was marketing StarLink, did not mention the issue at all, instead merely explaining: "As you are aware, it is AgrEvo's continued responsibility to assure that foods the firm markets are safe, wholesome, and in compliance with all applicable legal and regulatory requirements."[27]

The FDA official didn't know how right he was. By September 2000, the agency had been overwhelmed by frightened calls from around the United States. One of them came from Grace Booth, who said that at a business lunch where she ate enchiladas, she suddenly experienced hot flashes and violent diarrhea, her lips swelled, and she lost her voice. "I felt like I was going to die," she told CBS.[28] Immediately transported to a California hospital, she survived thanks to the quick administration of an anti-allergy medication. All the reports reaching the FDA spoke of violent reactions associated with the consumption of corn-based products. Interviewed by

CBS, Dr. Marc Rosenberg, an allergist advising the government in this sorry affair, confirmed that the symptoms "varied from just abdominal pain and diarrhea [and] skin rashes to a very small group having very severe life-threatening reactions."

As a very detailed report from Friends of the Earth pointed out in July 2001, "The StarLink debacle is a case study in the near total dependence of our regulatory agencies on the 'regulated' biotech and food industries, and . . . in [their] regulatory incompetence."[29] The group reported that the FDA took a week to confirm the presence of StarLink in the food chain, for a reason that it never would have suspected: "We later learned that this delay was due to the simple fact that after two years of StarLink cultivation on hundreds of thousands of acres across the country,* the FDA still did not have the expertise to even test for this potentially allergenic protein."[30] To be able to conduct laboratory tests, the FDA had had to ask for help from Aventis. Likewise, when the EPA was forced to establish a test to measure the allergenicity of the Bt protein, it had to turn to the manufacturer to supply it with a sample of the molecule. Finally, claiming that it could not isolate enough of the protein expressed in the plant, the company supplied a synthetic substitute from the *E. coli* bacterium. Experts pointed out that the test would be biased because "the same protein is not necessarily identical in different species."[31] After months of procrastination, the EPA cautiously concluded that there was "a medium likelihood that StarLink [was] an allergen."[32] The health authorities then buried the file, losing a perfect opportunity to understand why the consumption of corn products had made hundreds of Americans gravely ill and almost killed some of them.

No to GM Wheat

In the interim, the debacle had cost Aventis $1 billion. First, the company had to indemnify the food distributors that had withdrawn from their shelves 10 million corn-based products. Then it had to repurchase stocks of StarLink seeds from all the dealers, farmers, and millers. But the magnitude of the catastrophe exceeded the darkest predictions: tests conducted by

*It is estimated that StarLink at the time accounted for 1 percent of the corn crop in the United States, approximately 370,000 acres.

USDA found that 22 percent of American corn was contaminated by the incriminated protein.[33] This gave a fatal blow to exports that the monarch butterfly affair had already severely reduced. *Nature* reported that, according to a USDA representative, StarLink was found in bakery products in Taiwan and even Japan.[34] An irritated John Wichtrich, an Aventis executive, told a meeting of the North American Millers' Association in San Antonio: "I know you are wondering: Will there ever be an end to this? Unfortunately, as of now, the answer is 'No'—there will never be an 'end' as long as there is zero tolerance for Cry9C in food."[35]

It is therefore easy to understand why resistance was organized in the North American plains when, in the midst of the StarLink debacle, Monsanto announced its intention to market its Roundup Ready wheat. It should be noted that the company was in very bad shape. As previously reported, in late 2002 the CEO, Hendrik Verfaillie, had been forced to resign for "poor financial performance," namely, $1.7 billion in losses for the year. But this was not the problem for the Canadian Wheat Board, which, on June 27, 2003, declared war not only on Monsanto but on its loyal government ally. Adrian Measner, president of the CWB, declared: "We will do everything in our power to ensure that GM wheat is not introduced in Canada."[36]

A short time before, the Canadian House of Commons Standing Committee on Agriculture and Agri-Food had met to discuss the question. Excluded from the deliberations, Greenpeace Canada had circulated a letter it had sent to Paul Steckle, the committee chairperson, in which it criticized the "conflict of interest created by the partnership between Monsanto and the Canadian government."[37] The letter notes that Agriculture and Agri-Food Canada (AAC, under the Ministry of Agriculture and Food) "provided top quality genetic material that was public property to Monsanto so that it could develop its RR wheat," and that it was AAC that had "carried out under contract the field tests of Monsanto GM wheat so that it could be granted plant variety protection status." Finally, the same ministry "provided Monsanto with at least $800,000 of funding under the Matching Investment Fund Initiative."[38] Under these circumstances, it is indeed hard to see how the Ministry of Agriculture and its partner, the Canadian Food Inspection Agency (CFIA), which had functioned as co-developers of RR wheat, could independently exercise their regulatory authority in assessing "as required the safety of agricultural biotechnology for human health, agriculture, and the environment."[39]

In its letter, Greenpeace also discussed at length the problem of genetic contamination that might arise from the marketing of RR wheat. Its experts suggested that the committee ask Monsanto representatives three questions at the hearing:

> Is Monsanto prepared to issue a public and legally binding declaration that would hold it responsible in the event of the genetic contamination of conventional and organic wheat . . . by its RR wheat?
>
> If so, how much money is Monsanto prepared to set aside to compensate the victims of such damages?
>
> If not, according to Monsanto who should pay for those damages?

Ian McCreary, vice president of the Canadian Wheat Board, told me: "It's true that the question of genetic contamination weighed heavily in our decision to reject RR wheat. The specter of StarLink was haunting us, and besides, we already had the example of transgenic canola, which had practically eliminated conventional canola in Canada."

When Transgenic Canola Eliminates Organic Canola: Inevitable Contamination

The first victims of genetic contamination were organic farmers, who had to give up their canola crops because they could not guarantee their integrity. To confirm this, I met Marc Loiselle, one of the leading figures in the resistance to Monsanto's wheat, who has been an organic farmer for twenty-two years.[40] He and his wife, Anita, work the farm established by his grandparents, who emigrated from Aquitaine a century earlier and settled in Vonda, about 30 miles from Saskatoon, the territory of Percy Schmeiser, the man who stood up to Monsanto.

On the day in September 2004 that I met him, Loiselle was worried: an unusual cold spell with temperatures at well below freezing had hit the plains, threatening the wheat harvest. Wheat was his entire life—it was his livelihood, of course, but also it connected him to the family saga and to the human adventure beyond. This practicing Catholic did not grow just any wheat: every year he planted more than 100 acres with an old variety threatened with extinction: Red Fife, highly valued by traditional bakers. As we drove down a straight road running to the horizon over the flat landscape, he told me that when European settlers had come to Canada they'd brought wheat seeds that were not adapted to the extremely harsh climatic conditions of the prairies. Then in 1842, David Fife, a Scottish farmer who had settled in Ontario, began to plant seeds that a friend from Glasgow had gotten from a cargo of Ukrainian wheat shipped from Danzig. That variety of red wheat, known as Red Fife in honor of its discoverer, soon spread through the plains like wildfire, because it had strong rust resistance and ripened early enough to escape autumn frosts. Later a breeder decided to cross it with Hard Red Calcutta, a variety from India, to increase yields and flour quality. Thus was born Marquis, which in the early twentieth century conquered a vast territory stretching from southern Nebraska to northern Saskatchewan, considered today one of the world's breadbaskets.

"This history," Loiselle told me, "is a very good illustration of the great saga of wheat, which humanity was able to develop in the four corners of the world because the exchange of seeds was not yet blocked by patents and Terminator."

We were now in a huge field of Red Fife wheat, surrounded by plots of Roundup Ready canola drying on the ground. "Before," he told me, "I rotated crops of wheat and canola or mustard. But I had to stop because my field was contaminated by my neighbor's transgenic canola, probably transported by the wind. My organic certification agency asked me not to grow canola or any related plant for at least five years, because it is known that canola seeds can lie dormant in the soil for that length of time. In any case, I don't think I will go back to growing organic canola, because it's impossible to protect against contamination."

"You can't plant hedges or buffer zones, as agricultural authorities recommend?" I asked.

"It wouldn't do any good. You can't prevent all natural events: birds, bees, wind. Agriculture works with living things, which are not collections of genes set down on a piece of paper. Contrary to what Monsanto claims, I can tell you that once a GMO is introduced, the farmer loses the capacity to choose what sorts of crops he wants to grow, because GMOs colonize everything. They infringe my freedom as a farmer to plant what I want where I want. That's why we were prepared to do anything to preserve wheat from that calamity."

In January 2002, Loiselle joined a class action suit that included most of the organic farmers in Saskatchewan, requesting damages from Monsanto and Aventis for the loss of their canola crops.[41] On December 13, 2007, the Supreme Court of Canada finally rejected the claim on technical grounds, determining that the complaint, whose basis it did not challenge, could not be treated in a class action but only through individual cases.

In the interim, what Loiselle and his colleagues criticized had been confirmed by a scientific study conducted by René Van Acker, an agronomist from the University of Manitoba, at the request of the Canadian Wheat Board.[42] "We conducted tests in twenty-seven grain elevators of certified non-transgenic canola seeds and we found that 80 percent were contaminated by the Roundup Ready gene," Van Acker told me when I met him in Ottawa in September 2004. "That means that now almost all Canadian canola fields include Roundup Ready plants. As for organic canola, it has already disappeared in Canada, where it's hard to find three square miles with no GMOs."

"How were you able to use the experience with canola for wheat?"

"The Canadian Wheat Board asked us to determine whether the Roundup Ready gene was likely to move from one wheat crop to another. To answer that question, we constructed a model of the flow of genes, which in canola operates from what we call 'gene bridges.' We compared all the elements of the model, one by one, and we determined that the situation would be similar for wheat and that a flow of genes was also possible."

"Couldn't two separate channels be organized based on the segregation of seeds?" I asked, adopting the argument frequently put forth by the promoters of biotechnology.

"It's impossible. Contamination in the field is inevitable and it makes any attempt at segregation before planting ineffective."

This conviction is shared by the owners of grain elevators, confirmed by a survey conducted in 2003 by the Institute for Agriculture and Trade Policy in Minneapolis.[43] It showed that 82 percent of the owners contacted were "very concerned" by the possible marketing of RR wheat, because "it's impossible to have a segregation system with zero tolerance." Two years earlier, an internal memorandum (obtained by Greenpeace) from Agriculture and Agri-Food Canada to the Minister of Agriculture, Lyle Vanclief, revealed that the segregation argument didn't convince government officials themselves: "If transgenic wheat is registered, it will be difficult and costly to keep it segregated from non-transgenic wheat through the production, handling and transport chain," the memo says.[44]

It should be noted that this is also the opinion of European officials, who officially, however, speak an entirely different language, designed to reassure recalcitrant populations. For example, a secret report submitted to the European Union in January 2002, which Greenpeace also obtained, confirmed that the introduction of transgenic crops into Europe would be a fatal blow to "organic and small farming of oilseed rape as well as for intensive production of conventional maize," and that the "cultivation of GE and non-GE crops on the same farm might be an unrealistic scenario, even for larger farms." Aware of the "sensitivity" of these conclusions, Barry McSweeney, director of the research center of the European Union, saw fit to attach a letter to the report in which he wrote: "Given the sensitivity of the issue, I would suggest that the report be kept for internal use within the Commission only."[45]

"Is transgenic contamination reversible?" I asked Van Acker, a bit horrified by all this information.

"Unfortunately, I don't think so. There is no backtracking possible. Once a GMO has been released in nature, you can no longer call it back. If you wanted to eliminate transgenic canola in western Canada, you would have to ask all farmers to stop growing the plant for at least ten years. Which is impossible because canola is our second-largest national crop, with 11 million acres in cultivation."

"What are the consequences for biodiversity?"

"That's a very important question, particularly for Mexico, which is the original source of corn, or for the countries in the Fertile Crescent, where wheat comes from. Canada and the United States export to those regions of the world. If transgenes are introduced into wild and traditional species of corn or wheat, it would lead to a dramatic impoverishment of biodiversity. There is also the problem of intellectual property rights. The Percy Schmeiser case shows that Monsanto thinks any plant belongs to it whenever it contains a patented gene: if this principle is not challenged, it will mean in the end that the company could control the genetic resources of the world, which are common property. Look at what's happening in Mexico; we're already at a crossroads."

PART III

Monsanto's GMOs Storm the South

12

Mexico: Seizing Control of Biodiversity

Adventitious presence is part of the natural order.

—Monsanto, *Pledge Report*, 2001–2002

"The hope of the industry is that over time the market is so flooded that there's nothing you can do about it. You just sort of surrender."[1] So said Don Westfall, vice president of Promar International, a Washington consulting firm working for biotech companies, in early 2001. This statement was echoing in my head when I landed in Oaxaca, Mexico, in October 2006. Nestled in the heart of a lush landscape of green mountains, the city, considered one of the jewels of the country's tourist industry, was in the throes of a violent social conflict.

The Transgenic Conquest of Mexican Corn

Hundreds of strikers and their families occupied a tent camp flying the banners of the Popular Assembly of the People of Oaxaca (APPO) in the *zócalo*, the magnificent colonial plaza bordered by arcades. The streets in the historic center were blocked by barricades, while the governor's palace, the courthouse, the regional assembly, and all the schools in the state of Oaxaca, considered one of the poorest in the country, had been closed for weeks. Starting with a teachers' strike, the conflict had spread to all sectors of society, and people were demanding the resignation of Ulises Ruiz Ortiz, the state governor. This political boss of the Institutional Revolutionary Party (PRI), corrupt and a devotee of repressive measures, had finally been disowned by his own party.

"You've come to cover the events?" asked the receptionist in my hotel, who had seen a procession of reporters from around the world.

"No, I've come because of the contamination of corn." He obviously found this unexpected answer surprising.

On November 29, 2001, *Nature* had published a study that created a stir and drew heavy fire from Monsanto in St. Louis. Signed by David Quist and Ignacio Chapela, two biologists at the University of California, Berkeley, it found that *criollo* (traditional) corn in Oaxaca had been contaminated by Roundup Ready and Bt genes.[2] The news was particularly surprising because in 1998 Mexico had declared a moratorium on transgenic corn crops in order to preserve the extraordinary biodiversity of the plant, whose genetic cradle was Mexico. Grown since at least 5000 BC, corn was the basic food for the Maya and Aztec peoples, who worshiped it as a sacred plant. An Indian legend says that the gods created man from an ear of yellow and white corn.

As a European for whom corn is always golden yellow, I was fascinated by the unsuspected diversity of the numerous Mexican varieties. Traveling around the indigenous communities of Oaxaca, four or five hours over potholed roads from the capital, I encountered everywhere women in colorful skirts drying, in front of their hovels, magnificent ears of corn colored pale yellow, white, red, violet, black, or an astonishing midnight blue, some mixing together several colors because of cross-pollination.

"In the Oaxaca region alone, we have more than 150 local varieties," said Secundino, a Zapotec Indian who was harvesting white corn by hand. "This variety, for example, is excellent for making tortillas. Look at this ear: it has a very good size and fine kernels, so I'll save it to plant next year."

"You never buy seeds from outside?"

"No. When I have a problem, I exchange with a neighbor: I give him ears for him to eat and he gives me seeds. It's old-fashioned barter."

"Do you always make tortillas with local corn?"

"Yes, always," he said with a smile. "It's more nourishing, because it's of much better quality than industrial corn. Besides, it's healthier, because we farm without chemical products."

"Industrial corn" means the 6 million tons of corn that flood in every year from the United States, 40 percent of which is transgenic. Because of NAFTA, the 1992 free trade agreement with Canada and the United States, Mexico has been unable to prevent the massive importation of corn; heavily subsidized by the American administration, it threatens local production because it is sold at half the price.* It is estimated that between 1994 and 2002, the price of Mexican corn fell by 44 percent, forcing many small farmers to head for city slums.

"Look," said Secundino, holding out in his hand like a gift a magnificent violet ear. "This corn was my ancestors' favorite."

"It existed before the Spanish conquest?"

"Yes, and now there is another conquest."

"What's the new conquest?"

"The transgenic conquest, which wants to destroy our traditional corn so industrial corn can dominate. If that happens, we will become dependent on multinational corporations for our seeds. And we will be forced to buy their fertilizers and their insecticides, because otherwise their corn won't grow. Unlike ours, which grows very well without chemical products."

The Media Lynching of Ignacio Chapela

"Small Mexican farmers are very conscious of the stakes raised by transgenic contamination, because corn is not just their basic food but a cultural symbol," said Ignacio Chapela, one of the authors of the *Nature* study, who had agreed to meet me at Sproul Plaza on the Berkeley campus. This is where the anti–Vietnam War movement took off in 1964, which denounced among other things the spraying of Agent Orange and the "merchants of death," among which was Monsanto.

It was an October Sunday in 2006, and the huge campus, where more than thirty thousand students and nearly two thousand teachers usually bustle about, was deserted. Only a police car drifted by like a damned soul. "That's for me," said Chapela. "I've been closely watched since this affair

*In 2007, the United States exported 11 percent of its corn to Mexico, which was worth $500 million; 30 percent of the corn consumed in Mexico came from the United States.

started, especially when there's a camera." When I looked incredulous, he went on: "You want proof? Come with me." We drove to the top of a hill overlooking San Francisco Bay. As we walked toward the lookout point, we saw the same police car, which parked conspicuously at the side of the road and stayed there throughout our conversation.

"How did you find out that Mexican corn was contaminated?" I asked, rather disturbed.

"I worked for fifteen years with the Indian communities in Oaxaca teaching them to analyze their environment," answered the Mexican-born biologist, who had worked for the Swiss company Sandoz (which became Novartis, and then Syngenta) for several years. "David Quist, one of my students, went there to run a workshop on GMOs. To explain the principles of biotechnology, he suggested that they compare the DNA of transgenic corn, from a can of corn he brought from the United States, with that of a *criollo* variety meant to serve as a control, because we thought it was the purest in the world. Imagine our surprise when we discovered that the samples of traditional corn contained transgenic DNA. We then decided to conduct a study, which confirmed the contamination of *criollo* corn."

To conduct their research, the two scientists took ears of corn from two localities in the Sierra Norte of Oaxaca. They found that four samples had traces of 35S promoter, derived from the cauliflower mosaic virus; two samples revealed the presence of a fragment from the bacterium *Agrobacterium tumefaciens*; and another the trace of a Bt gene.[3] "As soon as we got the results, we alerted the Mexican government, which conducted its own study that confirmed the contamination."

On September 18, 2001, the Mexican environment minister announced that his experts had done tests in twenty-two farming communities and found contaminated corn in thirteen of them, with a level of contamination between 3 and 10 percent.[4] Oddly, this announcement went practically unnoticed, but a few months later Ignacio Chapela and David Quist became a focus of attention, probably because of the reputation of *Nature*, which published their article in late November. When they'd submitted the article to the journal eight months earlier, the two scientists had received compliments on the quality of their study, and the peer review process followed its normal course: the article was sent to four reviewers, who approved it. But as a local paper, the *East Bay Express*, pointed out in May 2002: "No one could have predicted the magnitude of the controversy to come."[5] The result was a veritable media lynching, largely organized from St. Louis.

"First," Chapela told me, "you have to understand why the study provoked the wrath of the unconditional promoters of biotechnology. It contained two revelations: the first concerned genetic contamination, which really surprised no one, because everyone knew it was bound to happen, including Monsanto, which always merely confined itself to minimizing the impact." Indeed, in its *Pledge Report*, the company approaches the thorny subject with infinite delicacy, not mentioning "contamination" but stating; "Adventitious presence is part of the natural order."[6] "But," Chapela went on, "the second point of our study was much more serious for Monsanto and similar companies. In investigating where the fragments of transgenic DNA were located, we found that they had been inserted into different places in the plant genome in a completely random way. That means that, contrary to what GMO producers claim, the technique of genetic engineering is not stable, because once the GMO cross-pollinates with another plant, the transgene splits up and is inserted in an uncontrolled way. The most virulent criticisms were particularly focused on that part of the study, denouncing our technical incompetence and our lack of expertise to evaluate this type of phenomenon."

The fact that "the transgenes were unstable" had "profound" implications, according to an article in *Science* in February 2002: "Because a gene's behavior depends on its place in the genome, the displaced DNA could be creating utterly unpredictable effects."[7] Three months later, the *East Bay Express* went further: " It undercut the very premise that genetic engineering is a safe and exact science."[8] "The study was nothing more than mysticism masquerading as science," retorted Matthew Metz, a former student of Chapela's at Berkeley.[9] Metz, who had become a microbiologist at the University of Washington, denigrated Chapela and Quist to the point of claiming that they had been taken in by "false positives" due to "laboratory contaminants."[10]

"Where did the attacks come from?" I asked Chapela.

"From two places. First from colleagues at Berkeley whom I had confronted in the past over a $25 million contract the biology department had signed with Novartis-Syngenta, my former employer, in 1998. This five-year contract gave the company the right to file patents on a third of our discoveries. The affair had created two camps in Berkeley representing two conflicting conceptions of science: on one side, those who, like me, wanted it to remain independent, and on the other, those who were prepared to sell their souls to obtain funding."

In June 2002, *New Scientist* identified these colleagues, who wrote an inflammatory letter to *Nature* in December 2001 asking the journal to retract the article, an unusual step. They were Mathew Metz, Nick Kaplinsky, Mike Freeling, and Johannes Futterer, a Swiss researcher whose boss was Wilhelm Gruissem, who worked at Berkeley, where he "was widely regarded as the man who brought Novartis to Berkeley."[11]

"But the worst campaign came from Monsanto," Chapela said. He concluded that it "had quite obviously received a copy of our article before it was published."

Monsanto's Dirty Tricks

Monsanto really did carry things to an extreme in this case, and the story I am about to tell is hard to believe. The very day Quist and Chapela's article was published in *Nature*, November 29, 2001, an obviously well-informed woman named Mary Murphy sent an e-mail to the pro-GMO science Web site AgBio World in which she wrote: "The activists will certainly run wild with news that Mexican corn has been 'contaminated' by genes from GM corn. . . . It should also be noted that the author of the *Nature* article, Ignacio H. Chapela, is on the Board of Directors of the Pesticide Action Network North America (PANNA), an activist group. . . . Not exactly what you'd call an unbiased writer."[12]

The same day, a person named Andura Smetacek posted on the same Web site a comment titled "Ignatio [*sic*] Chapela—activists FIRST, scientist second," in which she had no qualms about spreading lies: "Sadly the recent publication by Nature Magazine of a letter (not a peer-reviewed research article subject to independent scientific analysis) by Berkeley Ecologist Ignatio Chapela are being manipulated by anti-technology activists (such as Greenpeace, Friends of the Earth, and the Organic Consumers Association) with the mainstream media to falsely suggest some heretofore undisclosed ill associated with agricultural biotechnology. . . . Research into Chapela's history with these groups of [eco-radicals] demonstrates his willingness to collude with them to attack biotechnology, free-trade, intellectual property rights, and other politically motivated agenda items."[13]

At the time the "smear campaign" that derailed Chapela's career was getting under way, Jonathan Matthews came upon these strange posts by chance.[14] Matthews was the head of GMWatch, an information service on GMOs based in Norwich in southern England. "At the time I was looking into AgBio World," he told me when I met him in November 2006, sitting in front of his computer. "It was breathtaking: the two e-mails from Mary Murphy and Andura Smetacek were distributed to the 3,400 scientists on AgBio World's distribution list. The campaign spread from there. Some scientists, such as Professor Anthony Trewavas of the University of Edinburgh, called on *Nature* to retract the article or to have Ignacio Chapela fired."

"Who is behind AgBio World?"

"Officially it's a nonprofit foundation that claims 'to provide science-based information on agricultural biotechnology issues to various stakeholders across the world,' as its Web site declares," he answered, showing me the site.[15] "It's run by Professor Chanapatna S. Prakash, director of the Center for Plant Biotechnology Research at Tuskegee University in Alabama. Originally from India, he is an adviser to USAID, and in that capacity, he has intervened frequently in India and Africa to promote biotechnology. He became famous in 2000, when he launched a 'Declaration of Support for Agricultural Biotechnology,' for which he secured the signatures of 3,400 scientists, including twenty-five Nobel Prize winners.[16] AgBio World had no qualms about accusing environmentalists on its Web site of 'fascism, communism, and terrorism, including genocide.' One day, when I was consulting the AgBio World archives, I received an error message giving me the name of the server that hosts the site: apollo.bivings.com. The Bivings Group, based in Washington, is a communications firm, one of whose clients is Monsanto, and it specializes in Internet lobbying."[17]

Matthews showed me a 2002 article by George Monbiot in *The Guardian* revealing that the firm had presented its expertise in an article on its Web site entitled "Viral Marketing: How to Infect the World." "There are some campaigns where it would be undesirable or even disastrous to let the audience know that your organization is directly involved . . . it simply is not an intelligent PR move. In cases such as this, it is important to first 'listen' to what is being said online. . . . Once you are plugged into this world, it is possible to make postings to these outlets that present your position as an uninvolved third party. . . . Perhaps the greatest advantage of viral marketing is that your message is placed into a context where it is more likely to be considered seriously." A senior executive from Monsanto is quoted on the Bivings site, thanking the PR firm for its "outstanding work."[18]

"Do you know who Mary Murphy and Andura Smetacek are?" I asked, feeling as though I were in the midst of a detective novel.

"Well," the director of GMWatch said with a smile, "*The Guardian*, to which I sent my findings, summed it up well: they are 'phantoms' or 'fake citizens.'[19] I spent a lot of time trying to find out who these two 'scientists' who had launched the campaign against Ignacio Chapela were. As for Mary Murphy, she has posted at least a thousand e-mails on the AgBio World site. For example, she put online a forged Associated Press article criticizing 'anti-GMO activists.' When you trace back to find the address of the server hosting her e-mail address, you find: bw6.bivwood.com. So 'Mary Murphy' seems to be a Bivings employee. When it came to 'Andura Smetacek,' I thought it should be easy to find a scientist with such an unusual name, especially since she claimed to be writing from London. She was the one who had initiated a petition demanding that José Bové be incarcerated. I went through the electronic phone directory, the electoral registry, and the list of credit card holders, but it was impossible to find any trace of her. I hired a private detective in the United States, but he didn't find anything either. Finally, I examined the technical details at the bottom of her e-mails indicating the Internet protocol address: 199.89.234.124. When you type it onto a directory of Web sites, you come upon 'gatekeeper2.monsanto.com,' with the owner's name, 'Monsanto Corporation, St. Louis.'"

"Who do you think is hiding behind 'Mary Murphy'?"

Matthews responds, "George Monbiot of *The Guardian* and I think it's Jay Byrne, who was in charge of Monsanto's Internet strategy. At an industry meeting in late 2001, he stated that it was necessary to 'think of the Internet as a weapon on the table. Either you pick it up or your competitor does, but somebody is going to get killed.'"[20]

"Fake scientists and fake articles—it's incredible!"

"Yes, they're really dirty tricks that represent the exact opposite of the qualities Monsanto claims it stands for in its *Pledge*: 'dialogue, transparency, sharing.'[21] These methods reveal a firm that has no desire to persuade with arguments and is prepared to do anything to impose its products everywhere in the world, including destroying the reputation of anyone who might stand in its way."

An Absolute Power

Meanwhile, the "conspiracy," as *The Ecologist* called it, had borne fruit.[22] On April 4, 2002, after failing to persuade Quist and Chapela to retract their article, *Nature* published an "unusual editorial note,"[23] constituting an "unprecedented disavowal" in the 133-year history of the celebrated journal.[24] "The evidence available is not sufficient to justify the publication of the original paper," the journal wrote. "A unique event in the history of technical publishing," this rebuff created a stir in the international scientific microcosm.[25] In a letter to the journal, Andrew Suarez of Berkeley expressed his surprise, commenting that the statement "reflects poorly on *Nature*'s editorial policy and review process . . . Why has *Nature* refrained from releasing similar editorial retractions of earlier publications later found to be incorrect or open to alternative interpretations?"[26] The answer to this question was suggested by Miguel Altieri, another Berkeley researcher: "*Nature* depends on its funding from big corporations. Look at the last page of the journal and see who funds the ads for jobs. Eighty percent are technology corporations, paying anywhere from $2,000 to $10,000 per ad."[27]

Nature's "backpedal[ing]"[28] was particularly surprising because a month earlier *Science* had reported that "two teams of Mexican researchers had confirmed biologist Ignacio Chapela's explosive findings."[29] Directed by Exequiel Ezcurra, the highly respected director of the Mexican National Institute of Ecology, one of the studies had analyzed samples of corn taken from twenty-two communities in Puebla and Oaxaca. Genetic contamination ranging from 3 to 13 percent had been found in eleven of them, and with contamination levels of 20 to 60 percent in four others. Ezcurra submitted an article to *Nature*, which rejected it in October 2002. "This rejection is due to ideological reasons," he stated, pointing to the "contradictory explanations" of the reviewers, one of whom said that the results were "obvious," and the other that they were "incredible."[30]

Meanwhile, Chapela had paid a heavy price: in December 2003, the Berkeley administration informed him that it had denied him tenure despite the 32–1 vote in favor by his department; he would have to leave the university at the end of his contract six months later. In other words, he was fired. He filed suit and won in May 2005: "Since then," he told me, "I bear the burden of being known as a whistle-blower. I have no funding to conduct the research that interests me, because in the United States now you can't work in biology if you don't accept funding from biotechnology firms. There was a time when science and the university loudly proclaimed their independence from governmental, military, and industrial institutions. That's over, not only because scientists depend on industry to survive, but because they themselves are part of industry. That's why I say that we're living in a totalitarian world, ruled by the interests of multinational corporations who recognize their responsibility only to their shareholders. It is hard to resist this absolute power. Look at what happened to Exequiel Ezcurra."

Unfortunately, I was unable to meet the former director of the Mexican National Institute of Ecology, who, a few years after denouncing *Nature*'s rejection of his study of the contamination of *criollo* corn, was in 2005 appointed director of scientific research at the San Diego Natural History Museum, where he had headed the Biodiversity Research Center from 1998 to 2001. I was surprised to find that in August 2005 he had co-signed a study published in *Proceedings of the National Academy of Sciences*, a publication of the National Academy of Sciences. Conducted at Washington University in St. Louis,* the study found an "absence of detectable transgenes in local landraces of maize in Oaxaca."[31] But in October 2006 I did meet one of his colleagues, Dr. Elena Alvarez-Buylla, in her laboratory at the Mexican National Institute of Ecology.

"How do you explain the fact that Dr. Ezcurra signed a study that contradicts his previous work to such an extent?" I asked.

"Only he knows," the biologist answered cautiously. "What I can say is that we began that work together and that I was pushed out. I was replaced by an American, Allison Snow from the University of Ohio, who picked up the study in progress. They decided to publish preliminary results, which I don't consider scientifically very rigorous." She is not the only one who thinks so: five international researchers—including Paul Gepts, whom I had met in July 2004 at the University of California, Davis, to discuss the patenting of life—also found that the "conclusions [of the study] are not scientifi-

*It should be noted in passing that Monsanto has deposited its archives at Washington University in Saint Louis, but they are unfortunately not accessible for journalists.

cally justified."[32] Nonetheless, the study was reported in many international newspapers, including *Le Monde*.[33]

"Since then," Alvarez-Buylla told me, "my laboratory has carried out another study throughout the country that found that the national level of contamination is on average from 2 to 3 percent, depending on the type of transgene, with some much higher peaks."

"What do you think about this dispute?"

"I think it has nothing to do with scientific rigor and that it is masking other interests. What's important to me now is to find out the medium-term effects of the contamination on *criollo* corn. That's why my research team did an experiment on a very simple flower, *Arabidopsis thaliana*, which has the smallest genome in the plant world, into which we introduced a gene by genetic engineering.[34] We then planted the transgenic seeds and observed their growth. We found that two genetically completely identical plants—they had the same genome, the same chromosomes, and the same transgene—could produce very different phenotypes [floral forms]: some had flowers identical to the natural variety, with four petals and four sepals, but others had aberrant flowers with abnormal bristles or bizarre petals. And some were plainly monstrous. In fact, the only difference among all these plants was the location of the transgene, which was inserted completely at random, by modifying the plant's metabolism."

"What does that have to do with corn?" I asked, contemplating one of the "monstrous" flowers that the scientist was displaying on her computer.

"From this experimental model we can extrapolate what risks happening when transgenic corn cross-pollinates with local varieties. It's very worrying, because there is a fear that the random insertion of a transgene may affect the genetic inheritance of *criollo* corn in a totally uncontrolled way."

The Monsters of Oaxaca

"The monsters are already in our mountains," said Aldo González, one of the leaders of the Union of Organizations of the Sierra Juarez of Oaxaca, to whom I had just recounted my conversation with Alvarez-Buylla. It was a morning in October 2006, and we were on our way from Oaxaca to a Zapotec community in a remote mountain area. González had put a portable computer on the backseat of his car. "It contains my war chest," he said with a smile, "the fruit of three years' work." In 2003, peasants had contacted his organization because they were worried when they saw corn plants growing in their fields that "looked sick and deformed." Some were abnormally high; others had deformed ears or unusual leaves. González came and took photographs and plant samples that he had tested by a laboratory that used the kits that enable European customs agents to detect transgenes in soybeans or corn imported from North America. "Every test turned out positive. I now have about three hundred photographs that I've taken all over the Sierra Juarez."

We had reached the little village of Gelatao. After the required introduction to the head of the community, González picked up a loudspeaker that echoed loudly in the steep-walled basin in which the village was set. "We invite you to participate in a meeting about the new diseases attacking our corn because of transgenic contamination," he explained as a screen was set up on the village square. With machetes at their waists, the men streamed in, sometimes accompanied by their wives carrying brightly colored cloths in which they would soon wrap the ears of the harvest.

"I am going to show you photographs of corn plants taken in our region," González told the audience. "I would like to know if you've already come across this type of plant in your community. You see, some very strange things are happening. This plant, for example, has one stalk here and another one there. Normally, a corn plant is not like that: there is always one leaf out of which an ear grows, but look at this, there are three ears coming out of the same leaf. They're really monsters. In general, we've come across these plants on roadsides or in gardens. It's possible that someone bought corn in a grocery store and he lost a few kernels on the way. The kernels germinated and that's how traditional corn was contaminated."

"I had a plant that looked like that last year," a young peasant said. "I showed it to the old people and they told me they'd never seen that. It's a new disease?"

"Yes," González answered. "But the problem is that there is no treatment."

"If I understand," another Indian said, "if this proliferation isn't stopped in our fields we'll soon be forced to buy corn, because ours will not produce anything anymore. That's very worrying: what can we do?"

"The first recommendation is if you find a bizarre plant, you should immediately pull up the seedling to prevent it from emitting pollen and contaminating the rest of your field. Generally speaking, you have to be very vigilant and keep close watch on your corn."

"If contamination spreads, what might be the consequences?" I asked

"That will be the end of *criollo* corn and of the whole rural economy. But the more I think about it, the more I tell myself that it's all intentional, because finally contamination benefits only multinational corporations like Monsanto. Once everything is contaminated, the company will be able to take control of the most widely grown grain in the world and collect royalties, as in Argentina and Brazil."

Indeed, the ravages of GMOs are not limited to North America and Mexico. They have also affected South America, Argentina in particular, where over the course of just a few years transgenic soybeans have become both the country's primary economic resource and, probably, its primary curse.

13

In Argentina: The Soybeans of Hunger

> The consistent rise in global acreage is evidence of the benefits of herbicide tolerant crops, including positive environmental impacts.
>
> —Monsanto, *Pledge Report*, 2005

It was April 13, 2005, in Buenos Aires, and Miguel Campos was having trouble hiding his anger. For several weeks, Argentina's secretary of agriculture, livestock, fishing, and food had been involved in a dangerous struggle with Monsanto. Not that this agricultural engineer was opposed to biotechnology. On the contrary, he was appointed to his position, like all his predecessors for the previous ten years, precisely because he was an unconditional supporter of GMOs. Throughout the two hours of our conversation he constantly extolled the agricultural and financial benefits of RR soybeans while attempting to persuade me that Monsanto's conduct was as vile as it was inexplicable.

"Monsanto was never able to patent its RR gene in Argentina because our laws do not permit it," he explained, speaking forcefully. "So the company agreed to give up seed royalties and promised not to sue farmers who replanted part of their harvest, as they have always done, completely legally. Now Monsanto is going back on its promises, demanding $3 per ton of soybeans or soy flour leaving Argentine ports, or $15 when the cargo reaches European ports. That's unacceptable."

Taking Over Argentina

Miguel Campos looked crestfallen, like a good student unjustly accused by a teacher he adores. For if there was a country where Monsanto could do whatever it wanted without the slightest obstacle, that country was certainly Argentina. At the time Campos was talking to me, half the cultivated land in the country was planted with transgenic soybeans—35 million acres and 37 million tons harvested, 90 percent of which was exported, primarily to Europe and China. If Monsanto were to reach its goals, the company would take in $160 million annually for exports to Europe alone—a jackpot.

"You don't think it was a trap?" I asked.

It seemed to me that Campos pretended not to understand. "A trap?"

"Well, Monsanto first created conditions favoring the spread of RR soybeans throughout the country, then the company asked you to pay the bill."

"If that was the strategy, it was a mistake. You don't change the rules of the game ten years later."

"Will you pay?"

"The conflict is serious, because Monsanto is threatening to attack all Argentine exports." In a statement reported by the Dow Jones Newswire on March 17, 2005, Campos had been blunter, denouncing Monsanto's "hoodlum-like attitude."

Ten years earlier, however, the transgenic adventure had begun like a fairy tale in the country of cattle and gauchos. When the FDA authorized the sale of RR soybeans on the North American market in 1994, Monsanto had already had its eyes on the Southern Cone for some time. Its target was, of course, Brazil, the world's second-largest producer of soybeans. But the deal was hardly in the bag because the Brazilian constitution required that transgenic crops go through preliminary tests of their environmental impact before their release was authorized. So Monsanto turned to Argentina, where the government of Carlos Menem, following the lead of the first Bush administration, constantly spoke of deregulation. During his ten-year rule (1989–99), Menem, who went on trial in October 2008 for illegal weapons sales, did his utmost to complete the work begun under the military dictatorship (1976–83): he dismantled what remained of the Argentine welfare state, privatizing whatever he could and opening the country's gates wide to foreign capital. This policy had a devastating impact on the agricultural sector, whose protective barriers were annihilated in order to hand production over entirely to the laws of the market.

Monsanto was prepared and entered the breach in the early 1990s, becoming the privileged interlocutor of Conabia, the National Advisory Commission on Agricultural Biotechnology, established by Menem in 1991 to provide Argentina with the appearance of GMO regulation. The commission, under the jurisdiction of the Secretariat of Agriculture, had only advisory status and was made up exclusively of representatives from public bodies, such as the National Seed Institute (INASE) or the National Institute of Agricultural Technology (INTA), and private players in the biotechnology industry, such as Syngenta, Novartis, and, of course, Monsanto, whose persistent interventionism is not hard to imagine. The opinions expressed by Conabia were indeed based directly on North American models; from the outset it adopted the principle of substantial equivalence, as its Web site indicated: "The Argentine standard is based on the identified characteristics and risks of the biotechnological product and not on the process that made the product possible." Concretely, the commission did nothing but analyze the data supplied by the multinational corporations; if tests were conducted, their only purpose was to test the adaptability of transgenic seeds to Argentine agricultural conditions.

Beginning in 1994, Monsanto sold licenses to the principal seed companies in the country, such as Nidera and Don Mario, who took care of introducing the Roundup Ready gene into the varieties in their catalogue. By a lucky coincidence, the two major newspapers in the country, *La Nación* and especially *Clarín* (which had the largest national circulation), plunged into the promotion—some called it propaganda—of biotechnology, labeling all opponents, even the most moderate, anti-progress fanatics or Luddites, to adopt the expression of Bill Clinton's former Secretary of Agriculture, Dan Glickman.* Countless editorials praised the biotechnological revolution with arguments oddly reminiscent of those presented by a certain company in Missouri: "With GMOs, science has made a decisive contribution to the war against hunger," Carlos Menem, for example, declared in an agricultural journal.[1] William Kosinski, Monsanto's "biotechnology educator," asserted:

*The most determined defender of GMOs in Argentina is Héctor Huergo, who edits the supplement *Clarín Rural*.

"Biotechnologies make possible harvests of better quality, higher productivity, and sustainable agriculture protecting the environment."[2]

"The introduction of GMOs into Argentina came about with no public or even parliamentary debate," according to Walter Pengue, an agricultural engineer at the University of Buenos Aires who specializes in the improvement of plant genetics and whom I met in Buenos Aires in April 2005.[3] "There is still no law regulating their marketing, and civil society, which is not even represented on the Conabia, is kept out of any decisions. After they were authorized in 1996, RR soybeans spread through Argentina at an absolutely unprecedented speed in the history of agriculture: an average of more than two million acres a year. We now have a veritable green desert devouring one of the world's breadbaskets."

The Magic Seeds

As soon as you head north out of Buenos Aires you encounter a stunning sight: as far as the eye can see are soybeans and more soybeans, sometimes interrupted by pastures with large herds of grazing cows. In the southern autumn month I was there, the harvest was already well along and Ruta Nacional 9 was jammed with trucks shuttling between the silos of soybeans and the ports on the Río Parana. This is the heart of the pampas, the vast legendary plain of Argentina that covers 20 percent of the national territory, 250,000 square miles bordered on the north by the Chaco region, the east by the Río Parana, the south by the Río Colorado, and the west by the Andes. As fertile as the U.S. corn belt, the *llanura pampeana* is one of the best pasturelands in the world and since the nineteenth century has been an area of intense agricultural development where, until the arrival of GMOs, the crops were grains (corn, wheat, sorghum), oil-producing plants (sunflowers, peanuts, soybeans), and fruits and vegetables, not to mention milk production, which was so well developed that the area was known as the "milk basin." In the national imagery, the pampas were the country's pride, able to produce food for ten times its population and therefore for export. "Cultivating the soil is serving the country," says a poster at the entrance to the headquarters of the Argentine Rural Society.

The man who met me after I had driven for five hours was from a true peasant family filled with that nourishing vision of agriculture. About forty, Héctor Barchetta farmed 315 acres about thirty-five miles from Rosario, the capital of the transgenic empire. A member of the Argentine Agrarian Federation, an association of seventy thousand small and medium-sized farms, he confessed that he was "at a complete loss." As he walked through the fields of RR soybeans that now cover 70 percent of his farm, he told me the story of a miracle that was in the process of becoming a nightmare.

In the 1990s, he confronted a problem that affected all the farmers of the pampas: the erosion of the soil because of excessively intensive exploitation. According to INTA, yields had fallen by 30 percent. "We didn't know which way to turn," he told me, "and that's when RR soybeans arrived. At first, they were really magic seeds, because we returned to high yields, with reduced costs and less work." In fact, as in the United States, transgenic agriculture was developed with the technique of "direct sowing" (*siembra directa*), which permits direct planting, with no prior plowing, in the residue of the previous crop. Promotion and technical advice were provided by Aapresid, the Argentine Association of No-Till Farmers, which bears a strong resemblance to its North American counterpart, the American Soybean Association (ASA).

Grouping together fifteen hundred large producers, Aapresid is the principal promoter of RR soybeans and Monsanto's most loyal ally in Argentina. "The technique of *siembra directa* is an integral part of the model of transgenic farming," according to the agronomist Walter Pengue. "At first it did lead to the restoration of soil fertility through an increase in organic matter supplied by the surface residues, which retained water. The technique cannot be dissociated from what Monsanto calls the 'technological package,' transgenic seeds and Roundup sold together, and there the company demonstrated its great skill by launching its 'package' at one third the price it charged in the United States." The price was so low, in fact, that North American producers, even though they were heavily subsidized, howled in protest against this "unfair competition."

Barchetta, for one, took the bait enthusiastically. "Before," he told me, "to destroy weeds I had to apply four or five different herbicides, but with RR soybeans, two applications of Roundup were enough. And then, to top it off, the mad cow crisis made the price of soybeans take off, and I stopped producing corn, wheat, sunflowers, and lentils, like all my neighbors." The European prohibition on animal-based feeds did bring about increased demand for vegetable proteins, including soy cakes. The price of soybeans reached historic highs, bringing about a rush on the new green gold in the pampas. "Thanks to the soybean boom I was able to survive the crisis," he went on. "Everything was done to spare the producers. While interest rates were skyrocketing, we could get Monsanto's package and not pay for it until after the harvest."

In 2001, Argentina was at the brink of bankruptcy. The government of Fernando de la Rúa was forced to resign under popular pressure. While *piqueteros*—strikers in revolt—ruled the streets, poverty took hold all over the country, where 45 percent of the population was living below the poverty line. Strangled by a colossal external debt, the governments of Eduardo Duhalde and then Néstor Kirchner used soybeans as a life preserver. "It's the engine of our economy," according to Campos. "The state collects a 20 percent tax on the oil and 23 percent on the seeds, which amounts to $10 billion [annually], 30 percent of the national currency. Without soybeans, the country would simply have gone under."

Soybeans Take Over the Country

The Argentine crisis was a boon for Monsanto that exceeded its wildest dreams. RR soybeans spread from the pampas like wildfire, steadily heading north into the provinces of Chaco, Santiago del Estero, Salta, and Formosa. Covering only 90,000 acres in 1971, soybeans spread over 20 million acres in 2000, 24 million in 2001, 29 million in 2002, and reached more than 39 million acres in 2007, accounting for 60 percent of the cultivated land. The phenomenon was so striking that there was talk of the *sojisación* of the country, a neologism designating a profound reordering of the agricultural world whose disastrous effects soon became apparent.

At first, although the crisis had crippled the national economy, the price of land skyrocketed, because it had become a safe investment providing significant quick profits. "In my area," Héctor Barchetta told me, "the price of an acre went from $800 to $3,000. The weakest producers ended up selling out, which brought about a concentration of landholdings." In the course of a decade, the average size of farms on the pampas increased from 617 to 1,328 acres and the number of farms was reduced by 30 percent. According to an agricultural census conducted by the National Institute of Statistics and Census (INDEC), 150,000 farms went out of business between 1991 and 2001, 103,000 of them after the advent of transgenic soybeans. By the end of that period, 6,000 owners held half the cultivated land in the country, while 39 million acres were already in foreign hands, a process that has accelerated since then.

According to Eduardo Buzzi, president of the Argentine Agrarian Federation, "We have witnessed an unprecedented expansion of agribusiness, industrial agriculture directed toward exports, to the detriment of family farming, which is disappearing. Farmers who leave are replaced by people who do not come from the agricultural world: pension funds or investors placing their money in 'seed pools,' who plunge into the monoculture of RR soybeans, in cooperation with multinational corporations like Cargill and Monsanto, all at the expense of food crops."

As RR soybeans continued their irresistible advance, transforming what was once the breadbasket of the world into a producer of cattle feed for the European market, food producers shrank at a rapid rate. According to official figures, from 1996–97 to 2001–2, the number of *tambos* (dairy farms) decreased by 27 percent and, for the first time in its history, Argentina had to import milk from Uruguay. Similarly, the production of rice fell by 44 percent, corn by 26 percent, sunflowers by 34 percent, and pork by 36 percent. In tandem with this movement came a staggering rise in the prices of basic consumer products: in 2003, for example, the price of flour went up 162 percent, lentils—a major element in the national diet—by 272 percent, and rice by 130 percent. According to Pengue, "The average Argentine is eating much less well than he did thirty years ago. And the irony is that we are being encouraged to replace cow's milk and beef, which have always been part of the national diet, with soy milk and soy steaks."

Pengue's comment is not a joke but a simple description of reality. In a country where *dulce de leche* (milk caramel) and *carne de vaca* (beef) are essential ingredients of the cultural heritage, Secretary of Agriculture Miguel Campos himself is quick to give you the address of a good *sojero* restaurant in Buenos Aires. He goes on to praise the generosity of the Soja Solidaria program launched in 2002 by Aapresid, which decided to "help," in its way, the 10 million people suffering from malnutrition, including one of every six children. The idea was simple: "Give away a kilo of soybeans for every ton exported." The campaign was backed by the major media, quick to present Soja Solidaria as a "brilliant idea which is going to change history."[4] As for the unavoidable Héctor Huergo, the editor of *Clarín Rural*, he encouraged the government to "replace current social welfare programs with a chain of solidarity with no cost thanks to a soybean distribution network, one of the most complete foods, which just has to be introduced into our culture."[5]

GMO promoters participated generously in the program: thanks to free diesel provided by Chevron-Texaco, soybean shipments were delivered to hundreds of food banks and school cafeterias in poor and slum neighborhoods, to hospices, hospitals, and every variety of charitable institution in Argentina. Throughout the country, workshops were set up where volunteers—the Catholic University of Córdoba called them "soybean soldiers"—taught cooks how to make milk, hamburgers, and other meat substitutes out of soybeans. On the Web site nutri.com, one learns, for example, that in Chimbas, in the remote province of San Juan, a municipal program educated six thousand people in "soybean consumption," and that a thousand volunteers had been mobilized to distribute soy milk to twelve thousand children.

When Soja Solidaria celebrated its first anniversary, Victor Trucco, the president of Aapresid, did not hide his satisfaction. "In time," he wrote in *Clarín*, "we will look back on the year 2002 as the year when soybeans were incorporated into the Argentine diet."[6] He drew up a balance sheet: "We have contributed 700,000 tons of soybeans, representing more than 600,000 pounds of high-value protein, 8 million quarts of milk, 5 million pounds of eggs, or 3 million pounds of meat." The statistics were likely designed to conceal a purpose that the Soja Solidaria Web site summed up in a sentence: "The plan helped the spread of soybeans" in the country.[7]

Rebel Soybeans: Toward the Sterilization of the Soil

One day, Walter Pengue arranged a visit to Jésus Bello, a farmer in the pampas who had started planting RR soybeans in 1997. For seven years, Pengue had been following several farms in the region and carefully examining their farming records. "At the beginning," he said, "I was rather in favor of transgenic soybeans, because I thought that with crop rotation and a reasonable use of glyphosate, it could be good for the environment and for the producers' pocketbooks, since weed control amounted to 40 percent of production costs. But now I'm very worried, because every element is in the red."

Bello nodded in agreement: "We're headed into a wall. We're spending more and more and the soil is exhausted." Bello, like Héctor Barchetta two hundred miles away, was confronted with a problem that was growing worse every year: the resistance of weeds to glyphosate. "From an agronomic perspective, that was known beforehand," said Pengue. "Before the advent of transgenic soybeans, producers used four or five different herbicides, some of which were very toxic, like 2,4-D, atrazine, and paraquat.* But the alternation between the different products prevented weeds from developing resistance to any single one of them. Now, the exclusive use of Roundup at any time of year has led to the appearance of biotypes that were first tolerant of glyphosate; to get rid of those weeds, it was necessary to increase the herbicide dose.† After tolerance came resistance, which can already be observed in some areas of the pampas."

"So Monsanto's commercial argument that Roundup Ready technology reduces herbicide use is mistaken?"

"Completely," said Bello. "I apply glyphosate twice, once after planting, the other time two months before the harvest. At first, I used less than a liter of herbicide per acre; now I need twice as much."

Pengue added: "Before the advent of RR soybeans, Argentina used an average of 1 million liters of glyphosate annually. In 2005, we reached 150 million liters. Monsanto does not deny that there is a resistance problem and has announced a new, more powerful herbicide with a new generation of GMOs, but that's not a way out of the vicious circle."

The cost for producers has been heavy. The time has passed when, to prime the pump, Monsanto offered a two-thirds discount on the price of its herbicide. The price very soon returned to normal, which led producers to turn to generics (principally Chinese) as soon as the company's patent expired in 2000. But at the same time a new problem arose that further increased costs: what was known as "rebel soybeans" in Argentina ("volunteers" in Canada), indicating that, in South America as in North America, the same causes produced the same effects. And as in the United States, Syngenta, Monsanto's Swiss competitor, which manufactures atrazine and paraquat,

*It will be recalled that 2,4-D is one of the components of Agent Orange; it is now (theoretically) banned in Europe and the United States. Atrazine was banned in the European Union in 2003. As for paraquat, which was, like Roundup, one of the most widely sold herbicides in the world, it was banned in the European Union on July 10, 2007.

†They are *Parietaria debilis, Petunia axilaris, Verbena litoralis, Verbena bonariensis, Hybanthus parviiflorus, Iresine diffusa, Commelina erecta*, and *Ipomoea sp*.

seized the opportunity: in 2003, one of its major ads proclaimed, "Soybeans are weeds."

In addition, the intensive use of Roundup tends to make the earth sterile. "I use constantly increasing amounts of fertilizer," Bello acknowledged. "Otherwise yields would collapse." It is hard to see how a total herbicide able to eliminate every kind of plant would spare the microbial flora essential for soil fertility. According to Pengue, "The disappearance of certain bacteria makes the earth inert, which blocks the process of decomposition and attracts slugs and fungi such as fusarium."

To top everything, in 2004 the price of soybeans began a downward tendency, which continued in 2005, to the point of causing producers like Bello and Barchetta lasting anxiety.* "What are we in the process of doing?" Barchetta asked, his eyes fixed on the plot he was about to harvest. "Before, I produced fifteen different food crops; now I only do transgenic soybeans. Maybe we've fallen into a trap. Maybe we're in the process of sacrificing the Earth and our children's future."

A Public Health Disaster

Dr. Darío Gianfelici was annoyed as he drove down the road: "Look at that. They plant soybeans even in the ditches at the side of the road. When they apply pesticides, you can be completely doused in spray. This country's public health authorities are completely irresponsible." When I met him in April 2005, Gianfelici was a doctor in Cerrito, a small town of five thousand about thirty miles from Paraná, in the province of Entre Ríos, in the heart of the empire of soybeans. In this region of the pampas once noted for its agricultural diversity, soybean cultivation grew from 1.5 million acres in 2000 to nearly 3 million acres three years later. During the same period, rice production fell from 370,000 to 128,000 acres.[8] At least twice a year, crop-dusting planes or *mosquitos* (farm equipment towed by tractors that spray herbicides using long mechanical arms in the shape of wings) inundate the region with Roundup, often reaching as far as house doors, since RR soybeans have invaded the whole area.

*After reaching $230 a ton in 2003, the price fell to $200 in 2004, then to $150 in mid-2005. But in 2006, there was a spectacular recovery, and it reached a peak in 2007, particularly because of the fad for biofuels.

"It's like a fever, an epidemic," Gianfelici said, pointing to the notorious *chorizos*, the sausage-shaped silos that are now scattered along the roadsides because there is no other place to store the enormous quantities of soybeans. The doctor became an anti-GMO activist not out of ideology but because he was worried by the evolution of the illnesses he was encountering in his practice. "I don't know whether biotechnology is a danger for public health," he explains, "but I denounce the damages to health caused by the massive spraying of Roundup, as well as the excessive consumption of RR soybeans." He mentioned the toxicity of glyphosate and especially of the surfactants, the inert substances that enable glyphosate to penetrate into the plant, such as polyethoxylated tallowamine. In Argentina more than elsewhere, Monsanto's advertising assuring that Roundup is "biodegradable" and "good for the environment" resulted in many people failing to take any precautions during spraying, which means that the substance has contaminated the entire environment: air, soil, and water table. All the while the representative of the state, Miguel Campos, has claimed with complete confidence that "Roundup is the least toxic herbicide there is."

Gianfelici was certain: "Several colleagues in the region and I have observed a very significant increase in reproductive anomalies such as miscarriages and premature fetal death; malfunctions of the thyroid, the respiratory system—such as pulmonary edemas—the kidneys, and the endocrine system; liver and skin diseases; and severe eye problems. We are also worried by the effects that might be caused by Roundup residues ingested by consumers of soybeans, because we know that some surfactants are endocrine disruptors. We have observed in the region a significant number of cases of cryptorchism and hypospadias* in boys, and hormonal malfunctions in girls, some of whom have their periods as young as three."

There are few like Gianfelici who dare to raise their voices against the devastating effects of the all-soybean policy. Of course, organizations such as Greenpeace and the radical ecologists of the Grupo de Reflexión Rural had denounced the marketing of GMOs and pointed out the dangers of biotechnology, but they were preaching in the wilderness. "With the crisis, there were thousands of other problems," according to Horacio Verbitsky, a

*Cryptorchism is a birth defect characterized by undescended testicles; hypospadias is a malformation of the urethra (it does not reach the tip of the penis).

columnist for the left-wing daily *Página 12*, who never wrote a thorough article on transgenic soybeans. "I admit that even I know nothing about it."

Oddly, it was the Soja Solidaria program that provoked the first institutional warnings—not about GMOs as such, but about the risks posed to children by the excessive consumption of soy products. For example, in July 2002, the Consejo Nacional de Coordinación de Políticas Sociales organized a forum on the subject where it was pointed out that "soy juice should not be called 'milk' and it can in no case replace milk." Health professionals pointed out that soy is much less rich in calcium than cow's milk and its heavy concentration of phytates blocks the body's absorption of metals such as iron and zinc, increasing the risk of anemia.* Above all, they strongly advised against the consumption of soy products by children younger than five, for a commonsense reason: it is known that soy is very rich in isoflavones, which act as hormone substitutes for premenopausal women and can therefore cause significant hormonal imbalances in growing bodies.†

"We are sowing the seeds of a veritable public health disaster," according to Gianfelici, "but unfortunately the authorities haven't recognized what's at stake and anyone who dares to talk about it is considered a crazy person opposed to the country's welfare."

That day, the doctor had an appointment in a Catholic school run by German nuns. The imposing colonial-style building emerged in the midst of a vast spread of soybeans. "Last week," the headmistress explained, "they sprayed Roundup just before it rained. Then there was bright sunshine, which caused evaporation. Many students started to vomit and complained of headaches." She had asked the provincial health services to investigate, and they determined it was a "virus." "Still, they did analyze the water, but they didn't find anything."

"Did they consider the possibility of poisoning due to chemical products?" the doctor asked.

"No," answered Angela, one of the teachers. "When we mentioned that hypothesis, they rejected it out of hand."

Angela knew what she was talking about. She lived in a little house sur-

*Phytates are phosphorus compounds that bind with certain metals, for example iron, and prevent their absorption by the intestine.

†Often called "phyto-estrogens," isoflavones are similar to female estrogens.

rounded by soybean fields. Every time they were sprayed, she had violent migraines, nausea, eye irritation, and joint pains. "I have talked to the technicians," she said. "The only thing I got from them was that they warn me when they're going to spray herbicide and I leave my house, with my family, for two days. They suggested that I sell my house, but where would I go? Soybeans are worth more than our lives."

Banging One's Head Against the Wall

When I saw how Campos flew off the handle when I questioned him about the environmental and health consequences of Roundup spraying, I understood that the subject was not a government priority. "Coming from a European reporter, that question takes the cake," he said emphatically. "Our herbicide use is much lower than that of France. The truth is that we are the least polluted country in the world."

The secretary of agriculture had obviously not been reading his country's newspapers. When you go through them, you find, for example, that a judge has opened an investigation in Rosario, following a complaint filed by a couple whose house is surrounded by soybean fields. Their son Axel was born with no toes on his left foot and with severe testicular and kidney problems.[9] Similarly, in Córdoba, mothers in the Ituzaingó neighborhood conducted a community protest to stop spraying in the nearby fields after they observed abnormal rates of cancer, particularly among children and young women. The affair caused some stir in parliament before getting lost in the maze of the justice system. "It's always like that," said Luis Castellán, an agronomist working for an agricultural development organization in Formosa, in northern Argentina. "Whenever there is a serious environmental problem, you cannot find a single expert who dares to stand up to the powerful soy lobby."

Castellán knew what he was talking about: in February 2003, he had been contacted by farmers from Colonia Loma Senés, a rural community in the province of Formosa near the border with Paraguay. They were desperately looking for an expert to certify the damage done to their food crops by the spraying of Roundup and 2,4-D on a seventy-five-acre plot that had been invaded by "rebel soybeans." They belonged to a neighbor living in Paraná who leased his land to a company from the province of Salta, which subcontracted with another company for seeds and spraying.

Welcome to the kingdom of GMOs! "Technicians"—often day laborers without protective gear who poison themselves for miserable wages—turned up one Saturday morning and sprayed until Sunday morning. "It was very hot that day and there was a strong wind in the region," recalled Felipe Franco, who farmed about twenty-five acres. "The product is very volatile and it drifted for about 400 yards." Twenty-three families who had taken refuge in their cinder-block houses were contaminated. "When I got there," Castellán told me, "their eyes were red and they had large spots on their faces and torsos. Many of them had violent headaches and nausea and complained of hot flashes and dry throats." Some of them never recovered, such as an old woman who was treated for eight months in Buenos Aires and still complains of unbearable bone and joint pain. The community asked the provincial government health services to write a report, but they concluded that lack of hygiene was the cause of all the problems. The families filed suit in the court in El Colorado, but the case bogged down because there was no health report. Only Castellán agreed to prepare a scientific description of the damage caused to the crops.

"We lost everything," said Franco. "The manioc, sweet potatoes, and cotton were devastated. The chickens and ducks died; some sows aborted, and the others produced scrawny piglets. The day of the spraying the plow horses had diarrhea and threw themselves on the ground; some of them died."

Castellán took photographs and samples of the affected plants, which he had analyzed by a laboratory at the University of the Littoral in Santa Fé. "I thought about it a lot before taking on this work," he admitted, "because I knew that I was taking risks."

"All the agronomists in the Ministry of Economy and Production refused," Franco confirmed. "We had to confront the police and the politicians who wanted to keep us quiet. Some neighbors gave up filing a complaint and decided to leave and go to the slums of Formosa."

"Monsanto says that transgenic soybeans can coexist with food crops. What do you think?" I inquired.

"It's impossible," answered Castellán, "especially in areas like this one, where small producers are surrounded by large spreads of GMOs. If something like this were to happen again, I don't know how many small producers would stay on the land."

Franco went on: "The problem is also the purpose of this production model. People who grow transgenic soybeans have a purely commercial aim;

they don't live where they farm, so they don't have to suffer the collateral damages. But we produce in order to live. We pay attention to the environment and the quality of what we produce, because we consume it or sell it in the market. This transgenic technology does not serve the farmer but an economic enterprise whose promoters are prepared to do anything to get rich."

Expulsion and Deforestation

Milli is a small rural community of ninety-eight families living in an area of seven thousand semi-arid acres located about thirty-five miles from Santiago del Estero in northern Argentina. You get there over a red dirt potholed road that winds through a brush-covered plain from which spring a few *quebrachos*, trees whose wood is so valuable that they are threatened with disappearance. This landscape is typical of the Gran Chaco region, which stretches to the Bolivian border.

"Here it's simply called *el monte*," said Luis Santucho, the lawyer for the peasant organization Mocase, when I met him in April 2005. "Before GMOs came, no one coveted this poor land, where thousands of small farmers have led a self-sufficient life for several generations." Santucho was eager to have me meet the community leaders of Milli, whose survival was threatened by the appetite of soybean producers, who had constantly been extending the agricultural frontier further north. A year before my visit, a provincial judge had turned up with armed men and bulldozers. "This is community land, with no property deeds," Santucho explained, "but with soybean money all kinds of skulduggery is possible." That day the population of Milli was able to drive off the assailants by blocking the roads. The *sojeros* then changed tactics. They tried to divide the community by offering to pay cash for twenty-five acres to some families, who hesitated because they'd never imagined having so much money.

"That stirred up a lot of trouble," said Luis, "but we didn't accept, because this is community land, it doesn't belong to anyone in particular. And where would we have gone? Life is hard here, but we have enough to eat every day." Chickens, ducks, and a litter of black pigs were running around the beaten-earth farmyard. Near the creek behind the little hut a cow and a horse were grazing. Every family was growing manioc, potatoes, and a little rice or corn. "*El monte* is a way of life," said Santucho, "but it's also great vegetable and animal biodiversity that is now threatened."

Indeed, the province of Santiago del Estero has the sad distinction of having one of the highest rates of deforestation in the world. Every year on average 0.81 percent of the forest is cut down, compared to a worldwide average of 0.23 percent. For instance, between 1998 and 2002, 540,000 acres simply went up in smoke and were replanted with RR soybeans.* "Between 1998 and 2004, nearly 2 million acres were cut down in Argentina," explained Jorge Menéndez, director of forests in the Ministry of Environment and Sustainable Development. "The situation is so worrying that it keeps me from sleeping. All the *bosques nativos* [native woodlands] are threatened: they are forests of great biodiversity whose flora and fauna antedate the discovery of America. Some animal species, such as pumas, jaguars, Andean cats, and tapirs, cannot live outside this particular ecosystem. If we do not impose rules on soybean farming, the damage will be irreparable."

"Isn't it the function of your ministry to define those rules?"

"Yes, but we don't carry much weight."

To recognize the magnitude of the catastrophe, all one needs to do is travel down Ruta Nacional 16 toward Salta or Chaco. Tree trunks are frequently piled at the roadside. Sometimes black smoke reveals the activity of *carboneros*, or charcoal makers, usually small farmers who have given up their land and are selling their labor to survive.

The cynicism is absolute: driven off by the beast, they are reduced to feeding it. A guard controlled access to the site. I bargained with him, and he let me through. The four-wheel drive went down the desolate road. As far as the eye could see were unspeakable heaps of crushed shrubs, torn up trees, and disemboweled bushes. "The bulldozers," murmured Guido Lorenz. Lorenz is a German geographer at the University of Santiago del Estero. Along with Pedro Colonel, a forest engineer, he frequently travels around the region to measure the extent of the plague. We were approaching the charcoal ovens. Men blackened with soot were unloading carts of wood. There was the sound of a tango. The boss explained that he had been out of work and he found this job, which would last for two years. It was to clear a four-thousand-acre plot belonging to the son of the governor of the

*During the same period, 290,000 acres were cut down in the neighboring province of Chaco and 420,000 in Salta.

neighboring province of Tucumán. The governor himself owns several thousand acres not far from here. "We cut down, we burn, then we plant soybeans," the man said.

We went on. Lorenz and Colonel had heard rumors of a huge operation of illegal deforestation about a hundred miles away. It involved a sixty-thousand-acre parcel recently acquired by an investor. On paper, Argentine law is very strict: in order to be able to cut down trees, owners have to get a permit setting the percentage of deforestation authorized, depending on the type of soil. In this sector, classified "fragile," that figure could not exceed 15 percent. "But once again," said Colonel, "soybean money took care of making the arrangements." Corruption and the lack of sanctions gave free rein to the bulldozers spreading desolation as far as the eye could see.

"They say they're extending the frontiers of agriculture, but they're really leaving a desert behind," Colonel went on. "They'll grow soybeans for a year or two and then they'll be forced to leave. The fertility of this soil is tied to vegetation that is a thousand years old; when it disappears, the soil is soon impoverished."

"This is a fragile environment, because we are in an arid or semi-arid climate zone," Lorenz explained. "Deforestation leads to a decline in the reserves of organic matter, causing erosion of the soil, which loses its ability to hold water. At the level of an entire watershed, water from gullies causes flooding in other areas. Deforestation is the source of the unusual floods we experienced recently in the province of Santa Fé. In addition, with the technique of direct planting, the Roundup sprayed stays on the surface; when it rains the herbicide residue is washed away by the flow of water and pollutes other areas in the watershed. The water mammals drink contaminates them and therefore the cows' milk, and so on."

"Unfortunately, that's not the *sojeros'* problem," Colonel said. "They are large companies or businessmen who come from Santa Fé or Córdoba and treat soybeans like a raw material. A subcontractor sends a man with a machine that plants, another sends a man with a plane to spray, and a third one comes with a combine that harvests the beans and takes them away. They don't use any local labor, except when they cut down the trees."

"We're really in an emergency situation, but no one in an official position has understood that yet," Lorenz concluded. "It's very serious, because the damage is irreversible."

14

Paraguay, Brazil, Argentina: The "United Soy Republic"

The good news is that practical experience clearly demonstrates that the co-existence of biotech, conventional and organic systems is not only possible, but is peacefully occurring around the world.

—Monsanto, *Pledge Report*, 2005

A tiny woman with an infinitely gentle gaze whom I met in January 2007 had understood in her flesh that transgenic soybeans were mortal enemies. To reach her, you have to drive for eight hours from Asunción, the capital of Paraguay, toward the Argentine border. As far as Iguazú, Ruta 7 runs through green pampas with huge grazing herds of cattle, dotted with palm trees and wooded hills. Then you turn off toward Encarnación, in the department of Itapuá. As far as the eye can see, hundreds of acres of RR soybeans stretch north toward nearby Brazil and south to the Argentine province of Formosa.

Silvino, Age Eleven, Killed by Roundup in Paraguay

Petrona Talavera, age forty-six, invited me into her humble shack located at the end of a red dirt road winding through Monsanto's GMOs. "This is the road where my son died," she said, handing me a welcoming cup of maté. "I'll fight to the bitter end to keep Paraguayan children from being poisoned by the agriculture that is killing them." Her husband, Juan, with whom she had raised eleven children, listened in silence.

It was January 2, 2003. Silvino, who was eleven, was coming home on his bicycle after buying noodles and a piece of meat in the only shop in the area, several miles from the house. On the road, he was sprayed by a *mosquito* driven by a *sojero* named Herman Schlender. "He came home soaked, complaining of nausea and a violent migraine," Petrona said. "I told him to lie down and I prepared a meal with the noodles and the meat. I didn't know that the product was so dangerous. In the afternoon the whole family experienced vomiting and diarrhea. Silvino was feeling worse and worse and I had to take him to the hospital." The boy went home after three days in intensive care, but the next day another soybean producer, Alfredo Lautenschlager, decided to spray his field, located about fifteen yards from the family's shack. Silvino did not survive the second poisoning. He died in the hospital on January 7.

Petrona then began her hard battle, determined that the crime would not go unpunished. Supported by the Coordinadora Nacional de Organizaciones de Mujeres Trabajadoras Rurales e Indígenas (Conamuri, National Coordinating Committee of Rural and Indigenous Women Workers), she filed a claim with the court in Encarnación. In April 2004, the two *sojeros* were each sentenced to two years in prison and fined 25 million guaranis. This was a first in the nation. The court found that the child had died as a result of poisoning by a toxic agricultural product that he "had absorbed through his respiratory system and orally, as well as through the skin." The two *sojeros* appealed, backed by Capeco, the organization of large-scale soy producers, the Paraguayan counterpart to the ASA in the United Sates and Aapresid in Argentina. The sentence was upheld in July 2006, but the defendants appealed to the Supreme Court.

In November 2006, the appeal was denied, but when I went to see Petrona in January 2007, they were still at liberty. During the three years the case went on, a group of NGOs was created and frequently organized demonstrations so the affair would not be buried. "The *sojeros* are very powerful," Petrona said, "more powerful than the government. They threatened to kill me. They paid several of our neighbors to make our life impossible and force us to leave. But where would we go? To a slum? Silvino had a classmate who died recently from poisoning, but her family did not file a claim for fear of reprisals, and because they couldn't afford it. How many Paraguayan children have already died in the face of complete indifference?"

The question was hard to answer. At the Ministry of Health, Dr. Graciela Camarra acknowledged that Roundup pollution had become a real public health problem but that for now it was impossible to number the victims. "We are trying to set up a reporting system so that we can be informed as soon as a suspect case appears, but it's not simple. I know of a case of two children who died after eating fruit sprayed with herbicide. And then there was Antonio Ocampo Benítez, talked about in the press, who almost died after swimming in a polluted river. There was another tragedy in an indigenous community in the department of San Pedro, where three children succumbed to the effects of spraying. We in the Ministry of Health are trying to persuade our colleagues in the Ministry of Agriculture to enforce the rules for the proper use of herbicides, but no one can stand up to the *sojeros*. And yet we're all concerned, even here in Asunción, because the fruits and vegetables we buy all come from the countryside."

Smuggled Seeds

"We are first in the world in soy production per inhabitant, with an average of sixteen hundred pounds per person," Tranquillo Favero bluntly declared in a June 12, 2004, interview with the Argentine daily *Clarín*. The Paraguayan "soybean king" explained that he had substantially contributed to this record, since he himself cultivated 125,000 acres in the departments of Alto Paraná and Amambay.

Welcome to Paraguay, which in ten years had risen to the rank of sixth-largest soy producer in the world and the fourth-largest exporter! From 1996 to 2006, surfaces devoted to soybean cultivation went from less than 2.5 million acres to 5 million acres, an increase of 10 percent a year. For good measure, the *Clarín* reporter hastened to add that the Paraguayan boom was due to the cultivation model graciously provided by Aapresid, the Argentine association of large *sojeros* closely associated with Monsanto. If he had gone a little further, he could have added that the organization had transmitted to its counterparts in Capeco not just the technique of direct sowing but also illegal RR soybean seeds. In 2004, no Paraguayan law authorized the cultivation of GMOs, even though they covered nearly half the cultivated land (this was still true in 2007).

"How is that possible?" The question startled Roberto Franco, deputy minister of agriculture, whom I met in Asunción on January 17, 2007. He seemed delighted to see me, so infrequently do European reporters show any interest in his country, which had been stifled for nearly forty years by the dictatorship of Alfredo Stroessner (1954–89).

"The transgenic seeds entered illegally," he said with a nervous smile. "It's what we call *bolsa blanca*, because they came in white sacks with no indication of their source."

"But where did they come from?"

"Well, mainly from Argentina, but also a little from Brazil."

"Who organized the smuggling?"

"Large Paraguayan soy producers, who have close ties with their Argentine colleagues."

"Do you think Monsanto played a role in the smuggling?"

"Well, we don't have any evidence. But it's not impossible that firms involved in this technology helped promote their varieties. Faced with this situation, the government had to act, because we export almost all our grain, 23 percent of it to the European Union, which requires the labeling of agricultural products that contain GMOs. We have no way of knowing whether the soy was transgenic or not. To avoid losing our markets—soybeans account for 10 percent of our GDP—we had to legalize the illegal crops."

"Putting it bluntly, the government was confronted with a fait accompli?"

"Yes. We have the same problem today with Bt cotton, which is in the process of spreading with no official authorization and no law to govern it."

"You don't think it was a trap?"

"Well, we're not the only ones; Brazil went through the same thing."

A strange coincidence indeed. In 1998, when RR soybeans were invading the North American plains and the Argentine pampas, Monsanto seemed to be champing at the bit in Brazil, the world's second-largest soybean producer. A petition filed by Greenpeace and the Brazilian Institute for Consumer Defense (IDEC) secured a temporary suspension of the marketing of GMOs on the grounds that "with no prior study of the environmental impact and the health risk to consumers, it would violate the precautionary principle of the Convention on Biodiversity" signed in 1992 in Río de Janeiro.

By lucky chance, smuggling was organized in the Brazilian state of Rio Grande do Sul; seeds were clandestinely imported from nearby Argentina, which led them to being nicknamed "Maradona" after that country's famed soccer player. Backed by Aapresid, Apassul (Seed Producers Association of Rio Grande Do Sul) organized sumptuous *churrascadas* (barbecues) to promote transgenic crops, right before the eyes of the authorities, who did nothing. "It's not unusual to see Argentine technicians in Brazilian fields who have come to lend a hand to their local colleagues," according to a 2003 report by Daniel Vernet in *Le Monde*, which quoted a statement by Odacir Klein, the agriculture secretary of Rio Grande do Sul: "The federal police conducts inspections on farms and on the roads to charge violators, and transmits the charges to the justice system, which almost never follows up."[1]

The result was that in 2002, when Luiz Inácio Lula da Silva, known as Lula, ran for president for the fourth time and campaigned against GMOs, they had already spread throughout the state of Rio Grande do Sul, and also in the neighboring states of Parana and Mato Grosso do Sul. Nine months after the Workers Party candidate had entered the Palácio do Planalto in Brasilia, the European Commission adopted two rules on September 22, 2003, on the traceability and the labeling of GM foods intended for human and animal consumption. This decision directly threatened Brazil's exports, since it was unable to distinguish between conventional and transgenic soybeans, because the latter officially did not exist.

Three days later, Lula signed a decree authorizing—temporarily—the sale of RR soybeans for the 2003 crop, then their planting and marketing for the 2004 season.* It offered an amnesty for all GMO producers, inviting them to come out of the closet and identify their crops so that segregation could be organized. The decision caused an uproar among peasant and ecological organizations, but also inside the Workers Party, which had promised not to release transgenic seeds before their environmental, health, and social impact had been seriously evaluated.

Aware of the disastrous consequences that would inevitably follow from the onward rush of soybeans, João Pedro Stedile, leader of the Landless Peasants Movement (MST), called Lula a "transgenic politician," and the environment minister, Marina Silva, seriously considered resigning. For the opponents of GMOs, the presidential decree signaled the surrender of the new government to agribusiness, represented by agriculture minister Roberto Rodrigues, and above all to Monsanto.

*The decree was renewed in October 2004. Then in March 2005, the lower house of the Brazilian Congress passed a law definitively authorizing transgenic crops.

Time to Collect

Monsanto had been waiting in the starting blocks for a long time. Its entire strategy in Brazil demonstrates that it had largely anticipated the takeover of the country by soybeans, and transgenic crops more broadly. It had been marketing its herbicides in Brazil since the 1950s, and it opened its first glyphosate production facility in São Paulo in 1976. But in the 1990s, when its RR soybeans were spreading illegally, it launched the construction of a new plant that its Brazilian Web site presented with all the expected fanfare: "In December 2001 Monsanto inaugurated, at the Petrochemical Pole of Camaçari, the first plant of the company designed to produce raw materials for the herbicide Roundup in South America. The investment is equivalent to US $500 million. . . . The Camaçari Plant, the largest unit of Monsanto installed out of the United States, is also the only Monsanto plant manufacturing raw materials for the Roundup production line. The production is directed to Monsanto units installed in São José dos Campos (SP), Zarate (Argentina) and Antwerp (Belgium); in the past those units received raw materials from the United States."[2]

As it was adapting its Roundup production capacities to the huge market it was seeking to develop, the company took control of Brazilian seeds in 1997 by acquiring Agroceres, the largest seed company in Brazil, and through the Brazilian subsidiaries of American seed companies that had come under its control in the United States, such as Cargill Seeds, DeKalb, and Asgrow. In 2007, Monsanto was the largest supplier of corn seeds in Brazil, and the second largest for soybean seeds, just behind Embrapa, the National Institute for Agricultural Research, which was desperately fighting to survive.

The culmination of Monsanto's work was the collecting of royalties, first in Brazil, followed by Paraguay, and finally Argentina. Scarcely had Lula legalized the illegal crops when Monsanto began negotiations with producers, exporters, and processors of the precious grain, brandishing its intellectual property rights to the RR gene. Threatened with a cutoff of seed supplies, the Brazilians did not resist for long; by January 2004, they had signed an agreement providing that royalties would be collected when producers delivered their crops to the grain elevators of dealers and exporters of soybeans, such as Bunge and Cargill, the American giant whose foreign operations

Monsanto had just purchased. Royalties were set at $10 per ton for the first year and $20 for the 2004 harvest. When you consider that 30 percent of Brazilian soybeans in 2003 were transgenic, amounting to about 16 million tons harvested, the math is simple: for the first year alone, intellectual property rights brought in $160 million for Monsanto.

In October 2004, it was the Paraguayan producers' turn to pay up. They didn't offer much resistance either, because in the end the official payment of royalties confirmed their triumph. The agreement provided for an initial payment of $3 per ton of soybeans, which was supposed to double within five years. As in Brazil, the fee was collected by dealers when the harvest was delivered, and they transferred it to Monsanto after deducting a commission. One week later, on October 22, 2004, Agriculture Minister Antonio Ibáñez issued a circular authorizing the sale of four varieties of transgenic soybeans belonging to Monsanto.

"In this matter, in fact, the government just legalized the violation, didn't it?" I asked Roberto Franco.

"Well, let's say we went along with it," he said hesitantly. "The large producers were the ones who negotiated directly with Monsanto. It wasn't like in Argentina, where the government handled the issue of royalties from the very beginning."

That's true. And it can be said that in Argentina, Monsanto hit a snag that has poisoned its relations with its loyal ally since 2004. It will be recalled that when it launched its RR soybeans, the company demonstrated extraordinary generosity by agreeing that producers would not pay royalties on the seeds. Eight years later, it was estimated that only 18 percent of the seeds used were certified, that is, bought at the list price from dealers subservient to Monsanto through licenses; the rest were seeds that had been saved or bought on the black market. Monsanto did not move until January 2004, when it suddenly threatened to withdraw from Argentina if all the producers did not pay the "technology fee."

At first Agriculture Secretary Miguel Campos didn't bat an eyelid. He even offered to set up a royalty fund paid for by a tax the government would collect from producers and turn over to Monsanto, the trifling sum of $34 million a year. To enter into force, the measure had to be approved by the Congress, which dragged its feet for fear of antagonizing the agricultural sector. "There is no question of our paying anything at all," I was told in April 2005 by Eduardo Buzzi, president of the Argentine Agrarian Federation.

"First of all, Monsanto didn't patent its gene in this country; besides, farmers are protected by law 2247, which guarantees what is called the 'principle of the farmer's exception,' that is, his right to replant part of his harvest, even if the original seeds are certified by breeders. There is no reason why Monsanto should enjoy special status."

"But your organization at first encouraged the development of transgenic soybeans."

"That's right, and we were totally taken in. How could such cynicism be imagined? The company had planned everything for the long term, relying on Aapresid, an association it finances to promote its products, with the complicity of government officials and the media. Everything had been calculated, even smuggling to Paraguay and Brazil, and we fell right into the trap."

"It's war?"

"Yes, the seed war, except that we're not worried about collecting dividends to satisfy shareholders but simply about staying alive."

A few days after our meeting, Buzzi flew to Munich, headquarters of the European Patent Office, to plead his cause. On March 14, 2005, Monsanto had sent a letter to soybean exporters informing them that the company was going to "go after any shipment of soybeans, soy flour, or soy oil leaving Argentine ports headed for countries where the RR gene is patented." For that purpose, it would request "the assistance of the customs authorities to take samples to detect the presence of the gene." If the test was positive, it would sue the exporters in European courts, demanding a penalty of $15 a ton in addition to legal costs. At this moment, although the European Patent Office has granted a patent for the RR gene, only five countries recognize it: Belgium, Denmark, Italy, Holland, and Spain. In 2004, those five alone imported 144,000 tons of soybeans and 9 million tons of soy flour from Argentina. "Monsanto's demand is completely illegal," according to Campos. "The patent covers only seeds, not beans, flour, or oil. European law does not permit Monsanto to collect royalties on Argentine products."

That remains to be seen. Monsanto for its part asserts that the gene belongs to it wherever it may be found, in the plant as well as in the products derived from it. And once you agree to take the first step into the infernal system of the patenting of life, the reasoning seems logical. In the meantime, the multinational corporation wasted no time in carrying out its threats: in 2005, it had ships inspected in Holland and Denmark in connection with lawsuits, and in early 2006, three shipments of soy flour were inspected in Spain. The cases were brought before the European Court in Brussels and Monsanto lost. If successful, these maneuvers could be a serious threat to Argentine exports, because in order to avoid disputes with uncertain outcomes, European dealers have begun to turn to other sources of supply. "It's unjust," says Campos, "because Monsanto has greatly profited from the boldness of Argentina, which authorized its seeds when they were very controversial. And it's thanks to Argentina that the company was able to make inroads into other countries on the continent."

The New Conquistadores

Back in Paraguay, the inroads Campos spoke of rather euphemistically have assumed the shape of an ecological and social catastrophe. "It's a new conquest," according to Jorge Galeano, president of the Agrarian and Popular Movement (MAP). "Nothing seems to be able to stop the *sojeros*, who use the same brutality as the conquistadores to increase their empire." When I was in Paraguay in January 2007, the peasant leader was eager to show me the latest line of the "soy frontier," which was constantly progressing toward the interior of the country. We left in a four-by-four from Vaquería, a small town 125 miles northeast of Asunción, in the department of Caaguazú. We drove on red dirt roads through a hilly and forested landscape of astonishing beauty. Along the way we passed Guarani Indians carrying bundles of wood; here and there thatch-roofed houses were lost amidst luxurious vegetation, with naked children splashing in a river under the burning sun. "Everything grows here," Galeano said, "corn, cassava, sweet potatoes, all kinds of beans, sugarcane, citrus fruits, bananas, maté. Families feed themselves from a tiny plot of land, because we're still waiting for the agrarian reform that is permanently threatened by soybeans."

He spoke of the history of his country, one of the poorest in Latin America, where 2 percent of the population owns 70 percent of the land, a glaring injustice that goes back to the Spanish conquest but that was accentuated by the 1870 war against the Triple Alliance, in which Paraguay was defeated by Argentina, Brazil, and Uruguay. To pay for the reparations demanded by the victors, the Asunción government sold off public land, privatizing 57 million acres between 1870 and 1914 for the benefit of Brazilian and Argentine citizens and companies. Fabulous estates of 175,000 acres still survive from that time. Starting in 1954, the dictatorship of Alfredo Stroessner further intensified the concentration of land ownership to the detriment of small farmers: 25 million acres fell into the hands of allies of the bloodthirsty general, the son of a Bavarian brewer, who distributed them to local political bosses or foreign companies in return for large bribes. In the 1970s, during the first expansion of (non-transgenic) soybeans, another deviation of the agrarian reform that was always being postponed led to the sale of huge territories in the public domain to Brazilian producers from Rio Grande do Sul and Paraná, the "Braziguayans" who organized the smuggling of RR seeds twenty years later. It is now estimated that sixty thousand producers share the transgenic bounty, 24 percent of them Paraguayan and the rest foreigners from Brazil, Germany, and Japan, or "international investors who have placed their money in the new green gold," to adopt the expression of Deputy Agriculture Minister Roberto Franco.* Putting it bluntly, they are foreign companies that buy huge properties to plant GMOs and have no qualms about driving off, by any means possible, the small farmers in their way.

"Look," said Galeano, "this is where the soy frontier has reached today." The view was startling. We were now going down a straight path running for several miles. To our left, toward the east, were soybeans as far as the eye could see, from which tiny clumps of trees infrequently emerged. To our right lay the wooded landscape rich in biodiversity that we had been going through for the past two hours. "Less than two years ago, these huge areas were populated by peasant and indigenous communities, all of whom finally left," Galeano explained. "The technique of the *sojeros* is always the same. First they contact the families and offer them food and toys for the children's birthdays. Then they come back and offer to rent their plots of land with a three-year contract. The families keep living there, keeping a small space for food crops. But they are very soon affected by the spraying, so the *sojeros* offer to buy their land outright. Since title deeds usually don't exist for these properties, because they are supposed to be part of the agrarian reform that never happened, the producers bribe well-placed government officials in Asunción and they become the legal owners of these 'liberated' plots, as they call them. Then the bulldozers come and destroy the entire natural habitat of these very fertile lands, and the next year monoculture takes over. That's why I say it's a new conquest, because the expansion of

*The Japanese Agency for International Cooperation has encouraged the emigration of Japanese settlers.

soybeans is based on the pure and simple elimination of human communities and ways of life."

"Is the phenomenon reversible?"

"Unfortunately, no. Even if small farmers could one day recover the land, it would be so contaminated by chemical products that it would take years for the initial quality of the soil to be restored. Transgenic soy is really a deadly enterprise against which we've decided to fight, whatever it costs."

Soy's Musclemen and Repression

In contrast to Argentina, where transgenic expansion has met little organized resistance, in Paraguay collective action against RR soybeans proliferated starting in 2002. Joined together in the National Front for Sovereignty and Life, peasant organizations such as Galeano's MAP and the MCP (Movimiento Campesino Paraguayo) and civil society organizations such as Conamuri, to which Petrona Talavera belongs, have been conducting campaigns against the takeover of the country by soybeans. A week does not go by without a demonstration, a blocking of roads, or an occupation of land to slow the "advance" of Monsanto's GMOs.

The government of President Nicanor Duarte chose to respond to this situation with repression and the criminalization of the anti-soy movement. Hundreds of peasants have been incarcerated since 2002 and a dozen assassinated. In some cases, the local police openly conduct themselves as an armed militia in the pay of the *sojeros*, with no qualms about shooting opponents on sight. For example, one day in February 2004, a truck carrying fifty peasants that had blocked *mosquitos* in the department of Caaguazú was fired on by M16s, killing two and seriously wounding ten. All around the country, with the approval of President Duarte, armed thugs have been recruited to protect the spraying machines and the large soybean properties.

Confident in their impunity, some *sojeros* have revived the proven techniques of the long Stroessner dictatorship and simply eliminated peasant leaders who are too troublesome. For example, on September 19, 2005, two policemen attempted to assassinate Benito Gavilán in Mbuyapey in the department of Paraguari by shooting him in the head. He miraculously survived but lost an eye. Almost everywhere in the areas bordering the "soy frontier," violent actions have been conducted aimed at dislodging recalcitrant small farmers by force. On November 3, 2004, in the department of Alto Paraná, seven hundred policemen were mobilized to expel two thousand landless peasants who were camping with their families next to the 160,000 acres of RR soybeans recently acquired by Agropeco, a company belonging to a Paraguayan of German origin and an Italian investor.[3] The two men had bought the huge estate from Stroessner's son, who had obtained it through a manipulation of the agrarian reform. The families were cultivating a strip of land along Ruta 6. During the operation, in which thirteen peasants were incarcerated, the crops and the camp were destroyed.

But the symbol of the dictatorial methods that flow from the transgenic model was the rural community of Tekojoja, located 45 miles from Caaguazú, a few miles from the "soy frontier." Fifty-six families have been carrying on a desperate battle against the appetites of two powerful *sojeros* of Brazilian origin, Ademir Opperman, a local potentate, and Adelmar Arcario, who owns 125,000 acres in Paraguay and five important grain elevators in the region. On December 3, 2004, the two partners organized an attempt to evict the families by force, by burning houses and destroying fifty acres of crops.[4] But with the support of MAP, the families resisted and reoccupied their land.

On June 24, 2005, at five in the morning, 120 policemen backed by private militia recruited by Opperman took the community by storm accompanied by two lawyers who displayed an order of expulsion signed by a judge. "They were forged property deeds illegally obtained from Indert [Instituto Nacional de Desarollo Rural y de la Tierra]," explained Jorge Galeano, who had rushed to the scene as soon as he heard of the operation. "The Supreme Court in Asunción acknowledged that the acquisition was illegal in September 2006, but since then the families have been living in a very precarious state."

On that January day in 2007 they had left their plastic tents to meet at the scene of the tragic events that had shattered their lives, hoping that my reporting would protect them from another violent action. "It was terrible," said a toothless old woman. "The police arrested 160 people, 40 of them children. We spent several days in jail. When we were released our houses had been burned down, our crops destroyed, and our animals killed. And then we lost two companions."

In silence, the families had approached two tombstones covered with flowers in the midst of a clearing. "Here is where they assassinated Angel

Cristaldo, who was only twenty, and Leoncio Torres, a forty-nine-year-old father, who were trying to block the way of the bulldozers," Galeano explained. "The police first claimed that they had died during a confrontation between the police and armed peasants, but we have proof that they were in fact murdered." The day of the assault, a Canadian anthropologist, Kregg Hetherington, who was doing research in the community of Tekojoja, witnessed the whole operation and took photographs. Galeano gave me copies of the photos, which show police in uniform surrounding the trucks loaded with the furniture that Opperman's men had stolen from the modest wood huts before they were set on fire. Armed men are bustling around tractors destroying the crops, while peasants are trying to block their progress with their bare hands. A man in a blue T-shirt is lying on the ground, his chest bloody. Another, also wearing a blue T-shirt, has a shattered arm. Faces are ravaged by pain. "I was also wearing a blue T-shirt," Galeano murmured. "Opperman's men got the wrong man." Thanks to Hetherington's testimony, an arrest warrant has been issued for the *sojero*, who was on the run when I visited Tekojoja.

It was already time to leave, because a few miles away another community was expecting us, also wanting to tell us of its distress: Pariri, where several hundred families have more or less survived surrounded by fields of GMOs. I had traveled through North and South America where transgenic crops proliferated, but I had never seen so much soy. It was a green ocean that covered every inch of space up to the beaten-earth square in front of the little church where the inhabitants of Pariri had come together. A man approached Galeano with his ten-year-old son, whose legs were covered with burns. To get to school, the boy had to go through a soybean field that had just been sprayed with Roundup. One woman complained of persistent migraines, another of vomiting; a man said he no longer had the strength to work since spraying had resumed. "What can we do?" asked an old man. "Leave the way forty families have already? To pick through garbage pails in a slum? Help us."

Galeano was moved. I was angry. I lit a cigarette and listened to the speech he improvised in front of men and women who were dying so pigs and chickens in Europe could eat soy, because we can't take the trouble to feed them with locally produced food. "Don't leave!" declared Galeano. "We have to resist the model of transgenic production that multinational corporations like Monsanto want to impose on us, because it will lead to an agriculture without farmers. Family farming the way we do it provides work for five people on every two acres cultivated, while RR soybeans employs only one full-time worker for fifty acres.* In the long run, judging by its actions, Monsanto's aim is to control the production and the food of the world, and that's why it wants to prevent us from doing our job. We don't want the transgenic model, because it's criminal: it pollutes the environment, destroys natural resources, and creates unemployment, poverty, insecurity, and violence. It makes us dependent on the outside for something as basic as food. It kills life, but once it's settled in, it's very difficult to go back. That's why we have to struggle, for us and most of all for the future of our children."

The Soy Dictatorship

On January 23, 2007, Tomás Palau met me in a house about a hundred miles from Asunción, where he had adopted the habit of retiring to read and write far from the uproar of the capital. That day, the sociologist who specializes in agrarian questions was beginning an article on the "United Soy Republic," an advertising slogan launched in early 2004 by Syngenta, Monsanto's Swiss competitor. In the ad, which had been distributed throughout the Southern Cone, you could see a green map linking Bolivia, Paraguay, Brazil, and Argentina, whose outline formed a soybean with the title "República Unida de la Soja." "Soy knows no borders," explained the second page, which sang the praises of a technical assistance service of the company supplying fertilizer and phytosanitary products to producers of RR soybeans.

"You can really say soybeans have taken over the Southern Cone," said Palau, "because now Monsanto's GMOs cover 100 million acres in the four countries shown on the map. But this staggering expansion, that has come at the expense of small farmers in the region, represents more than a mere agricultural phenomenon; it is also a real hegemonic political program. And in that sense, Syngenta's advertising slogan is perfectly right; it's even a confession." Palau explained that, in his opinion, "Monsanto does now control the agricultural and trade policy of Brazil, Paraguay, Argentina, Bolivia, and

*In Argentina, figures supplied by the Agriculture Ministry indicate one paid position for five hundred acres cultivated.

soon of Uruguay, and its power greatly exceeds that of the national governments. It's the company that decides what seeds and what chemical products will be used in those countries, what crops will be suppressed, and in the end what people will eat and at what price. The recalcitrant are taken to court, because the patents are the final link in the totalitarian chain. All that is done with the assistance of producers' associations like Aapresid and Capeco, who maintain close relations with the ASA in St. Louis."

I had been able to observe myself the links that united Monsanto's three cooperating associations. At the end of our conversation, Deputy Agriculture Minister Roberto Franco had invited me to go with him to a reception being held the next day on the property of Jorge Heisecke, the president of Capeco. That evening a delegation of twenty members of the American Soybean Association was expected, led by John Hoffman, who had been so hospitable to me at his Iowa farm. I had reorganized my schedule to seize this opportunity. Unfortunately, after six hours on the road, I was never able to penetrate Heisecke's huge estate, which was protected by armed guards, despite the intervention of the deputy minister, who offered lame excuses. Long live the "United Soy Republic."

"Do you know how many small farmers in Paraguay have given up farming because of soybeans?" I asked Palau.

"At the last census, the nationwide statistics indicated that 100,000 people out of a total of 6 million inhabitants were leaving the countryside annually to settle in cities," he answered, not surprised by my misadventures. "That amounts to between 16,000 and 18,000 families. It's estimated that about 70 percent of the migrants leave because of soybeans. That's huge, when you consider that families usually end up in slums where they live in conditions of extreme poverty. But beyond the social problems GMOs cause, the greatest impact is the loss of food security. When they leave their land, small farmers stop producing for themselves, but also for others. Since 1965, Paraguay has shifted from a food surplus to a deficit, which means it now imports more food products than it exports. That's why I say that Monsanto and its allies, including in the end its competitors Syngenta and Novartis [who may eventually merge], are engaged in an imperialist, even dictatorial strategy intended to subject populations politically through tight control over food supplies. Recall the 'Santa Fé document' published in 1980, which constituted the basis of the Reagan Doctrine, in which national security advisers presented the food supply as a political weapon that had to be controlled to annihilate enemy governments. Well that's exactly what Monsanto is doing now."

On my way to Asunción, where I would board a plane back to France, I thought about the conversation I had had two years earlier with Walter Pengue, the Argentine agronomist who had become one of the world's best known specialists on the impacts of transgenic soybeans.

"The transgenic model is the latest incarnation of industrial agriculture," he had explained over a glass of Argentine cabernet sauvignon. "It's the last link in a model of intensive production, based on a technological package that includes not only seeds and herbicide but a whole series of inputs, such as fertilizer and insecticides, without which there is no yield, that are sold by the multinational corporations of the North to the countries of the South. That's why we can speak of the second agricultural revolution. The first, the one that came in the postwar years, was piloted by national agricultural organizations, like INTA in Argentina, and was aimed at developing countries' food-producing capacities relying on the peasant class. The second is driven by supranational interests and leads to an agricultural model turned toward exports, where there are no more active participants in the fields. This model is directed purely toward supplying low-cost fodder for the large industrial feedlots in the countries of the North, and leads to the development of monocultures that threaten the food security of the countries of the South. In ten years, the Argentine economy has gone back a century by becoming dependent on commodity exports whose prices are set in world markets where the power of multinational corporations is decisive. When the price of soybeans collapses, we can expect the worst."

"What are the consequences of RR soybeans for conventional and organic soybeans?"

"That's another very important point in transgenic agriculture, which leads to biouniformity, which is another danger for food security. GM soybeans have practically made conventional and organic soybeans disappear, because they are contaminated and their prices have declined drastically. But there is something more serious: if half a country is planted with a single variety, it creates a veritable highway for natural diseases that can annihilate a country's entire production. A threat is now hanging over soybeans that has no phytosanitary remedy, soybean rust. It began in Brazil, spread to Paraguay, and then Argentina. The lack of diversity of plant species prevents resistance to disease. Don't forget what happened in the nineteenth century to potatoes in Ireland. The great famine of 1845–49 that killed off a large part of the population and drove tens of thousands of people into exile was due particularly to the lack of biodiversity, which favored the development of the blight that no natural barrier was able to stop."

"What is Monsanto's long-term objective?"

"I think the company is seeking to control the food produced in the world. To do that, it has to get its hands on the seeds in the locations where they are used by farmers. First it appropriates the seeds, then the processing of grains, then the supermarkets, and in the end it controls the entire food chain. The seeds are the first link in the chain: whoever controls seeds controls the food supply and thereby controls mankind."

A month before I traveled to Paraguay in January 2007, I had been able to observe the effects of this terrible logic on the other side of the planet in an even more dramatic context, in India, where the cultivation of Monsanto's transgenic cotton had become associated with death—the subject of the next chapter.

15

India: The Seeds of Suicide

> Our products provide consistent and significant benefits to both large- and small-holder growers. In many cases farmers are able to grow higher-quality and better-yielding crops.
>
> —Monsanto, *Pledge Report*, 2006

It was December 2006, and we had barely arrived when a funeral procession came around the corner of an alley running between whitewashed walls, shattering the torpor of the little Indian village under the burning sun. Wearing the traditional costume—white cotton tunic and trousers—the drummers led the group toward the nearby river, where the funeral pyre had already been set up. In the middle of the procession, weeping women desperately held on to robust young men with somber faces carrying a stretcher covered with brilliantly colored flowers. Gripped by emotion, I glimpsed the face of the dead man: eyes closed, aquiline nose, brown moustache. I will never forget this fleeting vision, which stains Monsanto's great promises with infamy.

Three Suicides a Day

"Can we film?" I asked, seized by sudden doubt, as my cameraman questioned me with a motion of his head. "Of course," answered Tarak Kate, an agronomist who heads an NGO specializing in organic farming who was traveling with me through the cotton-growing region of Vidarbha, in the southwestern Indian state of Maharashtra. "That's why Kishor Tiwari brought us to this village. He knew there would be a funeral of a peasant who had committed suicide."

Kishor Tiwari is the leader of the Vidarbha Jan Andolan Samiti (VJAS), a peasant movement whose members have been harassed by the police because they have insistently denounced the "genocide" caused by Bt cotton in this agricultural region, formerly celebrated for the quality of its "white gold." Tiwari nodded when he heard Kate's answer. "I didn't say anything to you for security reasons. The villagers inform us whenever a farmer has committed suicide, and we come to all the funerals. Right now, there are three suicides a day in the region. This young man drank a liter of pesticide. That's how peasants kill themselves: they use the chemical products the transgenic cotton was supposed to enable them to avoid."

As the procession headed off toward the river, where the body of the young victim would soon be cremated, a group of men approached my film crew. They looked suspicious, but Tiwari's presence reassured them: "Tell the world that Bt cotton is a disaster," an old man said angrily. "This is the second suicide in our village since the beginning of the harvest. It can only get worse, because the transgenic seeds have produced nothing."

"They lied to us," the head of the village added. "They had said that these magic seeds would allow us to make money, but we're all in debt and the harvest is nonexistent. What will become of us?"

We then headed toward the nearby village of Bhadumari, where Tiwari wanted to introduce me to a young widow of twenty-five whose husband had committed suicide three months earlier. "She's already talked to a reporter from the *New York Times*," he told me, "and she's ready to do it again. This is very unusual, because usually the families are ashamed."[1] Very dignified in her blue sari, the young woman met us in the yard in front of her modest earthen house. The younger of her two sons, one three years and the other ten months old, was sleeping in a hammock that she rocked with her hand during the conversation, while her mother-in-law, standing behind her, silently showed us a photograph of her dead son. "He killed himself right here," said the widow. "He took advantage of my absence to drink a can of pesticide. When I got back he was dying. We couldn't do anything."

As I listened to her, I recalled an article published in the *International Herald Tribune* in May 2006 in which a doctor described the ordeal of the sacrificial victims of the transgenic saga. "Pesticides act on the nervous system—first they have convulsions, then the chemicals start eroding the stomach, and bleeding in the stomach begins, then there is aspiration pneumonia—they have difficulty in breathing—then they suffer from cardiac arrest."[2]

Anil Kondba Shend, the husband of the young widow, was thirty-five, and cultivated about three-and-a-half acres. In 2006, he had decided to try Monsanto's Bt cotton seeds, known as "Bollgard," which had been heavily promoted by the company's television advertising. In those ads, plump caterpillars were overcome by transgenic cotton plants: "Bollgard protects you! Less spraying, more profit! Bollgard cotton seeds: the power to conquer insects!" The peasant had had to borrow to buy the precious seeds, which were four times more expensive than conventional seeds. And he had had to plant three times, his widow recalled, "because each time he planted the seeds, they didn't resist the rain. I think he owed the dealers 60,000 rupees.* I never really knew, because in the weeks before his death he stopped talking. He was obsessed by his debt."

"Who are the dealers?" I asked.

"The ones who sell transgenic seeds," Tiwari answered. "They also supply fertilizer and pesticides, and lend money at usurious rates. Farmers are chained by debt to Monsanto's dealers."

"It's a vicious circle," Kate added, "a human disaster. The problem is that GMOs are not at all adapted to our soil, which is saturated with water as soon as the monsoon comes. In addition, the seeds make the peasants completely dependent on market forces: not only do they have to pay much more for their seeds, but they also have to buy fertilizer or else the crop will fail, and pesticides, because Bollgard is supposed to protect against infestations by the cotton bollworm but not against other sucking insects. If you add that, contrary to what the advertising claims, Bollgard is not enough to drive off the bollworms, then you have a catastrophe, because you also have to use insecticides."

"Monsanto says that GMOs are suitable for small farmers: what do you think?" I asked, thinking of the firm's claims in its 2006 *Pledge Report*.

"Our experience proves that's a lie." said the agronomist. "In the best case, they may be suitable for large farmers who own the best land and have the

*About $1,200 ($1 equals about 50 rupees). There is no minimum wage in India, but most workers earned less than 6,000 rupees a month in 2006.

means to drain or irrigate as the need arises, but not for small ones who represent 70 percent of this country's population."

"Look," Tiwari interrupted, spreading out a gigantic map he'd gone to get from the trunk of his car.

The vision was startling: every spot in what is known in Vidarbha as the cotton belt was marked with a death's head. "These are all the suicides we've recorded between June 2005, when Bt cotton was introduced into the state of Maharashtra, and December 2006," the peasant leader said. "That makes 1,280 dead. One every eight hours! But here, where it's blank, is the area where rice is grown: you see there are practically no suicides. That's why we say that Bt cotton is in the process of causing a veritable genocide."*

Kate showed me a small space where there were no death's heads. "This is the sector of Ghatanji in the Yavatamal district," he said with a smile. "That's where my association is promoting organic farming among five hundred families in twenty villages. You see, we don't have any suicides."

"Yes, but suicides of cotton farmers are nothing new; they existed before GMOs came on the scene."

"That's true. But with Bt cotton they've greatly increased. You can observe the same thing in the state of Andhra Pradesh, which was the first one to authorize transgenic crops, before it got into a battle with Monsanto."

According to the government of Maharashtra, 1,920 peasants committed suicide between January 1, 2001, and August 19, 2006, in the entire state. The phenomenon accelerated after Bt seeds came on the market in June 2005.[3]

Hijacking Indian Cotton

Before I flew off to the huge state of Andhra Pradesh in southeastern India, Kishor Tiwari was eager to show me the cotton market of Pandharkawada, one of the largest in Maharashtra. On the road leading to it, we crossed a column of carts loaded with sacks of cotton pulled by buffalo. "I warn you," Tiwari said, "the market is on the edge of exploding. The farmers are exhausted, the yields were catastrophic, and the price of cotton has never been

*From January to December 2007, the VJAS recorded 1,168 suicides.

so low. This is the result of the subsidies the American administration gives its farmers, which has a dumping effect on world prices."*

We had barely gone through the imposing gate into the market when we were assailed by hundreds of angry cotton farmers who surrounded us so we couldn't move. "We've been here several days with our harvest," one of them said, brandishing a ball of cotton in each hand. "The dealers are offering a price that's so low we can't accept it. We all have debts to pay."

"How much is your debt?" Tarak Kate asked.

"Fifty-two thousand rupees," the farmer answered.

What came next was an incredible scene in which dozens of peasants spontaneously declared, one after another, the amount of their debts: 50,000 rupees, 20,000 rupees, 15,000 rupees, 32,000 rupees, 36,000 rupees. Nothing seemed able to stop this litany running through the crowd like an irresistible tidal wave.

"We don't want any more Bt cotton!" yelled a man whom I couldn't even pick out from the crowd.

"No!" roared dozens of voices.

Kate, clearly very moved, asked: "How many of you are not going to plant Bt cotton next year?"

A forest of hands went up that, miraculously, the cameraman, Guillaume Martin, managed to film, even though we were literally crushed in the midst of this human tide, which made filming extremely difficult. "The problem," said Kate, "is that these farmers will have a lot of trouble finding nontransgenic cotton seeds, because Monsanto controls practically the entire market."

Beginning in the early 1990s—in fact, at the same time as it was setting its sights on Brazil, the world's largest soybean producer—Monsanto was carefully preparing the launch of its GMOs in India, the world's third-largest cotton producer after China and the United States. An eminently symbolic plant in the country of Mahatma Gandhi, who made the growing of cotton the spearhead of his nonviolent resistance to British occupation, cotton has been grown for more than five thousand years on the Indian subcontinent. It now provides a livelihood for 17 million families, mainly in southern states (Maharashtra, Gujarat, Tamil Nadu, and Andhra Pradesh).

*Subsidies to American farmers amounted to $18 billion in 2006. See Somini Sengupta, "On India's Farms, a Plague of Suicide," *New York Times*, September 19, 2006. Three days after we filmed, a riot broke out in the market and the police arrested several farmers, including Kishor Tiwari.

Established in India since 1949, Monsanto is one of the country's major suppliers of phytosanitary products. There is a large market for herbicides and especially insecticides, because cotton is very susceptible to a wide variety of pests, such as bollworms, wireworms, cotton worm weevils, mealybugs, spider mites, and aphids. Before the "green revolution," which encouraged the intensive monoculture of cotton with high-yield hybrid varieties, Indian farmers managed to control infestations by these insects through a system of crop rotation and the use of an organic pesticide derived from the leaves of the neem tree. The many therapeutic properties of this tree, venerated as the "free tree" in all the villages of the subcontinent, are so well known that it has been the subject of a dozen patents filed by international companies, obvious cases of biopiracy that have led to endless challenges in patent offices. For example, in September 1994, the American chemical company W.R. Grace, a competitor of Monsanto, secured a European patent for the use of neem oil as an insecticide, preventing Indian companies from marketing their products abroad except if they paid royalties to the multinational corporation, which was also flooding the country with chemical pesticides.[4]

These were the chemical pesticides that caused the first wave of suicides among indebted cotton farmers in the late 1990s. The intensive use of synthetic insecticides produced a phenomenon well known to entomologists: the development of resistance by the insects to the products intended to combat them. The result was that to get rid of the insects, farmers had to increase doses and turn to increasingly toxic molecules, so much so that in India, where cotton covers only 5 percent of the land under cultivation, it alone accounts for 55 percent of the pesticides used.

The irony of the story is that Monsanto was perfectly capable of benefiting from the deadly spiral that its products had helped create and which, in conjunction with the fall in cotton prices (from $98.20 a ton in 1995 to $49.10 in 2001), had led to the death of thousands of small farmers. The company praised the virtues of Bt cotton as the ultimate panacea that would reduce or eliminate the need to spray for bollworms, as its Indian subsidiary's Web site proclaims.

In 1993, Monsanto negotiated a Bt technology license agreement with the Maharashtra Hybrid Seeds Company (Mahyco), the largest seed company in India. Two years later, the Indian government authorized the importation of a Bt cotton variety grown in the United States (Cocker 312, containing the

Cry1Ac gene) so that Mahyco technicians could crossbreed it with local hybrid varieties. In April 1998, Monsanto announced that it had acquired a 26 percent share in Mahyco and that it had set up a 50-50 joint venture with its Indian partner, Mahyco Monsanto Biotech (MMB), for the purpose of marketing future transgenic cotton seeds. At the same time, the Indian government authorized the company to conduct the first field trials of Bt cotton.

"This decision was made outside any legal framework," said Vandana Shiva, whom I met in her offices at the Research Foundation for Science, Technology and Ecology in New Delhi in December 2006. The holder of a Ph.D. in physics, this internationally known figure in the antiglobalization movement received the "alternative Nobel Prize" in 1993 for her service to ecology and her efforts against the control of Indian agriculture by multinational agrichemical corporations. "In 1999," she told me, "my organization filed an appeal with the Supreme Court denouncing the illegality of the trials conducted by Mahyco Monsanto. In July 2000, although our petition had had not yet been considered, the trials were authorized on a larger scale, on forty sites spread over six states, but the results were never communicated, because we were told they were confidential. The Genetic Engineering Approval Committee had asked that tests be done on the food safety of Bt cotton seeds, used as fodder for cows and buffaloes, which thus might affect the quality of milk, as well as on cottonseed oil which is used for human consumption, but that was never done. In a few years, Monsanto had carried off a hijacking of Indian cotton with the complicity of government authorities, who opened the door to GMOs by sweeping away the principle of precaution that India had always upheld."

"How was that possible?" I asked.

"Well, Monsanto did considerable lobbying. For example, in January 2001, an American delegation of judges and scientists very opportunely met Chief Justice A.S. Anand of the Supreme Court and vaunted the benefits of biotechnology at the very time the court was supposed to issue a decision on our appeal. The delegation, led by the Einstein Institute for Science, Health, and the Courts, offered to set up workshops to train judges on GMO questions.[5] Monsanto also organized several trips to its St. Louis headquarters for Indian journalists, scientists, and judges. The press was also extensively used to propagate the good news. It's appalling to see how many personalities are capable of stubbornly defending biotechnology when they obviously know nothing about it."

It should be noted in passing that not only Indian personalities fell for Monsanto's line. A company press release on July 3, 2002, reported with obvious satisfaction that a European delegation had gone on a tour of Chesterfield Village, the biotechnology research center in St. Louis. "The delegation of visitors represented government agencies, nongovernmental organizations, scientific institutions, farmers, consumers, and journalists from 12 countries that are involved in biotechnology and food safety," according to the press release.[6]

"Do you think there was also some corruption?" I asked.

"Well," Shiva answered with a smile, searching for words, "I don't have any proof, but I can't exclude it. Look at what happened in Indonesia."

On January 6, 2005, the U.S. Securities and Exchange Commission (SEC) launched a two-pronged proceeding against Monsanto, accused of corruption in Indonesia. According to the SEC, whose findings can be consulted on the Web, Monsanto representatives in Jakarta had paid estimated bribes of $700,000 to 140 Indonesian government officials between 1997 and 2002 for them to favor the introduction of Bt cotton into the country.[7] They had, for example, offered $374,000 to the wife of a senior official in the Agriculture Ministry for building a luxury house. These generous gifts, it was claimed, had been covered by fake invoices for pesticides. In addition, in 2002, Monsanto's Asian subsidiary was said to have paid $50,000 to a senior official in the Environment Ministry for him to reverse a decree requiring an assessment of the environmental impact of Bt cotton before it was marketed. Far from denying these accusations, Monsanto signed an agreement with the SEC in April 2005 providing for the payment of a $1.5 million fine. "Monsanto accepts full responsibility for these improper activities, and we sincerely regret that people working on behalf of Monsanto engaged in such behavior."[8]

The Dramatic Failure of Monsanto's Transgenic Cotton

The fact remains that on February 20, 2002, much to the chagrin of peasant and ecological organizations, India's Genetic Engineering Approval Committee gave a green light to the cultivation of Bt cotton. Mahyco Monsanto Biotech pulled out all the stops: it hired a Bollywood star to promote GMOs on television (which enjoys a large audience in India), while tens of thousands of posters were put up throughout the country showing smiling farmers posing next to shiny new tractors, supposedly acquired with the profits from Bt cotton.

The first year, 55,000 farmers, 2 percent of India's cotton growers, agreed to join the transgenic adventure. "I heard it is a miracle seed that will free me from the bondage of pesticide spraying," a twenty-six-year-old farmer from Andhra Pradesh, one of the first states to authorize the marketing of GMOs (in March 2002), told the *Washington Post* in 2003. "Last season, every time I saw pests, I panicked, I sprayed pesticides on my cotton crop about 20 times. This season, with the new seed, I sprayed only three times."[9]

Regardless of this obvious advantage (which soon disappeared because insects developed resistance to Bt plants), the remainder of the picture was much less brilliant, as farmers interviewed by the *Washington Post* reported at the end of their first GM harvest. "I got less money for my Bt cotton because the buyers at the market said the staple fiber length was shorter," said one. The yield also did not improve. "The price of the seed is so high, now I wonder if it was really worth it."[10] In fact, because the patenting of seeds has (for now) remained prohibited in India, Monsanto could not apply the same system as in North America, that is, require that farmers buy their seeds every year under threat of legal action. To make up for its "losses," it decided to rely on a quadrupling of seed prices: while a 450-gram packet of conventional seeds sold for 450 rupees, the same amount of Bt seeds cost 1,850 rupees. Finally, the *Washington Post* reported, "the ruinous boll weevils have not disappeared." These less-than-stellar results did not keep Ranjana Smetacek, public relations director for Monsanto India* from declaring confidently: "Bt cotton has done very well in all the five states where it was planted."[11]

The accounts presented by the *Washington Post* were, however, confirmed by several studies. The first was commissioned in 2002 by the Andhra Pradesh Coalition in Defense of Diversity (CDD), which brought together 140 civil society organizations, including the Deccan Development Society (DDS), a very respected NGO that specializes in careful farming and sustainable development. The CDD asked two agronomists, Dr. Abdul Qayum, a former official in the state Agriculture Department, and Kiran Sakkhari, to

*The reader will recall the fake scientists under whose names the campaign of defamation against Ignacio Chapela over Mexican corn had been launched; one of them was named "Andura Smetacek," and Jonathan Matthews, the British researcher who uncovered the truth, had remarked on this unusual name. Perhaps the schemers in St. Louis had simply taken it from their Indian staff.

compare the agricultural and economic results of Bollgard with those of non-transgenic cotton in the district of Warangal, where 12,300 farmers had succumbed to Monsanto's promises.

The two scientists followed a very rigorous methodology consisting of monthly observation of the transgenic crops, from planting in August 2002 to the end of the season in April 2003, in three experimental groups. In two villages, where twenty-two farmers had planted GMOs, four were selected by random drawing. In midseason (November 2002), twenty-one farmers from eleven villages were questioned about the state of their transgenic crops, followed up by a visit to their fields. Finally, at the end of the season, in April 2003, a survey was conducted among 225 small farmers, chosen at random among the 1,200 Bt cotton producers in the district, 38.2 percent of whom owned less than five acres, 37.4 percent between five and ten acres, and 24.4 percent more than ten acres (the latter were considered large farmers in India). During the same period, they also recorded the performance of producers of conventional cotton (the control group). I provide all these details to emphasize that a scientific study worthy of the name requires this kind of effort, or else it's nothing but smoke and mirrors. The results of this large field investigation were conclusive: "The cost of cultivation for Bt cotton was Rs. 1092 more than that for non-Bt cotton because there was only a meager reduction in the pesticides consumption on Bt crop. On an average, there was a significant reduction (35%) in the total yield of Bt cotton, while there was a net loss of Rs 1295/ in Bt cultivation in comparison with non-Bt cotton, where the net profit was Rs 5368/-. Around 78 per cent of the farmers, who had cultivated Bollgard this year, said they would not go for Bt the next year."[12]

To put flesh on this scientifically irreproachable investigation, the DDS added to the initiative a group of "barefoot camerawomen," to use the expression of P.V. Satheesh, the founder and director of the ecological organization. The six women, all illiterate peasants and *dalits* (untouchables, on the bottom rung of the traditional social scale), were trained in video techniques in a workshop set up by the DDS in October 2001 in the little village of Patapur under the name of Community Media Trust. From August 2002 to April 2003, they shot film monthly of six small Bt cotton producers in the district of Warangal who were participants in the agronomists' study.

The result was a film that is an extraordinary document of the failure of transgenic crops. It shows first all the hope that farmers placed in Bt seeds.

Everything goes well for the first two months: the plants are healthy and there are no insects. Then disillusionment strikes. The plants are very small and cotton bolls less numerous than in the adjacent conventional cotton fields. In the October dry season, when parasites have deserted the traditional crops, the GM plants are besieged by cotton thrips and whiteflies. In November, when the harvest begins, anxiety can be seen in farmers' faces: the yields are very low, the bolls hard to pick, and the cotton fiber shorter, which means a 20 percent reduction in price.

I met the filmmakers in December 2006 in a cotton field in Warangal where they had come to film with the two agronomists. I was impressed by the professionalism of these extraordinary women, who, carrying sleeping babies on their backs, set up camera, stand, microphones, and reflector to interview a group of farmers who were desperate because of the failure of their Bt crops.

Since the first report published by the two agronomists, the situation had only gotten worse, triggering the second wave of suicides, which soon reached the state of Maharashtra. Worried by this tragic situation, the Andhra Pradesh government conducted a study that confirmed the conclusions reached by Qayum and Sakkhari.[13] Aware of the electoral consequences this disaster might have, the head of the Agriculture Department, Raghuveera Reddy, then demanded that Mahyco indemnify the farmers for the failure of their crops, a demand the company ignored.

Propaganda and Monopoly

In its defense, Monsanto brandished a study very opportunely published in *Science* on February 7, 2003.[14] The influence of scientific studies is extraordinary as long as they are backed by prestigious journals, which seldom or never verify the source of the data presented. Matin Qaim, then at the University of Bonn, and David Zilberman of the University of California, Berkeley, neither of whom "had ever set foot in India," as Vandana Shiva put it, found that according to field trials carried out in "different states in India," Bt cotton "substantially reduces pest damage and increases yields . . . as much as 88 percent." "What is really disturbing is that the article extolling the outstanding performance of Bt cotton is based exclusively on data supplied by the company that owns Bt cotton, Mahyco-Monsanto," commented the *Times of India*. "The data presented by the authors is . . . not based on the first Bt cotton harvest—as one would expect—but on the yield from a few selected trial plots belonging to the company. No data from farmers' fields has been included."[15] And yet the newspaper noted that "the same paper has been quoted extensively by several agencies as proof of the spectacular performance of GM crops"—which indeed was the purpose of the publication in *Science*.

The article was commented on at length in a 2004 FAO report titled *Agricultural Biotechnology Meeting the Needs of the Poor?*[16] This document caused a lot of ink to flow, because it was an argument in favor of GMOs. It was claimed they were capable of "increasing overall agricultural productivity" and that they "could help reduce environmental damage caused by toxic chemicals," according to the introductory note by Jacques Diouf, the director general of the UN organization. The report was in any event deeply satisfying to Monsanto, which hastened to put it online.[17]

Similarly in France, just before the *Science* article was published, Agence France-Presse distributed a laudatory presentation of it. I quote an excerpt, because it illustrates perfectly how disinformation stealthily makes its way through the media, although one would be hard pressed to attack the press agency, because after all it was only extrapolating from the carefully calculated unspoken suggestions of the original article: "Cotton genetically modified to resist a harmful insect could see yields increase as much as 80 percent, according to researchers who carried out trials in India," the dispatch stated. "The results of their work are surprising: before this, only a tiny increase in yields had been observed in similar trials conducted in China and the United States."[18] One can imagine the impact this information—widely picked up by the media, as, for example, *Le Bulletin des Agriculteurs* in Quebec—might have on small and midsize farmers who are constantly struggling for survival. This was especially the case because, disregarding all the data collected in the field, Qaim made so bold as to assert that "despite the higher cost of seeds, farmers quintupled their revenues with genetically modified cotton." His colleague David Zilberman had the virtue of clearly revealing the real purpose of the study in an interview with the *Washington Post* in May 2003: "It would be a shame if anti-GMO fears kept important technology away from those who stand to benefit the most from it."[19]

The *Times of India* was more prosaic. "Who will pay for the failure of Bt cotton?" the newspaper asked, pointing out that a law passed in 2001, the

Protection of Plant Varieties and Farmers' Rights Act, required breeders to indemnify farmers who had been "deceived" about the seeds they were sold with respect to "yield, quality, pest resistance," and so on.[20]

This was precisely the law that the Andhra Pradesh Commissioner of Agriculture intended to apply. When he was unable to do so, he decided in May 2005 to ban from the state three varieties of Bt cotton produced by Mahyco Monsanto (which were introduced a short time later in the state of Maharashtra).[21] In January 2006, the conflict with Monsanto reached a new stage: Agriculture Commissioner Raghuveera Reddy filed a complaint against Mahyco Monsanto with the Monopolies and Restrictive Trade Practices Commission (MRTPC), the Indian body charged with regulating commercial practices and antitrust laws, denouncing the exorbitant price of transgenic seeds as well as the monopoly established by the GMO giant on the Indian subcontinent. On May 11, 2006, the MRTPC found in favor of the commissioner and required that the price of a 450-gram packet of seeds be reduced to the price Monsanto charged in the United States and China, a maximum of 750 rupees (as opposed to 1,850 rupees). Five days later, the company appealed the decision to the Supreme Court, but the appeal was dismissed on the grounds that the decision was entirely a state matter.[22]

That was the situation when I got to Andhra Pradesh in December 2006. Mahyco Monsanto had finally lowered its seed price to the level demanded by the state government, but the conflict was far from over, because the thorny problem of financial compensation remained. "In January 2006," Kiran Sakkhari told me, "the Agriculture Department threatened to cancel the company's marketing licenses if it did not indemnify farmers for their last three harvests."

"But I thought Andhra Pradesh had banned three Bt cotton varieties in 2005."

"That's right. But Mahyco Monsanto immediately replaced them with new transgenic varieties. The government was unable to stop it, short of asking New Delhi to totally prohibit GMOs. And the result was just as catastrophic, as we showed in a second study.[23] This year there is a chance that it will be even worse, because, as you can see in this field of Bollgard cotton, the plants have been attacked by a disease known as rhizoctonia, which causes rot in the section of the plant between the root and the stalk. The plant eventually dries out and dies.

"Farmers say they've never seen that," Abdul Qayum said. "In our first study, we saw the disease in only a few Bt cotton plants. But it spread over time, and now it can be observed in many Bt cotton fields that are beginning to contaminate non-transgenic fields. Personally, I think there is a bad interaction between the receiving plant and the gene introduced into it. It causes weakness in the plant so that it is no longer resistant to rhizoctonia."

"Generally," Sakkhari went on, "Bt cotton is not resistant to stress conditions such as drought or heavy rains."

"But," I said, "according to Monsanto, sales of transgenic seeds are constantly rising in India."[24]

"That's what the company claims, and overall it's true, even if the figures it presents are hard to verify. But the situation can in large part be explained by the monopoly it was able to establish in India, where it has become very difficult to find non-transgenic cotton seeds. And this is very worrying, because, as we found in our second study, the promise that Bt cotton would reduce the use of pesticides has not been kept; quite the reverse."

Insect Resistance to Bt Plants: A Time Bomb

The agronomist showed me the results of the second study, covering the 2005–6 season. While in 2002–3, the year following the introduction of Bt seeds, the use of insecticides was slightly lower for transgenic plants than for conventional cotton, three years later the "great promise" had been definitively buried: pesticide expenditures were on average 1,311 rupees per acre for conventional cotton growers and 1,351 rupees for their Bt counterparts. "This result did not surprise us, and it can only get worse," Qayum explained, "because any serious agronomist or entomologist knows very well that insects develop resistance to chemical products designed to fight them. The fact that Bt plants constantly produce the insecticide toxin is a time bomb that we will pay for one day, and the cost may be very high, both from the economic and the environmental point of view."

In fact, the prospect that cotton (or corn) parasites would mutate by developing resistance to the Bt toxin was raised even before Monsanto put its GMOs on the market. In the mid-1990s, the strategy the company adopted, in agreement with the EPA, was to have growers of Bt plants agree by contract to preserve plots of non-Bt crops, called "refuges," where normal insects were supposed to proliferate so that they would crossbreed with their cousins that had become resistant to *Bacillus thuringiensis*, thereby causing genetic dilution. When insects are constantly confronted with a theoretically fatal dose of poison, they are all exterminated, except for a few specimens endowed with a gene resistant to the poison. The survivors mate with their fellows, possibly transmitting the gene in question to their descendants, and so on for several generations. This is known as "co-evolution," which, over the long course of the history of life, has enabled species threatened with extinction to adapt in order to survive a fatal disease. To keep this phenomenon from developing among Bt plant parasites, the sorcerer's apprentices imagined that they just had to maintain a population of healthy insects on the non-transgenic plots—the refuges—so they could mate with their cousins that had become resistant to Bt, thereby preventing the resistant insects from reproducing among themselves.

Once that was established, it remained to determine the size the refuges should have so that the plan would work. The subject was a matter of intense negotiations between Monsanto and the scientists, with the EPA merely recording the outcome. At first, some entomologists argued that the surface area of the refuges should be at least equivalent to that of the transgenic plots. Monsanto, of course, protested, suggesting at first that the surface area of the refuge should equal 3 percent of that of GMOs. In 1997, a group of university researchers working in the Midwest corn belt courageously jumped into the arena with a recommendation that refuges should be equivalent to 20 percent of the transgenic plots, and twice that if the plots were treated with pesticides other than Bt.

This was still too much for Monsanto, as Daniel Charles reports in *Lords of the Harvest*. "'Monsanto looked at the recommendations and said, "We can't live with that,"' says Scott McFarland, a young lawyer who was working for Pioneer at the time." The company contacted "the National Corn Growers Association, which also had its headquarters in St. Louis. Monsanto's representatives convinced the leadership of the NCGA that large refuges were a threat to farmers' free use of Bt."[25] This went on until September 1998, when the parties met in Kansas City to come to an agreement. As the discussions were getting bogged down in battles over arbitrary percentages, an agricultural economist from the University of Minnesota convincingly demonstrated that, according to his estimates, if the refuges were only 10 percent the size of the transgenic plots, then corn borers—the target parasite of Bt corn—would have a 50 percent chance of developing resistance in the short term and that it would cost farmers a good deal. With their wallets directly affected, the farmers joined the camp of the entomologists.

This is why around the world, Bt growers' manuals since then have required that refuges be equivalent to at least 20 percent of the GMO surface area. But it must be acknowledged that this amounts once again to tinkering and improvisation, because no serious study has been conducted to verify that this compromise—worked out in one corner of Missouri—has any scientific validity. And when Michael Pollan questioned Monsanto representatives on the issue for the *New York Times* in 1998, they answered: "If all goes well, resistance can be postponed for 30 years," which can only be called a short-term policy.[26] When Pollan persisted with Jerry Hjelle, Monsanto vice president for regulatory affairs, attempting to find out what would happen after that crucial period, "the response [was] more troubling. . . . 'There are a thousand other Bt's out there. . . . We can handle this problem with new products. . . . The critics don't know what we have in the pipeline. . . . Trust us.'"

In the meantime, ten years after the inauguration of Bt crops, it is possible to draw up a preliminary assessment of the shiny bureaucratic edifice. First, as an Associated Press dispatch pointed out in January 2001, according to a survey conducted in 2000 "30 percent of [American] Bt corn growers do not follow the published recommendations for the management of resistance," because they found them too restrictive.[27] To tell the truth, I understand them. But they should, of course, stop supporting such an absurd system, which will sooner or later collapse like a house of cards, as a 2006 study conducted by Cornell University researchers in cooperation with the Chinese Academy of Science showed.[28] Considered "the first to look at the longer-term economic impact of Bt cotton," the study covered 481 of the 5 million GM producers in China. It found that "the substantial profits they have reaped for several years by saving on pesticides have now been eroded." According to the authors, while for the first three years after the introduction of Bt crops, farmers had "cut pesticide use by more than 70 percent and had earnings 36 percent higher than farmers planting conventional cotton," in 2004 "they had to spray just as much as conventional farmers, which resulted in a net average income of 8 percent less than conventional cotton farmers because Bt seed is triple the cost of conventional seed." Finally, after seven years, "insects have increased so much that farmers are now having to spray their crops up to 20 times a growing season to control them." The researchers' conclusion, despite their support for GMOs, is devastating:

"These results should send a very strong signal to researchers and governments that they need to come up with remedial actions for the Bt cotton farmers. Otherwise, these farmers will stop using Bt cotton, and that would be very unfortunate."

The argument made Abdul Qayum and Kiran Sakkhari smile. "In India, where the majority of farmers cultivate between two and five acres, the strategy of refuges is frankly ridiculous. It all shows that GMOs, which are the latest version of the green revolution, were invented for large farmers in the North."

16

How Multinational Corporations Control the World's Food

> Through dialogue with many people, Monsanto has learned to appreciate that agricultural biotechnology raises some moral and ethical issues that go beyond science. These issues include choice, democracy, globalization, who has the technology, and who will benefit from it.
>
> —Monsanto, *Pledge Report*, 2005

If anyone in India knows the subject of the "green revolution" well, it is Vandana Shiva, one of whose books, published in 1989, is titled *The Violence of the Green Revolution: Ecological Degradation and Political Conflict in Punjab*.[1] In this fundamental book, the feminist antiglobalization figure dissects the misdeeds of the agricultural revolution, launched in the wake of World War II, that was later called "green" because it was supposed to slow the expansion of the "red revolution" in "underdeveloped" countries, particularly in Asia, where the rise of Mao Zedong to power in China in 1949 threatened to create imitators.

The "Only" Goal of the Second Green Revolution Is to Increase Monsanto's Profits

"I'm not saying that the green revolution did not begin with good intentions, namely, to increase food production in Third World countries," Shiva told me, "but the perverse effects of the industrial agriculture model that underlies it have had tragic environmental and social consequences, particularly for small farmers." For our second meeting, in December 2004, the militant Indian intellectual had invited me to the farm of Navdayana (Nine Grains), an association for the preservation of biodiversity and the protection of farmers' rights that she had established in 1987, located in the state of Uttarakhand, in northern India, on the border of Tibet and Nepal. Here, a few miles from Dehradun in the foothills of the Himalayas, where she was born, she has established a center for agricultural education intended to promote the growing of traditional wheat and rice crops that the green revolution almost caused to disappear, replacing them with high-yield varieties imported from Mexico.

The agroindustrial concept that in 1968 was labeled the green revolution was born in 1943 in the capital of Mexico.* That year, Henry Wallace, vice president of the United States (and co-founder of Pioneer Hi-Bred, which invented hybrid corn), offered to his Mexican counterpart a "scientific mission" designed to increase national wheat production. Sponsored by the Rockefeller Foundation, under the auspices of the Mexican Agriculture Ministry, this pilot project was set up in a Mexico City suburb, where in 1965 it adopted the name of the International Maize and Wheat Improvement Center (CIMMYT, Centro Internacional de Mejoramiento de Maíz y Trigo).

In October 2004, I visited this renowned research center, which still operates as a nonprofit organization and now employs a hundred highly qualified international researchers, as well as five hundred associates from forty countries. In the entrance hall, a huge painting pays tribute to the father of the green revolution, Norman Borlaug, born on an Iowa farm in 1914, who was hired by the Rockefeller Foundation in 1944 and won the Nobel Peace Prize in 1970 "in recognition of his important contribution to the green revolution."[2] For twenty years, this agronomist, who is now an ardent supporter of GMOs, had a single obsession: to increase wheat production by creating varieties permitting a tenfold increase in yields. To reach the goal he came up with the idea of crossing CIMMYT's varieties with a Japanese dwarf variety, Norin 10. Increasing yields involves forcing the plant to produce larger and more numerous kernels at the risk of causing the stem to break. Hence

*The expression was used for the first time on March 8, 1968, by William Gaud, administrator of the United States Agency for International Development (USAID), in a speech delivered in Washington.

the trick of "stem shortening," in breeders' jargon, through the introduction of a gene for dwarfism.*

This is how, in the space of a century, wheat yields increased from about four hundredweight per acre in 1910 to an average of thirty-two hundredweight, while the height of wheat stalks decreased by three feet. But this exploit was accompanied by a side effect criticized by opponents of the green revolution: an increased use of phytosanitary products, without which the "miracle seeds," as the CIMMYT varieties were called, were of absolutely no use. In order to produce such a large quantity of kernels, the plant had to be stuffed with fertilizers (nitrogen, phosphorus, potassium), which eventually brought about a decline in the natural fertility of the soil. In addition, it had to be watered copiously, which depleted aquifers. Furthermore, extreme vegetal density was manna for insect pests and fungi, which meant the massive use of insecticides and fungicides. Finally, the obsession with yields brought about a general decline in the nutritional quality of the kernels and a reduction in the biodiversity of wheat, a number of varieties of which simply disappeared.

In the 1960s, aware of the irremediable nature of the losses associated with its promotion of high-yield varieties, CIMMYT opened a "germoplasm bank," which now stores at –3°C some 166,000 varieties of wheat. To supply it, its associates comb countrysides around the world in search of rare grains, such as the wild wheat specimens found at the Iranian edge of the Fertile Crescent, which its technicians were in the process of labeling when I visited the center.

Nonetheless, CIMMYT's dwarf varieties have spread around the world. In the North, including the Communist countries, breeders used them in their cross-pollination programs. The countries of the South, led by India, sent technicians to be trained at the center, nicknamed the "School of the Wheat Apostles." In 1965, an unusual drought devastated the wheat crop in the Indian subcontinent, and there was a threat of famine. The government of Indira Gandhi decided to buy eighteen thousand tons of high-yield seeds imported from Mexico, the largest transfer of seeds in history. Trained by CIMMYT, Indian agronomists propagated the green revolution in the re-

*Today, to further shorten stems, large wheat growers have no hesitation in applying to their crops a hormone, euphemistically named "plant growth regulator."

gions of Punjab and Haryana, considered India's breadbasket. They were given financial support by the Ford Foundation, which was in a good position to supply tractors and farm machines. At the same time high-yield varieties of rice were introduced into the country, at the initiative of the International Rice Research Institute (IRRI), established in 1960 by the Rockefeller and Ford Foundations on the model of CIMMYT.

"It is always said that thanks to the green revolution, India achieved self-sufficiency in food supplies and that in the five years from 1965 to 1970 its wheat production increased from 12 to 20 million tons," I was told by Vandana Shiva, whose last book is titled *Seeds of Suicide*.[3] "The country is now the world's second-largest wheat producer, with a production of 74 million tons, but at what cost? Exhausted soil, a worrying decline in water reserves, widespread pollution, the spread of monocultures at the expense of food crops, and the exclusion of tens of thousands of small farmers who have moved to slums because they could not adapt to an extremely costly model of farming. The first wave of suicides signaled the failure of the first green revolution. Unfortunately, the second green revolution, the GMO revolution, has been even more deadly, even though it was directly in line with the first."

"Why? How are they different?"

"The difference between the two is that the first green revolution was led by the public sector: government agencies controlled agricultural research and development. The second green revolution is led by Monsanto. The other difference is that although the first green revolution did have the hidden aim of selling more chemical products and farm machines, its principal motive even so was to provide more food and to guarantee food security. In the end, even though it was done at the expense of other crops, such as leguminous plants, the country produced more rice and wheat to feed people. The second green revolution has nothing to do with food security. Its only aim is to increase Monsanto's profits, and the company has succeeded in imposing its law around the world."

"What is Monsanto's law?"

"Patent law. The company has always said that genetic engineering was a way of getting patents, and that's its real aim. If you look at the research strategy it is now pursuing in India, you'll see that it is testing twenty plants into which it has introduced Bt genes: mustard, okra, eggplant, rice, cauliflower, and so on. Once it has established ownership of genetically modified seeds as the norm, it will be able to collect royalties; we will depend on the company for every seed we plant and every field we cultivate. If it controls seeds, it controls food; it knows that, and that's its strategy. It's more powerful than bombs or weapons; it is the best way to control the people of the world."

"But it's illegal to patent seeds in India," I said, a bit staggered by the picture she had painted.

"Sure. But for how long? Monsanto and the American government have been pressuring the Indian government for ten years to apply the TRIPS [Trade Related Aspects of Intellectual Property Rights] agreement of the WTO, and I'm afraid that the barriers will finally collapse."

Patents on Life, or Economic Colonization

Before explaining the TRIPS agreement, a headache for the WTO since its founding in 1995, I have to come back to the question of patents, which is of capital importance for the future of the planet. After listening to Shiva, one might think that she was exaggerating and that the patenting of seeds is just a gimmick of little concern to us. Skeptics should take a second look: the patenting of living things, particularly of seeds, is the tool through which Monsanto could appropriate the most lucrative of markets, the world's food. And the company has done everything it can to bring this about.

Shiva was quick to take an interest in this colossal challenge "because of the Bhopal disaster," as she told me the first time we met, in Bhopal, which was then commemorating the twentieth anniversary of the tragedy.[4] During the night of December 2, 1984, a cloud of toxic gas descended on the city: within a few hours, ten thousand people had died after suffering terribly, and twenty thousand died in the following weeks. The deadly gas came from a factory belonging to the American multinational corporation Union Carbide, a competitor of Monsanto's that manufactured chemical pesticides.

"It was the Bhopal tragedy that convinced me we had to promote organic farming, hence the neem tree, as an alternative to the multinational corporations' deadly pesticides," Shiva recalled. Remember that the Office of European Patents granted a patent on the use of neem oil to W.R. Grace in September 1994. From that point on, the patenting of life became the Indian activist's great cause; with the support of Greenpeace, she succeeded in having the patent rejected ten years later, along with an American patent held by a Texas company, RiceTec, on a variety of basmati rice. Since then, she has been fighting against a European and American patent held by Monsanto on a variety of wheat prized for the making of chapatis and cookies because of its low gluten content.[5] According to the terms of the patents, Monsanto holds a monopoly on the growing, crossbreeding, and processing of this variety, which originated in northern India.

"The patenting of life is a continuation of the first colonization," Shiva said. "The word 'patent' itself comes from the age of conquest. 'Letters patent' was the name given to an official public document—in Latin, *patens* means 'open' or 'obvious'—bearing the seal of European sovereigns [and] granting to adventurers and pirates the exclusive right to conquer foreign countries in their name. At the time Europe was colonizing the world, letters patent were directed at territorial conquest, whereas today's patents are aimed at economic conquest through the appropriation of living organisms by the new sovereigns, the multinational corporations like Monsanto. The same principle was operative in both cases, namely, the patents then and now were based on a denial of the life that existed before the arrival of the white man. When the Europeans colonized America, the land of the New World was declared *terra nullius*, 'empty land,' meaning devoid of white men. In the same way, the patenting of life and of the biosphere is based on an allegation of 'empty life,' because as long as the genes of living organisms have not been dissected in a laboratory, the organisms have no value. This is a denial of the labor and knowledge of millions of people who have maintained the biodiversity of life for millennia and who, moreover, live from it."

"What are the consequences of patents on life for the peoples of the South?" I asked, fascinated by the clarity of her thinking.

"They are huge, because patents are playing the same role as enclosures in sixteenth-century England. This movement, originating before the Industrial Revolution, privatized by enclosing common land that had been used communally, where the poorest villagers, for example, could graze their animals. The patent similarly encloses living things, such as plants that feed and heal people, and finally contributes to the exclusion of the poorest from the means of livelihood and even survival. As can be seen with food and medicine, as soon as a patent is filed, it means royalties and consequently an increase in price. This is why food, crop maintenance products, and medicines are excluded from Indian patent law, so that they remain accessible to everyone. The extension of the Western system of patents, as advocated by the World Trade Organization, and before that by the final round of GATT, directly undermines the economic rights of the poorest."

Monsanto and the Multinational Corporations Behind the WTO Agreement on Intellectual Property Rights

The General Agreement on Tariffs and Trade (GATT) was put in place in 1947 by the major capitalist powers of the time with the purpose of regulating customs duties on international trade. The 1986 ministerial conference of Punta del Este, inaugurating what became known as the "Uruguay Round," marked a decisive turning point in the history of GATT, in effect signing its death warrant. It was in the course of the eighth and final session of these intergovernmental trade negotiations in 1994 that the American government won agreement for the inclusion of four areas that had until then been under exclusively national political jurisdiction: agriculture, investments, services (telecommunications, transportation, and the like), and intellectual property rights (IPR). The U.S. trade representative justified the inclusion of this last area, with which I am particularly concerned, by the fact that "nearly 200 American transnational companies were deprived of 24 billion dollars of copyright earnings because of the weakness or absence of protection for intellectual property in some countries, primarily in the countries of the South," as a study by the University of Quebec reported.[6]

The inclusion of these new areas under GATT's jurisdiction, which had at first been a simple customs union, was the focus of intense negotiations, because they "raised questions that went beyond trade," namely, "fundamental rights" such as the "rights to employment, health, food, and self-determination," as Shiva has pointed out.[7] In December 1989, Arthur Dunkel, director-general of GATT, submitted a proposed final document, but it was not until April 1994 that the definitive agreement was signed by the 123 member countries in Marrakesh, ratifying the creation of the World Trade Organization, which officially replaced GATT on January 1, 1995.

The founding document of the WTO, which meets in Geneva, contains twenty-nine sectorial agreements making possible the subjection of any good or service to the laws of the market, and therefore the transfer to private companies (over which governments and citizens have no means of control) of areas that traditionally were a matter of public policy. The association of these sectors with trade is so far from obvious that the drafters of the agreements got around the problem by using the expression "trade-related," thereby pointing to the subterfuge.

This was notably the case with the TRIPS agreement, which, it turns out, "was largely designed by a coalition of companies gathered under the name of Intellectual Property Committee (IPC)," including the "major players in the area of biotechnology," as the researchers from Quebec pointed out.[8] Established in the United States in March 1986, the IPC brought together thirteen multinational corporations, principally from the chemical, pharmaceutical, and computer industries: Bristol-Myers, DuPont, FMC Corporation, General Electric, General Motors, Hewlett-Packard, IBM, Johnson and Johnson, Merck, Pfizer, Rockwell International, Warner Communications, and Monsanto.

As soon as it was established, the committee contacted the Union of Industrial and Employers' Confederations of Europe (UNICE), official organ of the European business world, and the Keidanren, the Japanese employers' confederation, to draft a common document, which was submitted to GATT in June 1988. Titled "Basic Framework of GATT Provisions on Intellectual Property: Statement of Views of the European, Japanese, and United States Business Communities," this document, which formed the basis for the TRIPS agreement, was aimed at extending to the rest of the world the patent system that already existed in the industrialized countries, which all told, through the offices in Washington, Munich, and Tokyo, registered 97 percent of the patents filed by companies (the vast majority from the North). The document framed the issue in these terms: "Disparities among systems for the protection of intellectual property result in excessive loss of time and resources in the acquisition of those rights. Holders find that the exercise of their rights is hindered by laws and regulations limiting market access and the repatriation of profits." There followed a short paragraph: "Biotechnology, or the use of microorganisms in production, is a sector in which patent protection has fallen behind the rapid progress of medicine, agriculture, pollution reduction, and industry. . . . This protection should apply to the processes as well as the products of biotechnology, whether they be microorganisms, parts of microorganisms (plasmids and other vectors), or plants."[9]

Seemingly convinced that what might be considered a hijacking of GATT was within its rights, Monsanto proudly asserted in June 1990: "Once created, the first task of the IPC was to repeat the missionary work we did in the U.S. in the early days, this time with the industrial associations of Europe and Japan, to convince them that a code was possible. . . . It was not an easy task but our Trilateral Group was able to distill from the laws of the more advanced countries the fundamental principles for protecting all forms of intellectual property. Besides selling our concepts at home, we went to Geneva where [we] presented [our] document to the staff of the GATT Secretariat. We also took the opportunity to present it to the Geneva-based representatives of a large number of countries. What I have described to you is absolutely unprecedented in GATT. Industry has defined a major problem for international trade. It crafted a solution, reduced it to a concrete proposal, and sold it to our own and other governments. The industries and traders of world commerce have played simultaneously the role of the patients, the diagnosticians, and the prescribing physicians."[10]

Despite this masterfully conducted collective lobbying, among the many sectors covered by the TRIPS agreement (copyright, trademarks, label of origin, industrial designs and models, and confidential information, including trade secrets), the sector opportunely suggested by Monsanto is the one that has stymied the implacable machine of the WTO since 1995. The controversy swirls around Article 27, paragraph 3(b), relating to "patentable subject matter." The official text provides: "Members may . . . exclude from patentability plants and animals other than micro-organisms, and essentially biological processes for the production of plants or animals other than non-biological and microbiological processes. However, Members shall provide for the protection of plant varieties either by patents or by an effective *sui generis* system or by any combination thereof. The provisions of this subparagraph shall be reviewed four years after the date of entry into force of the WTO Agreement."

The language of this article is so abstruse that it was partially responsible for the paralysis of the third ministerial conference of the WTO, held in Seattle in December 1999. After reading and rereading it, one can figure out that animals and plants but not microorganisms may be excluded from the patent system. But it also stipulates that "plant varieties" shall be protected "either by patents or by an effective *sui generis* system." This apparent contradiction is in fact intended directly for transgenic seeds: they may now, backed by sanctions, be "protected" (that is, manufacturers can collect royalties) at a minimum by the system set up by the UPOV agreements. It is precisely because the "protection" of seeds also brings about the protection of the foods derived from them that many countries of the South, led by South Africa, India, and Brazil, have demanded that Article 27, paragraph 3(b), be revised. They are also worried about the consequences of the patenting of microorganisms (theoretically including genes), which can only encourage biopiracy, that is, the theft of genetic resources and the traditional knowledge associated with them, to the detriment of the rural and indigenous communities that have maintained those resources for millennia.

The WTO: A Veritable Nightmare

To get a clear picture, I went to Geneva on January 13, 2005, to meet with Adrian Otten, director of intellectual property for the WTO, and I asked at the outset a basic question that suddenly made him tense up: "What is the goal of the TRIPS agreement?" Stammering a bit, he finally answered, "Well, I suppose that one of the fundamental objectives is to establish common international rules for member governments of the WTO to protect the intellectual property rights of certain member countries of the WTO, as well as those of their citizens and companies."

"And which article has caused a problem?" I asked, to see if I had understood the WTO's gibberish.

"Well, it's Article 27, paragraph 3(b), which adds a clause to the TRIPS agreement according to which inventions connected to plants and animals should be subject to patenting."

Put like that, it was as clear as spring water.

"The goal of the TRIPS agreement is that a patent obtained in the United States—for example, by Monsanto—will be automatically applicable everywhere in the world," I had been told a month earlier in New Delhi by Devinder Sharma. Chairman of the Forum for Biotechnology and Food Security, this noted Indian journalist is a fierce opponent of the WTO. "If you observe the international evolution of the patent system, you can see that it follows exactly that of the Patent Office in Washington. With the TRIPS agreement, every country has to follow the model of the United States or else suffer severe commercial penalties, because the WTO has absolutely extraordinary powers of coercion and reprisal. That means that if a country doesn't enforce respect for Monsanto's intellectual property rights, for example on a patented seed, the company will inform the American government, which will file a complaint with the WTO Dispute Settlement Body. The TRIPS agreement was also designed by multinational corporations to seize the genetic resources of the planet, chiefly in Third World countries, which have the greatest biodiversity. India is a particular target, because it is a megadiverse country where there are 45,000 plant species and 81,000 animal species. That's why so many of us say the world of the living is no concern of the WTO, but rather of the Biodiversity Convention signed under the auspices of the UN in Río de Janeiro in 1992. Signed by two hundred countries, this treaty says that genetic resources are the exclusive property of states, who must commit themselves to preserving them and organizing an equitable sharing of the exploitation of the traditional knowledge associated with those resources."

"Can the TRIPS agreement be reconciled with the Biodiversity Convention?"

"Absolutely not, because the two documents contradict one another. And that's why the United States didn't sign the convention. The problem is that the TRIPS agreement takes precedence over the convention, because it is under the jurisdiction of the WTO, which obeys the orders of multinational corporations like Monsanto, which, under cover of the globalization of trade, in fact rule the world."

For those who think these words are excessive, I will quote a UN report published in June 2000 by the Sub-Commission on the Promotion and Protection of Human Rights: "The greater percentage of global trade is controlled by powerful multinational enterprises. Within such a context, the notion of free trade on which the rules [of the WTO] are constructed is a fallacy. . . . The net result is that for certain sectors of humanity—particularly the developing countries of the South—the WTO is a veritable nightmare."[11]

Conclusion

A Colossus with Feet of Clay

> The people of this company are poison: like the god of death, they take away life.
>
> —A member of the Community Media Trust in Pastapur, Andhra Pradesh

The scene took place at the headquarters of TIAA-CREF (Teachers Insurance and Annuity Association, College Retirement Equities Fund) in an upscale neighborhood of Manhattan in July 2006. Established ninety years earlier, this prestigious pension fund is one of the most important financial institutions in the United States, holding $437 billion in assets. Ranked eightieth on *Fortune*'s list of the five hundred largest companies in the country, TIAA-CREF has a special characteristic figuring prominently on all its official documents: the company provides "financial services for the greater good." The only people eligible to join the pension fund are those who serve the "greater good," including "professors, nurses, deans, hospital and university administrators, doctors," and the like, amounting to 3.2 million members. Since 1990, TIAA-CREF has had a department specializing in "responsible investment," which 430,000 clients have joined. The reason I had asked to meet representatives of the venerable company was that I had discovered it was one of the twenty largest shareholders in Monsanto, 1.5 percent of whose shares it held at the time.*

*According to the SEC, in June 2006 Monsanto's principal shareholders were: Fidelity Investment (9.1 percent), Axa (6.1 percent), Deutsche Bank (3.6 percent), Primecap Management (3.6 per-

Reputation Is a Risk Factor for Companies

I met that day with John Wilcox, head of the company's corporate governance practice, and Amy O'Brien, director of the Socially Responsible Investing group. "Considering the special nature of your clientele, are there companies in which you refuse to invest?" I asked, under a little stress because the public relations director was also there, sitting behind me and taking notes.

"Of course," O'Brien said. "For instance, our investors don't want us to invest their money in tobacco companies, because of the burdens they impose on society. And, generally speaking, they are sensitive to the conduct of companies in social and environmental matters."

"That means you take into account a company's reputation?"

"Absolutely," Wilcox answered without hesitation. "Reputation is increasingly considered a risk factor. Until recently, non-financial criteria for a company's performance, such as its reputation or its environmental practices, were of no interest to Wall Street analysts, probably because they are hard to quantify and they involve the long term. But this is very clearly changing. There are increasing numbers of citizens who demand that the companies they invest their savings in share their values."

"I have read that TIAA-CREF holds 1.5 percent of Monsanto's shares."

"It's possible," Wilcox said. "I really don't know."

"That company's reputation is very controversial. How do you explain that investment?"

"I don't think we offered it in our portfolio of shares for responsible investing," O'Brien said hesitantly, visibly embarrassed. "I'm not sure, but in any case the company is especially controversial in Europe because of genetically modified organisms, but not in the United States."

"But Agent Orange, PCBs, bovine growth hormone, aren't they American stories? Did you inform your clients about the litigation Monsanto has had to deal with in the last several years?"

"No," Wilcox answered. "I'm going to examine Monsanto's risk factors and ask for advice from the people managing our stock portfolios."

---

cent), State Street Corp (3 percent), Barclays (3 percent), Morgan Stanley (2.9 percent), Goldman Sachs Group (2.7 percent), Vanguard Group Inc. (2.5 percent), Lord Abbett & Co (2.4 percent), American Century Investment Management Inc. (2.4 percent), and General Electric (2.3 percent).

A Risky Company for Investors

Still in Manhattan, not far from the offices of TIAA-CREF, I went to see Marc Brammer, who works for Innovest Strategic Value Advisors, the leader in what is known as "extrafinancial analysis," which consists of evaluating the social and environmental performances of companies on a scale ranging from AAA for the best to CCC for dunces. The grades are used to advise investors so that they may reduce their financial risks and increase the yield of their investments. With offices in New York, London, Tokyo, and Paris, Innovest has taken on the task of developing a clientele with portfolios focused on sustainable development. In January 2005, Brammer published a report titled "Monsanto and Genetic Engineering: Risks for Investors," in which he presented a summary of the company's activities and noted its "management and strategy" in the area of biotechnology.[1] The result was a grade of CCC. "It's the worst environmental grade," he told me. "We have observed that in almost every industrial sector the companies with above-average environmental grades generally do better in the stock market than those with below-average grades, by a range of three hundred to three thousand points annually. This means that Monsanto is a risky business for shareholders in the medium and long term."

"Who are Monsanto's shareholders?"

"They are very dispersed, but the principal investors are pension funds and banks, which represent tens of thousands of small shareholders."

"How do you explain the fact that a fund like TIAA-CREF has invested in Monsanto?"

"It's surprising, because it's an institution that really encourages responsible investing. On the other hand, it's rather typical of the way pension funds operate: they make short-term calculations and are very sensitive to market rumors. In the case of Monsanto, it's clear that it is overvalued because of unconditional support on Wall Street."

"What are the principal risk factors for investors?"

"The primary one is market rejection, which is a real time bomb for Monsanto. GMOs are some of the most strongly rejected products that have ever existed. More than thirty-five countries have adopted or proposed legislation limiting the imports of GMOs or requiring the labeling of food containing biotech ingredients. Most European food distributors have established measures to guarantee that no biotech ingredients are used in their products. This is true for Nestlé, Unilever, Heinz, ASDA [British subsidiary of Wal-Mart], Carrefour, Tesco, and many others. Outside of Europe, there is also strong consumer opposition to GMOs in Asia and Africa.

"Even in the United States, Monsanto was forced, for example, to withdraw its Bt potatoes from the market after companies such as McDonald's, Burger King, McCain, and Pringles refused to buy them. I'm sure that if the FDA decided to label GMOs, Monsanto would lose 25 percent of its market overnight. In fact, twenty surveys conducted between 1997 and 2004 clearly indicate that more than 80 percent of Americans want biotech products to be labeled.* So much so that one of the consequences of the lack of labeling of GMOs is the absolutely exponential development of the market for organic products in the United States."

Monsanto has fully understood the danger labeling represents for its biotech business. When a citizen initiative in 2002 succeeded in getting a referendum on the ballot in Oregon on the labeling of GMOs, Monsanto quickly organized a campaign under the name of the Coalition Against the Costly Labeling Law, with its "allies from biotechnology and the food industry," which cost the tidy sum of $6 million. "The general feeling," argued Monsanto spokesman Shannon Troughton, "is that the measure, if passed, would create a new set of bureaucratic rules and regulations and provide meaningless information at a considerable cost to consumers."[2] The initiative, the first of its kind in the United States, was finally rejected by 73 percent of the voters, on the grounds that the labeling would be too costly.

"The other risk factor that threatens Monsanto's performance are the flaws in the regulatory system, perfectly illustrated by the StarLink disaster," Brammer went on. "We have calculated that if it were faced with a similar case, the company would lose $3.83 a share. The fundamental problem with GMOs is that only Monsanto benefits from them; risks are for the others, and regulatory agencies have abdicated their role of assessment and super-

*The surveys referred to in the Innovest report were done by ABC News (in which 93 percent of those surveyed wanted GMOs labeled as such), Rutgers University (90 percent), Harris Poll (86 percent), *USA Today* (79 percent), MSNBC (81 percent), Gallup Poll (68 percent), Grocery Manufacturers of America (92 percent), *Time* magazine (81 percent), Novartis (93 percent), and Oxygen/Market-Pulse (85 percent).

vision. The opacity of the regulatory process feeds rejection by consumers in the United States because they don't have the right to choose what they eat, but also in Europe, as the example of MON 863 corn shows."

The Flaws of the Regulatory System: The Example of MON 863 Corn

While the French government announced in January 2008 that it was implementing the "safeguard clause" for MON 810 corn, suspending the cultivation of this Monsanto Bt variety until the European Union has reconsidered its authorization, I would like to recall the history of MON 863, a close cousin of MON 810. MON 863 contains a toxin (Cry3Bb1) intended to protect it against the corn root worm, while MON 810 has been engineered (Cry1Ab) to resist infestation by the corn borer.* The MON 863 affair is a perfect illustration of the worrisome way, to put it mildly, in which GMOs are regulated in Europe.

In August 2002, Monsanto filed a request for marketing approval of MON 863 with the German authorities, submitting a technical file including a toxicological study conducted on rats for twenty-nine days. In conformity with European regulations, the authorities examined the data supplied by Monsanto and transmitted a negative opinion to the Brussels Commission, on the grounds that the GMO contained a marker for resistance to an antibiotic, which infringed directive 2001/18 strongly advising against its use. The commission was then required to distribute the file to the member states to solicit their opinions, which would be examined by the European Food Safety Authority (EFSA), the European scientific committee charged with evaluating the food safety of GMOs.

In France, the Commission du Génie Biomoléculaire (CGB) received the file in June 2003. Five months later, on October 28, 2003, the CGB issued an unfavorable opinion, not because of the presence of the antibiotic resistance marker, but because, as Hervé Kempf explained in *Le Monde*, it was "very troubled by the deformities observed in a sample of rats fed with 863 corn."[3] "What struck me in this case was the number of anomalies," ex-

*According to Greenpeace, the corn root worm, a very harmful insect, was introduced into Europe during the Balkan War; it is said to have come in American air force planes.

plained Gérard Pascal, director of research at INRA and member of the CGB since its creation in 1986. "There are too many areas here where one observes significant variations. I've never seen that in another case. It has to be reconsidered."[4]

The "variations" included a "significant increase in white blood cells and lymphocytes in males in the sample fed with MON 863; a reduction in reticulocytes (immature red blood cells) in females; and a higher frequency of anomalies (inflammation, regeneration) of kidneys in males," as well as a reduction in weight in the test animals.[5] As Kempf points out, "no one would have known anything about it" if the lawyer Corinne Lepage, former environment minister in the government of Alain Juppé and president of CRIIGEN,* "had not forced entry into the CGB" to obtain, after a legal battle lasting a year, the transcripts of discussions leading to the CGB's negative opinion, which was "unusual for a commission that had always tended to support the authorization of GMOs." The deliberations of the scientific committees of member states of the European Union, like those of EFSA, are indeed confidential, which gives one an idea of the transparency of the process for the evaluation of GMOs.

There was a new development in the affair on April 19, 2004, when EFSA issued an opinion in favor of marketing MON 863. According to the authority, the anomalies observed by the CGB "are part of the normal variation of control populations"; as for the kidney deformities, they were "of minimal importance."[6]

How could two scientific committees express such different opinions on the same case? The answer to this question was provided by the European branch of Friends of the Earth, which in 2004 published a very detailed (and very troubling) report on the operations of EFSA.[7] Established in 2002 under the authority of European directive 178/2002 on the safety of food products, this institution includes eight scientific committees, one of which is charged exclusively with the evaluation of GMOs. It is precisely this committee, known as the GMO panel, that the report considers.

Friends of the Earth begins by observing: "In just over a year [the GMO panel] has published twelve scientific opinions, virtually all favourable to the biotechnology industry. These opinions have been used by the European

*CRII-GEN is the Committee for Independent Research and Information on Genetic Engineering, of which Professor Gilles-Éric Séralini is a member.

Commission, which is under increasing pressure from the biotechnology industry and the United States, to force new GM products onto the market. . . . They are being used to create a false impression of scientific agreement when the real situation is one of intense and continuing debate and uncertainty. Concerns about the political use of their opinions have been expressed by members of the EFSA themselves."

According to the report, this situation is due to the close ties between "certain members" of the GMO panel and the giants of biotechnology, led by the panel's chairman, Harry Kuiper. Kuiper is the coordinator of Entransfood, a project funded by the European Union to "facilitate market introduction of GMOs in Europe, and therefore bring the European industry in[to] a competitive position." In that capacity, he is a participant in a working group that includes Monsanto and Syngenta. Similarly, Mike Gasson works for Danisco, a partner of Monsanto; Pere Puigdomenech was co-chair of the Seventh International Plant Molecular Biotechnology Congress, sponsored by Monsanto, Bayer, and DuPont; Hans-Yorg Buhk and Detlef Bartsch are "well-known for their pro-GM views and have even appeared in promotional videos produced by the biotechnology industry." One of the few outside experts consulted by the panel is Dr. Richard Phipps, who had signed a petition in favor of biotechnology circulated by AgBio World[8] and turned up on Monsanto's Web site to support bovine growth hormone.[9]

Friends of the Earth then considered several cases, including MON 863. It appears that the reservations expressed by the German government about the presence of an antibiotic resistance marker were dismissed out of hand by the GMO panel, which relied on an opinion it had issued in a press release on April 19, 2004: "The Panel has confirmed that ARMs [antibiotic resistance markers] are in the majority of cases still required in order to ensure the efficient selection of transgenic events in plants," Kuiper stated. Friends of the Earth commented: "The Directive does not ask for confirmation of whether ARMs are an efficient tool for the biotech industry; the assessment required is whether they could have adverse effects on the environment and human health."

The end of the story was equally exemplary. After EFSA's favorable opinion was issued, Greenpeace asked the German Agriculture Ministry to have the technical file supplied by Monsanto (1,139 pages) made public, so that it could be independently analyzed. The ministry replied that this was impossible; Monsanto refused to make the data public because they were cov-cred by the "trade secret" privilege. After a legal battle that lasted several months, Monsanto was finally compelled to make the data public by a decision of the court of appeals in Munich on June 9, 2005.

"It's really unbelievable, when it's a question of verifying the safety of a pesticidal plant designed to enter into the food chain, that Monsanto could first claim 'trade secret,' then file two lawsuits to deny access to the raw data in its study," according to Gilles-Éric Séralini, who had followed the case very closely. At the request of Greenpeace, and simultaneously with Arpad Pusztai, the University of Caen scientist first did an assessment of the toxicological file that had been dragged out of Monsanto, which confirmed the anomalies observed by the French CGB.[10] Then, under the auspices of CRII-GEN, he conducted an independent analysis of the raw data, in which he applied a more refined statistical method, considering in particular organs, dosage, and exposure time to GMOs. This analysis revealed that the effects of 863 corn on rats were much more significant than those observed initially, "which indicates the need to conduct further tests."[11]

"Indeed," Séralini commented, "the story of MON 863 corn shows the inadequacy of the process for the approval of GMOs, which ought to be assessed in the same way as any pesticide or medicine is, by testing them on three mammalian species over two years, which would permit an assessment of their long-term toxicity, not only of their possible acute toxic effects." In the meantime, faced with these disturbing revelations, the European Commission discreetly swept MON 863 corn under the rug by banning its cultivation but not its importation and therefore not its consumption.

What If GMOs Were Tomorrow's Agent Orange?

"Contrary to what Monsanto claims, it is not an agricultural company, but a chemical company," argues Marc Brammer. "The proof is that the only GMOs it has succeeded in getting on the market are plants resistant to its star herbicide, Roundup, which still accounts for 30 percent of its revenue, or insecticidal plants.* Those plants are of no interest to consumers who are still waiting for the miracle GMOs that the company has constantly

*According to its 10-K form, Monsanto had revenues of $7.3 billion in 2006, $2.2 billion of which were from Roundup.

promised them, such as the golden rice that it announced with great public fanfare."

To be precise, Monsanto did not invent golden rice, which was cobbled together, with the best intentions in the world, by two European researchers, Ingo Potrykus from Zurich and Peter Beyer from Freiburg. This GM rice was supposed to produce beta-carotene, the precursor to vitamin A, deficiency of which leads to the death of a million children annually in the Third World and causes blindness in 300,000 others. Published in *Science* in 2000, the laboratory results seemed so promising that golden rice made headlines in many newspapers as the embodiment of the great promise of biotechnology.[12] Funded by the Rockefeller Foundation, the two researchers decided to launch their creation on the market, but they were confronted by an inextricable problem of patents: to make their golden rice, they had used genes and procedures covered by no fewer than seventy patents belonging to thirty-two companies or research centers. That meant that unless they sold the precious grains at astronomical prices, the business was doomed to failure. At that point a philanthropic association by the name of Monsanto intervened. At an agricultural conference in India in 2000, the company announced that it would "give away certain patent rights to speed [the] use of a genetically modified rice that could save millions of malnourished children."[13] The development of the rice, said Hendrik Verfaillie, who would soon succeed Robert Shapiro, "clearly demonstrates that biotechnology can help not only countries in the West, but in the developing world as well."[14]

However, as soon as golden rice was grown in real conditions, it produced such a pathetic amount of beta-carotene that it did absolutely no good. "We never found out why," said Brammer, "but this story is a good illustration of the unknowns surrounding the process of genetic engineering. They represent a medium- and long-term risk for Monsanto; we have no guarantee that GMOs will not be tomorrow's Agent Orange."

I will not enumerate all the surprises held in store over the years by products derived from genetic tinkering, such as, for example, the discovery by a Belgian scientist of an "unknown DNA fragment" in Monsanto Roundup Ready soybeans.[15] I merely advise the reader to consult a European Commission Web site listing the scientific studies it sponsors on the safety of GMOs. One example is a research study with the title "The Mechanisms and Control of Genetic Recombination in Plants."[16] In their presentation of the project, the authors note: "A major problem with present day technology is the non-predictability of the integration of . . . transgenes," which "may induce unpredictable and undesirable mutations in the host genome." The researchers proposed to ascertain the facts of the situation, proof that GMOs have entered the food chain before this important question was resolved.

Another example is "Effects and Mechanisms of Bt Transgenes on Biodiversity of Non-Target Insects: Pollinators, Herbivores, and Their Natural Enemies."[17] I think the monarch butterfly would have appreciated it if this study had been conducted *before* Bt corn was put on the market. A final example is "Safety Evaluation of Horizontal Gene Transfer from Genetically Modified Organisms to the Microflora of the Food Chain and Human Gut."[18] The findings of this British study have since been published, and the least that can be said is that they are not reassuring. The researchers gave seven volunteers each a hamburger and a milk shake containing RR soy, then analyzed the bacteria in their intestines. In three cases out of seven, they "detected very low levels of the gene for resistance to the herbicide."[19] It would certainly be useful, in the name of the principle of precaution, if the experiment were repeated over a period of two years with a daily intake of Monsanto soy products (a normal diet in the United States).

Genetic Contamination Is a Major Risk Factor

When one dissects Monsanto's activity reports (contained in 10-K forms) since 1997, one is struck by the place taken up by litigation. First there are suits filed by victims of the company's chemical activities, such as the residents of Anniston or the Vietnam War veterans.

"If the veterans' second class action were successful, it could lead to bankruptcy for Monsanto," Marc Brammer told me when I met him in the summer of 2006. "Not forgetting PCBs, bovine growth hormone, and Roundup, which may lead to more suits. In addition to the risks incurred by its past and present chemical activities, there are those associated with genetic contamination, which is an inexhaustible source of potential litigation. So far the StarLink disaster has cost Aventis $1 billion. But contamination is continuing, and so it is impossible to estimate the final cost to the company."

The reader may recall the uproar provoked in 2006 by the discovery of traces of unauthorized GMOs in American rice.[20] Produced by Bayer CropScience, one of Monsanto's competitors, the GM rice had never been approved for consumption or planting; the contamination, which came from field tests conducted on a Louisiana farm between 1998 and 2001, affected thirty countries, led to a collapse in American rice exports, and resulted in "losses of up to $253 million from food-product recalls in Europe."[21]

"We are involved in various intellectual property, biotechnology, tort, contract, antitrust, employee benefit, environmental, and other litigation, claims, and legal proceedings and government investigations."[22] This is according to Monsanto's 2005 10-K form, under the heading "Litigation and Other Contingencies." Under the heading of "Legal Proceedings," the firm enumerates, in a catalogue worthy of Prévert or perhaps Kafka, all the lawsuits in which it is a party, either as plaintiff or defendant.[23] Some proceedings pit it against its competitors, the Swiss Syngenta, the German Bayer, or the American Dow Chemical Company, over "who is the first to have discovered one or another gene or active principle." Similarly, the University of California filed a complaint against Monsanto for violation of a patent covering bovine growth hormone. One also finds that Syngenta has filed an antitrust claim asserting that Monsanto holds a monopoly on glyphosate-tolerant corn seeds. Reuters has wondered: "Monsanto Co.'s domination of the biotech crop market is indisputable, but is it illegal?"[24]

Says Marc Brammer: "The danger hanging over Monsanto is the same that once threatened Microsoft. It's not impossible that the company will one day be found guilty of violating American antitrust and antiracketeering laws. If that were to happen, it would be very costly." In 1999, a class action had been brought by farmers in federal court in St. Louis, claiming that the company had conspired with Pioneer Hi-Bred to fix the prices of seeds at a very high level. But the claim was dismissed in 2003 by Judge Rodney Sippel, the same judge who was so severe against farmers accused of violating Monsanto patents.[25]

One year later, the *New York Times* published a very detailed investigation in which, after interviewing dozens of executives of seed companies, the paper confirmed the suspicions of conspiracy hanging over the world leader in GMOs. Among other things, Monsanto had asked Mycogen, a California seed company, "not to compete with Monsanto and its partners on the price of biotech seeds in exchange for access to some of Monsanto's patented technologies, according to former executives" of the company (since acquired by Dow Chemical).[26] These allegations were later repeated in fourteen class action suits filed in fourteen different U.S. courts, as the company acknowledged in its 2005 10-K form.

"We are attacking the monopoly on seeds that Monsanto has acquired by what we consider to be illegal means," Adam Levitt, one of the plaintiffs' attorneys, who works for a well-known Chicago firm, told me in October 2006, "namely, the abuse of patent rights, such as the prohibition on farmers keeping their seeds or the requirement to buy only Roundup and not a generic glyphosate; also the obligation imposed on licensed dealers to sell a high percentage of Monsanto products. We also accuse the company of having stifled competition with unfair trade practices and having conspired to fix seed prices at an exorbitant level. All of that seems to us to amount to a violation of U.S. laws."

"Do you think you'll win?"

The question made Levitt smile, and he reminded me that he was paid on contingency. He concluded with obvious pleasure: "The fact that Monsanto has hired the largest law firms in the country for its defense makes us think that the company takes the matter seriously."

I will add that for us too, the citizens of the Earth, the matter is serious. After tracking the company for four years, I think I am in a position to state that we can no longer say we didn't know, and that it would be irresponsible to allow the food of humanity to fall into Monsanto's hands. For if there is one thing I'm certain I do not want, for myself and even less for my three daughters and my future grandchildren, it is the world according to Monsanto.

Notes

Introduction: The Monsanto Question

1. The program was broadcast on November 15, 2005.

2. Available on DVD in the Alerte Verte collection (www.alerte-verte.com), this film was awarded the grand prize at the Festival International du Reportage d'Actualité et du Documentaire de Société (FIGRA–Le Touquet), the Buffon Prize at the International Scientific Film Festival of Paris, and the best reportage prize, the grand prize, and the Ushuaïa TV prize at the International Ecological Film Festival in Bourges.

3. *Trans. note:* The U.S. Patent Office finally rejected the patent in April 2008.

4. This report was broadcast on Arte on October 18, 2005. It is available on DVD in the Alerte Verte collection.

5. In the words of the International Service for the Acquisition of Agri-Biotech Applications (ISAAA), a pro-GMO organization that provides these figures (www.isaaa.org).

6. Monsanto, *Pledge Report*, 2005 (www.monsanto.com/who_we_are/our_pledge/recent_reports.asp), 12.

7. Ibid., 3.

8. Ibid., 30.

9. Ibid., 9.

10. Ibid., inside front cover.

1. PCBs: White-Collar Crime

1. See Dennis Love, *My City Was Gone: One American Town's Toxic Secret, Its Angry Band of Locals, and a $700 Million Day in Court* (New York: William Morrow, 2006).

2. "Technical Report Evaluation of Monsanto's Polychlorinated Biphenyl (PCB). Process for PCB Losses at the Anniston Plant," United States Environmental Protection Agency, March 2005, www.epa.gov/Region4/waste/sf/annistonsf/10302197.PDF.

3. www.chemicalindustryarchives.org/dirty-secrets/annistonindepth/toxicity.asp.

4. Soren Jensen, "Report of a New Chemical Hazard," *New Scientist* 32 (1966): 612.

5. The story was told by the resident during a trial hearing. Trial transcript, *Owens v. Monsanto*, CCV-96-J-440-E, N.D. Alabama, April 5, 2001, 551.

6. *San Francisco Chronicle*, September 24, 1969.

7. *Le Dauphiné libéré*, Isère Nord edition, August 17, 2007.

8. Directive 96/59/CE. See Marc Laimé, "Le Rhône pollué par les PCB: un Tchernobyl français?" http://blog.mondediplo.net/-Carnets-d-eau-, August 14, 2007.

9. *Industrie-Déchets*, February 2007.

10. U.S. Public Health Service and U.S. Environmental Protection Agency, "Public Health Implications of Exposure to Polychlorinated Biphenyls (PCBs)," www.epa.gov/waterscience/fish/files/pcb99.pdf.

11. The two studies are presented in Ruth Stringer and Paul Johnston, *Chlorine and the Environment: An Overview of the Chlorine Industry* (Dordrecht: Kluwer Academic Publishers, 2001).

12. "Whales in Sound Imperiled," *Anchorage Daily News*, July 22, 2001.

13. *Chemical and Engineering News*, January 14, 2002, http://pubs.acs.org/cen/topstory/8002/8002notw1.html.

14. I recommend reading this very thorough article, which can be found at www.washingtonpost.com/ac2/wp-dyn?pagename=articleandcontentId=A46648-2001Dec31.

15. *Anniston Star*, February 23, 2002.

16. *Anniston Star*, August 8, 2003; *Wall Street Journal*, August 21, 2003.

17. " US: General Electric Workers Sue Monsanto Over PCBs," Reuters, January 4, 2006.

28. *The Ecologist*, March 22, 2007; *Sunday Times*, June 3, 1973.

2. Dioxin: A Polluter Working with the Pentagon

1. Renate D. Kimbrough, "Epidemiology and Pathology of a Tetrachlorobenzodioxin Poisoning Episode," *Archives of Environmental Health*, March–April 1977; *The Lancet*, April 2, 1977, 748.

2. *New York Times*, August 28, 1974.

3. Coleman D. Carter, "Tetrachlorobenzodioxin: An Accidental Poisoning Episode in Horse Arenas," *Science*, May 16, 1975.

4. See Robert Reinhold, "Missouri Now Fears 100 Sites Could Be Tainted by Dioxin," *New York Times*, January 18, 1983.

5. *New York Times*, August 13, 1983, November 18, 1983, November 29, 1983, and December 1, 1983.

6. James Troyer, "In the Beginning: The Multiple Discovery of the First Hormone Herbicides," *Weed Science* 49 (2001), 290–97.

7. Raymond R. Suskind et al., "Progress Report. Patients from Monsanto Chemical Company, Nitro, West Virginia," unpublished report, July 20, 1950.

8. J. Kimmig and Karl-Heinz Schulz, "Berufliche Akne (Sog. Chlorakne) durch Chlorierte Aromatische Zyklische Äther [Occupational Acne (So-Called Chloracne) Due to Chlorinated Aromatic Cyclic Ether]," *Dermatologia* 115 (1957): 5404–46.

9. Peter Downs, "Cover-up: Story of Dioxin Seems Intentionally Murky," *St. Louis Journalism Review*, June 1, 1998. See also Robert Allen, *The Dioxin War: Truth and Lies about a Perfect Poison* (London: Pluto Press, 2004).

10. "The Monsanto Files," *The Ecologist*, September/October 1998, http://web.archive.org/web/20000902182550/www.zpok.hu/mirror/ecologist/SeptOct. Required reading!

11. Brian Tokar, "Agribusiness, Biotechnology, and War," http://www.social-ecology.org/2002/09/agribusiness-biotechnology-and-war.

12. Richard H. Kohn, "Foreword," in William Buckingham Jr., *Operation Ranch Hand: The Air Force and Herbicides in Southeast Asia, 1961–1971* (Washington, DC: Office of Air Force History, 1982), iv.

13. Buckingham, *Operation Ranch Hand*, 1, 4, 3.

14. Ibid., 14–15.

15. Ibid., 33.

16. The most reliable estimates are those published by James Mager Stellman, "The Extent and Patterns of Usage of Agent Orange and Other Herbicides in Vietnam," *Nature*, April 17, 2003.

17. *Le Monde*, April 26, 2005.

18. GAO, "U.S. Ground Troops in Vietnam Were in Areas Sprayed with Herbicide Orange," FPCD 80-23, November 16, 1979, 1.

19. Written September 9, 1988, this letter was read by Tom Daschle to a Senate committee on November 21, 1989.

20. Diane Courtney et al., "Teratogenic Evaluation of 2,4,5-T," *Science*, May 15, 1970.

21. In 1978, the EPA ordered a halt to the spraying of 2,4,5-T in national forests after recording a "statistically significant increase in miscarriages" in women living near the forests sprayed. *Bioscience* 454 (August 1979).

22. Joe Thornton, *Science for Sale: Critique of Monsanto Studies on Worker Health Effects Due to Exposure to 2,3,7,8-Tetrachlorodibenzo-P-Dioxin (TCDD)*, Greenpeace, November 29, 1990. The study was presented at the National Press Club in Washington (*Washington Post*, November 30, 1990).

23. Plaintiffs' brief, October 3, 1989; see also Allen, *The Dioxin War*.

24. Judith A. Zack and Raymond R. Suskind, "The Mortality Experience of Work-

ers Exposed to Tetrachlorodibenzodioxin in a Trichlorophenol Process Accident," *Journal of Occupational Medicine* 22, no. 1 (1980): 11–14; Judith A. Zack and William R. Gaffey, "A Mortality Study of Workers Employed at the Monsanto Company Plant in Nitro, West Virginia," *Environmental Science Research* 26, no. 6 (1983): 576–91; Raymond R. Suskind and Vicki S. Hertzberg, "Human Health Effects of 2,4,5-T and Its Toxic Contaminants," *Journal of the American Medical Association* 251, 18 (1984): 2372–80.

25. Peter Schuck, *Agent Orange on Trial: Mass Toxic Disasters in the Courts* (Cambridge, MA: Harvard University Press, 1987), 86–87, 155–64. Monsanto had produced 29.5 percent of the Agent Orange used in Vietnam, compared to 28.6 percent for Dow Chemical, but some of its supplies contained forty-seven times as much dioxin as those of Dow.

3. Dioxin: Manipulation and Corruption

1. *Wall Street Journal*, January 1987.

2. 492 N.E. 2d 1327, 1340 (Ill. 1986), Clark C.J., concurring (opinion on appeal concerning a procedural motion).

3. *Kemner v. Monsanto*, plaintiff's brief, October 3, 1989.

4. Marilyn Fingerhut, "Cancer Mortality in Workers Exposed to 2,3,7,8-Tetrachlorodibenzo-P-Dioxin," *New England Journal of Medicine* 324, no. 4 (January 24, 1991): 212–18.

5. Anthony B. Miller et al., *Environmental Epidemiology*, vol. 1: *Public Health and Hazardous Waste* (Washington, DC: National Academy Press, 1991), 207.

6. Joe Thornton, *Science for Sale*, Greenpeace, November 29, 1990.

7. Raymond R. Suskind, testimony and cross examination, *Boggess v. Monsanto*, Civil No. 81-2098-265 (S.D. W. Va., 1986).

8. Alastair Hay and Ellen Silberberg, "Assessing the Risk of Dioxin Exposure," *Nature* 315 (May 9, 1985), 102–3.

9. Judith A. Zack and William R. Gaffey, "A Mortality Study of Workers Employed at the Monsanto Company Plant in Nitro, West Virginia," *Environmental Science Research* 26, no. 6 (1983): 576–91.

10. Alastair Hay and Ellen Silberberg, "Assessing the Risk of Dioxin Exposure."

11. Report of proceedings: testimony of Dr. George Roush, *Kemner v. Monsanto*, Civil No. 80-L-970, Circuit Court, St. Clair County, Ill., July 8–9, 1985.

12. *Kemner v. Monsanto*, plaintiff's brief, October 3, 1989.

13. *Harrowsmith*, March–April 1990.

14. EPA, *Drinking Water Criteria Document for 2,3,7,8-Tetrachlorodibenzo-P-Dioxin*, Office of Research and Development, Cincinnati, ECAO-CIN-405 (April 1988).

15. Cate Jenkins, "Memo to Raymond Loehr: Newly Revealed Fraud by Mon-

santo in an Epidemiological Study Used by the EPA to Assess Human Health Effects from Dioxins," February 23, 1990.

16. Dick Carozza, "Sentinel at the EPA: An Interview with William Sanjour," *Fraud Magazine*, September–October 2007, http://pwp.lincs.net/sanjour/Fraud%20%Magazine%209-o7htm.

17. The court decision is online at www.whistleblowers.org/sanjourcase.htm.

18. William Sanjour, *The Monsanto Investigation*, July 20, 1994, pwp.lincs.net/sanjour/monsanto.htm.

19. "Key Dioxin Study a Fraud, EPA Says," *Charleston Gazette*, March 23, 1990.

20. Case opening, EPA no. 90-07-06-101 (10Q), August 20, 1990; Cate Jenkins, "Cover-up of Dioxin Contamination in Products, Falsification of Dioxin Health Studies," November 15, 1990, EPA, www.mindfully.org/Pesticide/Monsanto-Coverup-Dioxin-USEPA15nov90.htm.

21. *Cate Jenkins v. EPA*, case no. 92-CAA-6 before the Department of Labor Office of Administrative Law Judges, complainant's post-hearing brief, November 23, 1992.

22. U.S. Department of Labor, Washington, May 18, 1994 (case no. 92-CAA-6).

23. *Jenkins v. EPA*, transcript, September 29, 1992.

24. *Washington Post*, May 17, 1990.

25. Elmo R. Zumwalt Jr., "Report to the Secretary of the Department of Veterans Affairs on the Association Between Adverse Health Effects and Exposure to Agent Orange," May 5, 1990, www.gulfwarvets.com/ao.html.

26. "A Cover-up on Agent Orange?" *Time*, July 23, 1990.

27. Thomas Daschle, "Agent Orange Hearing," *Congressional Record*, S. 2550, November 21, 1989.

28. Alfred M. Thiess, R. Frentzel-Beyme, and R. Link, "Mortality Study of Persons Exposed to Dioxin in a Trichlorophenol-Process Accident that Occurred in the BASF AG on November 17, 1953," *American Journal of Industrial Medicine* 3, no. 2 (1982): 179–89.

29. Stephanie Wachinski, "New Analysis Links Dioxin to Cancer," *New Scientist*, October 28, 1989. The fraud was also revealed by Friedmann Rohleder at a conference on dioxin held in Toronto September 17–22, 1989.

30. R.C. Brownson, J.S. Reif, J.C. Chang, and J.R. Davis, "Cancer Risks Among Missouri Farmers," *Cancer* 64, no. 11 (December 1, 1989): 2381–86.

31. Aaron Blair, "Herbicides and Non-Hodgkin's Lymphoma: New Evidence from a Study of Saskatchewan Farmers," *Journal of the National Cancer Institute* 82 (1990): 544–45.

32. Pier Alberto Bertazzi et al., "Cancer Incidence in a Population Accidentally Exposed to 2,3,7, 8-Tetrachlorodibenzo-PARA-Dioxin," *Epidemiology* 4 (September 1993): 398–406.

33. Lennart Hardell and A. Sandstrom, "Case-Control Study: Soft Tissue Sarco-

mas and Exposure to Phenoxyacetic Acids or Chlorophenols," *British Journal of Cancer* 39 (1979): 711–17; Mikael Eriksson et al., "Soft Tissue Sarcoma and Exposure to Chemical Substances: A Case Referent Study," *British Journal of Industrial Medicine* 38 (1981): 27–33; Lennart Hardell et al., "Malignant Lymphoma and Exposure to Chemicals, Especially Organic Solvents, Chlorophenols, and Phenoxy Acids," *British Journal of Cancer* 43 (1981): 169–76; Lennart Hardell and Mikael Eriksson, "The Association between Soft Tissue Sarcomas and Exposure to Phenoxyacetic Acids: A New Case Referent Study," *Cancer* 62 (1988): 652–56.

34. Royal Commission on the Use and Effects of Chemical Agents on Australian Personnel in Vietnam, *Final Report*, 9 vols. (Canberra: Australian Government Publishing Service, 1985).

35. "Agent Orange: The New Controversy: Brian Martin Looks at the Royal Commission That Acquitted Agent Orange," *Australian Society* 5, no. 11 (November 1986): 25–26.

36. Monsanto Australia Limited, "Axelson and Hardell: The Odd Men Out," Submission to the Royal Commission on the Use and Effect of Chemical Agents on Australian Personnel in Vietnam, Exhibit 1881, 1985.

37. Quoted in Lennart Hardell, Mikael Eriksson, and Olav Axelson, "On the Misinterpretation of Epidemiological Evidence, Relating to Dioxin-Containing Phenoxyacetic Acids, Chlorophenols, and Cancer Effects," *New Solutions*, spring 1994.

38. Richard Doll and Richard Peto, "The Causes of Cancer: Quantitative Estimates of Avoidable Risks of Cancer in the United States Today," *Journal of the National Cancer Institute* 66, no. 6 (June 1981): 1191–308.

39. "Renowned Cancer Scientist Was Paid by Chemical Firm for 20 Years," *The Guardian*, December 8, 2006.

40. *American Journal of Industrial Medicine*, November 3, 2006.

41. Arnold Schechter et al., "Food as a Source of Dioxin Exposure in the Residents of Bien Hoa City, Vietnam," *Journal of Occupational and Environmental Medicine* 45, no. 8 (August 2003): 781–88.

42. Le Cao Dai et al., "A Comparison of Infant Mortality Rates between Two Vietnamese Villages Sprayed by Defoliants in Wartime and One Unsprayed Village," *Chemosphere* 20 (August 1990): 1005–12.

43. *New Scientist*, March 20, 2005.

44. *New York Times*, March 10, 2005.

45. *Corpwatch*, November 4, 2004.

4. Roundup: A Massive Brainwashing Operation

1. www.roundup-jardin.com/page.php?rup=service_roundup_roundup.

2. Institute for Agriculture and Trade Policy, Minneapolis, *Sustainable Agriculture Week*, April 11, 1994.

3. *Problems Plague the EPA Pesticide Registration Activities*, U.S. Congress, House of Representatives, House Report 98-1147, 1984.

4. EPA, Office of Pesticides and Toxic Substances, "Summary of the IBT Review Program," Washington, DC, July 1983.

5. EPA, "Data Validation: Memo from K. Locke, Toxicology Branch, to R. Taylor, Registration Branch," Washington, DC, August 9, 1978.

6. EPA, Communications and Public Affairs, "Note to Correspondents," Washington, DC, March 1, 1991.

7. *New York Times*, March 2, 1991.

8. Ibid.

9. "Testing Fraud: IBT and Craven Laboratories," June 2005, www.monsanto.com/pdf/products/roundup_ibt_craven_bkg.pdf. *Trans. note:* This site is no longer accessible.

10. Caroline Cox, "Glyphosate Factsheet," *Journal of Pesticide Reform* 108, no. 3 (1998), www.mindfully.org/Pesticide/Roundup-Glyphosate-Factsheet-Cox.htm. This very thorough article provides an excellent summary of all the questions raised by Roundup.

11. www.mindfully.org/Pesticide/Monsanto-v-AGNYnov96.htm.

12. Attorney General of the State of New York, Consumer Frauds and Protection Bureau, Environmental Protection Bureau, *In the Matter of Monsanto Company, Respondent. Assurance of Discontinuance Pursuant to Executive Law § 63 (15)*, New York, April 1998.

13. Isabelle Tron, Odile Picquet, and Sandra Cohuet, *Effets chroniques des pesticides sur la santé: État des connaissances*, Observatoire Régional de Santé de Bretagne, January 2001.

14. Sheldon Rampton and John Stauber, *Trust Us, We're Experts! How Industry Manipulates Science and Gambles with Your Future* (New York: Tarcher/Putnam, 2002).

15. Fabrice Nicolino and François Veillerette, *Pesticides: Révélations sur un Scandale Français* (Paris: Fayard, 2007).

16. Julie Marc, "Effets Toxiques d'Herbicides à Base de Glyphosate sur la Régulation du Cycle Cellulaire et le Développement Précoce en Utilisant l'Embryon d'Oursin," Université de Biologie de Rennes, September 10, 2004.

17. Helen H. McDuffie et al., "Non-Hodgkin's Lymphoma and Specific Pesticide Exposures in Men: Cross-Canada Study of Pesticides and Health," *Cancer Epidemiology Biomarkers and Prevention* 10 (November 2001): 1155–63.

18. Lennart Hardell, Michael Eriksson, and Marie Nordström, "Exposure to Pesticides as Risk Factor for Non-Hodgkin's Lymphoma and Hairy Cell Leukemia: Pooled Analysis of Two Swedish Case-Control Studies," *Leukemia and Lymphoma* 43 (2002): 1043–9.

19. Anneclaire J. De Roos et al., "Integrative Assessment of Multiple Pesticides

as Risk Factors for Non-Hodgkin's Lymphoma among Men," *Occupational and Environmental Medicine* 60, no. 9 (2003).

20. Anneclaire J. De Roos et al., "Cancer Incidence among Glyphosate-Exposed Pesticide Applicators in the Agricultural Health Study," *Environmental Health Perspectives* 113, no. 1 (January 2005): 49–54.

21. Marc, "Effets Toxiques d'Herbicides à base de glyphosate."

22. A report entitled "Étude Phyto Air," financed by the Nord-Pas-de-Calais region and conducted by the Institut Pasteur of Lille, is a good source of information on the problems posed by the additives contained in herbicides. www.pasteur-lille.fr/images_accueil/Rapport%20Phytoair.pdf.

23. Institute for Science in Society, press release, March 7, 2005.

24. Julie Marc, Odile Mulner-Lorillon, and Robert Bellé, "Glyphosate-Based Pesticides Affect Cell Cycle Regulation," *Biology of the Cell* 96, no. 3 (April 2004): 245–49.

25. Tye E. Arbuckle, Zhiqiu Lin, and Leslie S. Mery, "An Exploratory Analysis of the Effect of Pesticide Exposure on the Risk of Spontaneous Abortion in an Ontario Farm Population," *Environmental Health Perspectives* 109, no. 8 (August 2001): 851–57.

26. John F. Acquavella et al., "Glyphosate Biomonitoring for Farmers and Their Families: Results from the Farm Family Exposure Study," *Environmental Health Perspectives* 112, no. 3 (March 2004): 321–26.

27. Lance P. Walsh et al., "Roundup Inhibits Steroidogenesis by Disrupting Steroidogenic Acute Regulatory (StAR) Protein Expression," *Environmental Health Perspectives* 108, no. 8 (August 2000): 769–76.

28. Eliane Dallegrave et al., "The Teratogenic Potential of the Herbicide glyphosate-Roundup® in Wistar Rats," *Toxicology Letters* 142, no. 1 (April 2003): 45–52.

29. Sophie Richard et al., "Differential Effects of Glyphosate and Roundup on Human Placental Cells and Aromatase," *Environmental Health Perspectives* 113, no. 6 (June 2005): 716–20; Nora Benachour et al., "Time- and Dose-Dependent Effects of Roundup on Human Embryonic and Placental Cells," *Archives of Environmental Contamination and Toxicology* 53, no. 1 (July 2007): 126–33.

30. Christian Ménard, "Rapport Fait au Nom de la Mission d'Information sur les Enjeux des Essais et de l'Utilisation des Organismes Génétiquement Modifiés," Assemblée Nationale, April 13, 2005, www.assemblee-nationale.fr/12/rap-info/12254-tl.asp. *Trans. note:* This site appears no longer accessible.

31. Marc, "Effets Toxiques d'Herbicides à base de glyphosate."

32. Rick A. Relyea et al., "The Impact of Insecticides and Herbicides on the Biodiversity and Productivity of Aquatic Communities," *Ecological Applications* 15, no. 2 (April 2005): 618–27.

33. University of Pittsburgh, press release, April 1, 2005.

34. Hsin-Ling Lee et al., "Clinical Presentations and Prognostic Factors of a

Glyphosate—Surfactant Herbicide Intoxication: A Review of 131 Cases," *Academic Emergency Medicine* 7, no. 8 (August 2000): 906–10.

35. *Pesticides News*, September 1996, 28–29.

36. Earthjustice Legal Defense Fund, "Spraying Toxic Herbicides on Rural Colombian and Ecuadorian Communities," January 15, 2002, www.mindfully.org/Pesticide/2002/Roundup-Human-Rights24jan02.htm.

5. The Bovine Growth Hormone Affair, Part One

1. *Los Angeles Times*, August 1, 1989. At the same time, Epstein wrote a scientific article: "Potential Public Health Hazards of Biosynthetic Milk Hormones," *International Journal of Health Services* 20, no. 1 (1990): 73–84.

2. Samuel Epstein also published another article: "Questions and Answers on Synthetic Bovine Growth Hormones," *International Journal of Health Services* 20, no. 4 (1990): 573–82.

3. The terms used by Congress were "knowing acts of non-disclosure" and "reckless acts." Samuel S. Epstein, Testimony on White Collar Crime, H.R. 4973, before the Subcommittee on Crime of the House Judiciary Committee, December 13, 1979.

4. "FDA Accused of Improper Ties in Review of Drug for Milk Cows," *New York Times*, January 12, 1990.

5. Judith C. Juskevich and C. Greg Guyer, "Bovine Growth Hormone: Human Food Safety Evaluation," *Science* 249 (August 24, 1990): 875–84.

6. Frederick Bever, "Canadian Agency Questions Approval of Cow Drug by US," Associated Press, October 6, 1998.

7. *Le Monde*, August 30, 1990.

8. Depending on the source, the level of IGF-1 present in milk from injected cows can be two to ten times higher than that found in natural milk. In the request for approval that Monsanto submitted to the British authorities, the company speaks of a level "as much as five times higher." T. Ben Mepham et al., "Safety of Milk from Cows Treated with Bovine Somatotropin," *The Lancet* 344 (November 19, 1994): 1445–6.

9. C. Xian, "Degradation of IGF-1 in the Adult Rat Gastrointestinal Tract Is Limited by a Specific Antiserum or the Dietary Protein Casein," *Journal of Endocrinology* 146, no. 2 (August 1, 1998).

10. June M. Chan et al., "Plasma Insulin-like Growth Factor-1 [IGF-1] and Prostate Cancer Risk: A Prospective Study," *Science* 279 (January 23, 1998): 563–66.

11. Susan E. Hankinson et al., "Circulating Concentrations of Insulin-like Growth Factor-1 and Risk of Breast Cancer," *The Lancet* 351 (1998): 1393–6.

12. *The Milkweed*, August 2006. This article surveys all the available scientific literature on the links between IGF-1 and breast cancer.

13. *Journal of Reproductive Medicine*, May 2006; *The Milkweed*, June 2006; *New York Times*, May 30, 2006. The number of twins in the United States has increased from 1.89 per 100 births in 1977 to 3.1 in 2002 (twice that in the United Kingdom).

14. "NIH Technology Assessment Conference Statement on Bovine Somatotropin," *Journal of the American Medical Association* 265, no. 11 (March 20, 1991): 1423–5.

15. Council on Scientific Affairs, American Medical Association, "Biotechnology and the American Agricultural Industry," *Journal of the American Medical Association* 265, no. 11 (March 20, 1991): 1429–36.

16. Eliot Marshal, "Scientists Endorse Ban on Antibiotics in Feeds," *Science* 222 (November 11, 1983): 601.

17. Barry R. Bloom and Christopher J. L. Murray, "Tuberculosis: Commentary on a Reemergent Killer," *Science* 257 (August 21, 1992): 1055–64.

18. Sharon Begley, "The End of Antibiotics," *Newsweek*, March 28, 1994, 47–52.

19. The GAO wrote a special report on the question of antibiotic residues in milk. It noted that there were few available tests to measure these residues—the FDA had only four, one of which was for penicillin—although thirty drugs were authorized for dairy herds, and reportedly seventy-two were used illegally. GAO, *Food Safety and Quality: FDA Strategy Needed to Address Animal Drug Residues in Milk*, GAO/PMED-92-96, 1992.

20. Erik Millstone, Eric Brunner, and Ian White, "Plagiarism or Protecting Public Health?" *Nature* 371 (October 20, 1994): 647–48.

21. Jeremy Rifkin, *The Biotech Century* (New York: Tarcher/Putnam, 1999).

22. Samuel Epstein had already expressed similar anger in the *Los Angeles Times*, March 20, 1994.

6. The Bovine Growth Hormone Affair, Part Two

1. 59 *Federal Register* 28 (February 10, 1994), 6279.

2. www.cfsan.fda.gov/~lrd/fr940210.html.

3. This thirty-two-page document was signed by Richard A. Merrill, Jess H. Stribling, and Frederick H. Degnan.

4. *Capital Times*, February 19–20, 1994.

5. *Washington Post*, May 18, 1994.

6. *New York Times*, July 12, 2003.

7. "Oakhurst to Alter Its Label," *Portland Press Herald*, December 25, 2003.

8. Associated Press, February 18, 2005.

9. Mark Kastel, "Down on the Farm: The Real BGH Story: Animal Health Problems, Financial Troubles," www.mindfully.org/GE/Down-On-The-Farm-BGH1995.htm.

10. *Metroland* (Albany), August 11, 1994.

11. *St. Louis Post-Dispatch*, March 15, 1995.

12. Their story is the subject of a chapter in Kristina Borjesson, ed., *Into the Buzzsaw: Leading Journalists Expose the Myth of a Free Press* (New York: Prometheus Books, 2002).

13. They can be consulted on www.foxbghsuit.com.

7. The Invention of GMOs

1. Edward L. Tatum, "A Case History in Biological Research," Nobel Lecture, December 11, 1958, http:nobelprize.org/nobel_prizes/medicine/laureates/1958/tatum-lecture.html.

2. Arnaud Apoteker, *Du poisson dans les fraises* (Paris: La Découverte, 1999).

3. Quoted by Robert Shapiro, "The Welcome Tension of Technology: The Need for Dialogue about Agricultural Biotechnology," Center for the Study of American Business, CEO Series, 37, February 2000.

4. Quoted by Hervé Kempf, *La Guerre secrète des OGM* (Paris: Seuil, 2003), 23.

5. Ibid., 25.

6. Susan Wright, *Molecular Politics: Developing American and British Regulatory Policy for Genetic Engineering, 1972–1982* (Chicago: University of Chicago Press, 1994), 107.

7. Daniel Charles, *Lords of the Harvest: Biotech, Big Money, and the Future of Food* (Cambridge, MA: Perseus, 2002), 24.

8. Quoted by Kempf, *La Guerre secrète des OGM*, 57.

9. Charles, *Lords of the Harvest*, 38.

10. Ibid., 37.

11. Luca Comai et al., "Expression in Plants of a Mutant *aroA* Gene from *Salmonella typhimurium* Confers Tolerance to Glyphosate," *Nature* 317 (October 24, 1985): 741–44.

12. Charles, *Lords of the Harvest*, 67.

13. Stephanie Simon, "Biotech Soybeans Plant Seed of Risky Revolution," *Los Angeles Times*, July 1, 2001.

14. Ibid.

15. *CropChoice News*, November 16, 2003, http://www.organicconsumers.org/ge/72803_ge_soybeans.cfm.

16. Charles, *Lords of the Harvest*, 75.

17. Simon, "Biotech Soybeans Plant Seed of Risky Revolution."

18. Apoteker, *Du poisson dans les fraises*, 36–37.

19. Kurt Eichenwald, "Redesigning Nature: Hard Lessons Learned; Biotechnology Food: From the Lab to a Debacle," *New York Times*, January 25, 2001.

20. *Coordinated Framework for Regulation of Biotechnology*, Office of Science and Technology Policy, 51 FR 23302, June 26, 1986, http://usbiotechreg.nbii.gov/CoordinatedFrameworkForRegulationOfBiotechnology.

21. Eichenwald, "Redesigning Nature."

22. Ibid.

23. Charles, *Lords of the Harvest*, 28.

24. Eichenwald, "Redesigning Nature."

25. Quoted by the Center for Regulatory Effectiveness, www.thecre.com/omb papers/1999-0129-F.htm.

26. Food and Drug Administration, "Statement of Policy: Foods Derived from New Plant Varieties," 57 FR 22983 (May 29, 1992).

27. 57 FR 22985 (emphasis added).

28. Charles, *Lords of the Harvest*, 143.

29. FAO, *Genetically Modified Organisms, Consumers, Food Safety, and the Environment*, www.fao.org/docrep/003/x9602e/x9602e00.htm.

30. Jeffrey M. Smith, *Seeds of Deception: Exposing Industry and Government Lies about the Safety of the Genetically Engineered Foods You're Eating* (Fairfield, IA: Yes! Books, 2003), 107–27; Jeffrey M. Smith, *Genetic Roulette: The Documented Health Risks of Genetically Engineered Foods* (Fairfield, IA: Yes! Books, 2007), 60–61. See the Web site www.seedsofdeception.com/Public/Home/index.cfm.

31. House of Representatives, *FDA's Regulation of the Dietary Supplement L-Tryptophan*, Human Resources and Intergovernmental Subcommittee of the Committee on Government Operations, U.S. House of Representatives, Washington, DC, 1991.

32. Arthur N. Mayenno and Gerald J. Gleich, "Eosinophilia-Myalgia Syndrome and Tryptophan Production: A Cautionary Tale," *Trends in Biotechnology* 12, no. 9 (September 1994): 346–52.

33. Quoted by Smith, *Genetic Roulette*, 61.

34. See, for example, "Information Paper on L-Tryptophan and 5-Hydroxy-L-Tryptophan," U.S. Food and Drug Administration, Office of Nutritional Products, Labeling, and Dietary Supplements, February 2001, www.cfsan.fda.gov/~dms /ds-tryp1.html.

35. Quoted in Steven Druker's statement to the FDA on November 30, 1999, "Why FDA Policy on Genetically Engineered Foods Violates Sound Science and US Law," Panel on Scientific Safety and Regulatory Issues, www.psrast.org/drukeratfda .htm.

36. Food and Drug Administration, "Statement of Policy: Foods Derived from New Plant Varieties," 57 FR 22991.

37. Smith, *Genetic Roulette*, 61.

8. Scientists Suppressed

1. www.biointegrity.org.

2. See Steven Druker's statement to the FDA on November 30, 1999, "Why FDA Policy on Genetically Engineered Foods Violates Sound Science and

US Law," Panel on Scientific Safety and Regulatory Issues, www.psrast.org/drukeratfda.htm.

3. *Alliance for Bio-Integrity v. Shalala*.

4. *New York Times*, October 4, 2000.

5. "Genetically Engineered Foods," *FDA Consumer*, January–February 1993, 14.

6. www.biointegrity.org/list.html.

7. Memorandum from the Division of Food Chemistry and Technology and Division of Contaminants Chemistry. Subject: "Points to Consider for Safety Evaluation of Genetically Modified Foods. Supplemental Information," November 1, 1991. www.safe-food.org/-issue/fda.html.

8. Samuel I. Shibko, Memorandum to Dr. James Maryanski, FDA Biotechnology Coordinator. Subject: "Revision of Toxicology Section of the Statement of Policy: Foods Derived from Genetically Modified Plants." Dated January 31, 1992. www.safe-food.org/-issue/fda.html.

9. Gerald B. Guest, DVM, Director of the Center for Veterinary Medicine, in a memorandum to Dr. James Maryanski, Biotechnology Coordinator. Subject: "Regulation of Transgenic Plants—FDA Draft Federal Register Notice on Food Biotechnology." Dated February 5, 1992. www.safe-food.org/-issue/fda.html.

10. Louis Pribyl, comments on "Biotechnology Draft Document, 2/27/92." Dated March 6, 1992. www.mindfully.org/GE/Louis-J-Pribyl-Comments-27feb92.htm.

11. Letter of James Maryanski to Dr. Bill Murray, Chairman of the Food Directorate, Canada, October 23, 1991, www.safe-food.org/-issue/fda.html.

12. Linda Kahl, Memorandum to James Maryanski, FDA Biotechnology Coordinator, January 8, 1992, www.biointegrity.org/FDAdocs/01/view1.html.

13. Statement of Policy: Foods Derived from New Plant Varieties, 57 FR 23000 (point 17d).

14. Jean Halloran and Michael Hansen, "Why We Need Labeling of Genetically Engineered Food," Consumers International, Consumer Policy Institute, April 1998; "Compilation and Analysis of Public Opinion Polls on Genetically Engineered Foods," Center for Food Safety, February 11, 1999.

15. *Time*, February 11, 1999.

16. "Citizen Petition before the United States Food and Drug Administration," March 21, 2000, www.fda.gov/ohrms/dockets/dailys/00mar00/032200/cp00001.pdf.

17. Douglas Gurian-Sherman, "Holes in the Biotech Safety Net: FDA Policy Does Not Assure the Safety of Genetically Engineered Foods," Center for Science in the Public Interest, Washington, DC, 2001.

18. "Flavr Savr Tomato: Pathology Branch's Evaluation of Rats with Stomach Lesions from Three Four-Week Oral (Gavage) Toxicity Studies," Memorandum from Dr. Fred Hines to Dr. Linda Kahl, June 16, 1993, www.mindfully.org/GE/FlavrSavr-Pathology-Review.htm.

19. www.biointegrity.org/FDAdocs/19/view1.html.

20. www.ilsi.org.

21. Sarah Boseley, "WHO 'Infiltrated by Food Industry,'" *The Guardian*, January 9, 2003.

22. "Biotechnologies and Food: Assuring the Safety of Foods Produced by Genetic Modification," *Regulatory Toxicology and Pharmacology* 12, no. 3 (1990), www.ilsi.org/AboutIlsi/IFBiC.

23. "Statement of Policy: Foods Derived from New Plant Varieties," 57 FR 23003.

24. Smith, *Seeds of Deception*; Smith, *Genetic Roulette*.

25. "Monsanto Employees and Government Regulatory Agencies Are the Same People!" *Green Block*, December 8, 2000, www.purefood.org/Monsanto/revolvedoor.cfm. See also *Agribusiness Examiner Newsletter*, June 16, 1999, and *Washington Post*, February 7, 2001.

26. "How Agribusiness Has Hijacked Regulatory Policy at the US Department of Agriculture," released at the Food and Agriculture Conference of the Organization for Competitive Markets, Omaha, Nebraska, July 23, 2004.

27. *St. Louis Post-Dispatch*, May 30, 1999.

28. Federal News Service, "Remarks of Secretary of Agriculture before the Council for Biotechnology Information," April 18, 2000.

29. Dan Glickman, "How Will Scientists, Farmers, and Consumers Learn to Love Biotechnology and What Happens If They Don't?" July 13, 1999, www.usda.gov/news/releases/1999/07/0285. Emphasis added.

30. www.ratical.org/co-globalize/MonsantoRpt.html.

31. Judith C. Juskevich and C. Greg Guyer, "Bovine Growth Hormone: Human Food Safety Evaluation," *Science* 249 (August 24, 1990): 875–84.

32. Erik Millstone, Eric Brunner, and Sue Mayer, "Beyond Substantial Equivalence," *Nature* 401 (October 7, 1999): 525–26.

33. Stephen Padgette et al., "The Composition of Glyphosate-Tolerant Soybean Seeds Is Equivalent to That of Conventional Soybeans," *Journal of Nutrition* 126, no. 3 (March 1996): 702–16.

34. Barbara Keeler and Marc Lappé, "Some Food for FDA Regulation," *Los Angeles Times*, January 7, 2001.

35. Marc Lappé et al., "Alterations in Clinically Important Phytoestrogens in Genetically Modified, Herbicide-Tolerant Soybeans," *Journal of Medicinal Food* 1, no. 4 (July 1, 1999).

36. www.soygrowers.com/newsroom/releases/documents/isobkgndr.htm.

37. Marc Lappé and Britt Bailey, *Against the Grain: Biotechnology and the Corporate Takeover of Your Food* (Monroe, ME: Common Courage Press, 1998).

38. *New York Times*, October 25, 1998.

39. Ian Pryme and Rolf Lembcke, "*In Vivo* Studies on Possible Health Consequences of Genetically Modified Food and Feed—with Particular Regard to Ingre-

dients Consisting of Genetically Modified Plant Materials," *Nutrition and Health* 17 (2003): 1–8.

40. Bruce Hammond et al., "The Feeding Value of Soybeans Fed to Rats, Chickens, Catfish, and Dairy Cattle Is Not Altered by Genetic Incorporation of Glyphosate Tolerance," *Journal of Nutrition* 126, no. 3 (March 1996): 717–27.

41. Manuela Malatesta et al., "Ultrastructural Analysis of Pancreatic Acinar Cells from Mice Fed on Genetically Modified Soybean," *Journal of Anatomy* 201, no. 5 (November 2002): 409–15; Manuela Malatesta et al., "Fine Structural Analyses of Pancreatic Acinar Cell Nuclei from Mice Fed on Genetically Modified Soybean," *European Journal of Histochemistry* 47, no. 4 (October–December 2003), 385–88. See also "Nouveaux soupçons sur les OGM," *Le Monde*, February 9, 2006.

9. Monsanto Weaves Its Web, 1995–1999

1. See François Dufour, " Les Savants Fous de l'Agroalimentaire," *Le Monde diplomatique*, July 1999. It is worth noting that for the 1996–97 growing year, France was able to supply 22 percent of its needs.

2. "Scientist's Potato Alert Was False, Laboratory Admits," *The Times* (London), July 13, 1998.

3. "Doctor's Monster Mistake," Scottish *Daily Record*, October 13, 1998.

4. *Daily Telegraph*, June 10, 1999.

5. "Le Transgénique, la Pomme de Terre, et le Soufflé Médiatique," *Le Monde*, August 15, 1998.

6. "Genetically Modified Organisms: Audit Report of Rowett Research on Lectins," press release, Rowett Institute, October 28, 1998.

7. *The Guardian*, February 12, 1999; "Le Rat et la Patate, Chronique d'un Scandale Britannique," *Le Monde*, February 17, 1999; "Peer Review Vindicates Scientist Let Go for 'Improper' Warning about Genetically Modified Food," *Natural Science Journal*, March 11, 1999.

8. *The Scotsman*, August 13, 1998.

9. "Testimony of Professor Philip James and Dr. Andrew Chesson," Examination of Witnesses, Question 247, March 8, 1999, www.parliament.the-stationery-office.co.uk/pa/cm199899/cmselect/cmsctech/286/9030817.htm.

10. "Loss of Innocence: Genetically Modified Food," *New Statesman*, February 26, 1999, 47.

11. "Furor Food: The Man with the Worst Job in Britain," *The Observer*, February 21 1999.

12. Quoted by Jeffrey Smith, *Seeds of Deception: Exposing Industry and Government Lies about the Safety of the Genetically Engineered Foods You're Eating* (Fairfield, IA: Yes! Books, 2003), 24.

13. "People Distrust Government on GM Foods," *Independent on Sunday*, May 23, 1999.

14. "Labour's Real Aim on GM Food," *Independent on Sunday*, May 23, 1999.

15. Memorandum Submitted by Dr. Stanley William Barclay Ewen, Department of Pathology, University of Aberdeen, February, 26, 1999, www.parliament.the-stationery-office.co.uk/pa/cm199899/cmselect/cmsctech/286/9030804.htm.

16. Laurie Flynn and Michael Sean Gillard, "Pro-GM Food Scientist 'Threatened Editor,'" *The Guardian*, November 1, 1999.

17. Stanley Ewen and Arpad Pusztai, "Effects of Diets Containing Genetically Modified Potatoes Expressing Galanthus Nivalis Lectin on Rat Small Intestines," *The Lancet* 354 (October 16, 1999): 1353–54.

18. Steve Connor, "Scientists Revolt at Publication of 'Flawed GM Study,'" *The Independent*, October 11, 1999.

19. Flynn and Gillard, "Pro-GM Food Scientist 'Threatened Editor.'"

20. Andrew Rowell, "The Sinister Sacking of the World's Leading GM Expert—and the Trail That Leads to Tony Blair and the White House," *Daily Mail*, July 7, 2003.

21. Monsanto 1997 annual report, quoted in *Washington Post*, November 1, 1999.

22. *New Yorker*, April 10, 2000.

23. *The Ecologist*, September–October 1998.

24. Hervé Kempf, *La Guerre secrète des OGM* (Paris: Seuil, 2003), 110.

25. *New Yorker*, April 10, 2000.

26. "Growth Through Global Sustainability: An Interview with Monsanto's CEO, Robert B. Shapiro," *Harvard Business Review*, January–February 1997.

27. Ibid.

28. "Interview, Robert Shapiro: Can We Trust the Maker of Agent Orange to Genetically Engineer Our Food?" *Business Ethics*, January–February 1997.

29. I recommend this fascinating article: Michael Specter, "The Pharmageddon Riddle," *New Yorker*, April 10, 2000.

30. "Interview, Robert Shapiro."

31. *The Ecologist* 28, no. 5 (September–October 1998).

32. Daniel Charles, *Lords of the Harvest* (Cambridge, MA: Perseus, 2002), 119.

33. Ibid., 120.

34. Ibid., 179.

35. Ibid., 151.

36. Ibid., 177.

37. Ibid., 200.

38. *Chemistry and Industry*, July 20, 1998.

39. *Daily Telegraph*, June 8, 1998.

40. Associated Press, June 7, 1998.

41. Reuters, August 11, 1998. In February 1999, the company was finally con-

demned for deceptive advertising (*The Guardian*, February 28, 1999). In all, thirty complaints had been filed.

42. *The Ecologist* 28, no. 5 (September–October 1998). This is essential reading. In addition, on July 1, 1999, the French weekly *Courrier international* published a translation, to which Monsanto replied in the July 29 edition. The reply contained the following: "With respect to Agent Orange, the writers for *The Ecologist* forgot to mention that detailed studies conducted for several years by the US Air Force and other bodies have shown that there are no major harmful health effects associated with this defoliant."

43. *The Guardian*, September 29, 1998.

44. Justin Gillis and Anne Swardson, "Crop Busters Take on Monsanto: Backlash against Biotech Foods Exacts a High Price," *Washington Post*, October 27, 1999. On October 26, Monsanto stock was quoted at $39.18 on the New York Stock Exchange, compared to $62.72 in August 1998.

45. Véronique Lorelle, "L'Arrogance de Monsanto a mis à mal son rêve de nourrir la planète," *Le Monde*, October 8, 1999.

46. Gillis and Swardson, "Crop Busters Take on Monsanto."

47. Michael D. Watkins and Ann Leamon, "Robert Shapiro and Monsanto," Harvard Business School case, rev. January 2, 2003.

48. Véronique Lorelle, "Le Patron de Monsanto, prophète des OGM, démissionne pour cause de mauvais résultats," *Le Monde*, December 20, 2002. In 2002, the company recorded a net loss of $1.7 billion.

10. The Iron Law of the Patenting of Life

1. For more details on the patenting of life, see my documentary *Les Pirates du vivant*, broadcast on Arte on November 15, 2005.

2. Monsanto, *Pledge Report*, 2005, 42 (emphasis added). The document can be found at http://www.monsanto.com/who_we_are/our_pledge/recent_reports.asp.

3. Monsanto, *Technology Use Guide*, art. 19 (emphasis added). Quoted by the report of the Center for Food Safety, *Monsanto vs. U.S. Farmers*, November 2005, 20, www.centerforfoodsafety.org/Monsantovsusfarmersreport.cfm.

4. Daniel Charles, *Lords of the Harvest* (Cambridge, MA: Perseus, 2002), 185.

5. Ibid., 155.

6. Ibid., 187.

7. Rick Weiss, "Seeds of Discord: Monsanto's Gene Police Raise Alarm on Farmers' Rights, Rural Tradition," *Washington Post*, February 3, 1999.

8. Center for Food Safety, *Monsanto vs. U.S. Farmers*.

9. *Chicago Tribune*, January 14, 2005.

10. "Lawsuits Filed against American Farmers by Monsanto," Administrative Office of the U.S. Courts, http://pacer.uspci.uscourts.gov.

11. Quoted by Charles, *Lords of the Harvest*, 187.

12. Ibid.

13. Weiss, "Seeds of Discord."

14. Associated Press, April 28, 2004.

15. Quoted by Center for Food Safety, *Monsanto vs. U.S. Farmers*, 44.

16. Interview with Robert Schubert, *CropChoice News*, April 6, 2001.

17. The story is reported in *Monsanto vs. U.S. Farmers*, 23ff. In addition, I talked to Scruggs's lawyer, James Robertson, who has film of the arrangement set up by Monsanto agents.

18. Associated Press, May 10, 2003.

19. "Monsanto 'Ruthless' in Suing Farmers, Food Group Says," *Chicago Tribune*, January 14, 2005. According to this article, of the ninety suits filed by Monsanto so far, forty-six had been heard in St. Louis.

20. *St. Louis Business Journal*, December 21, 2001.

21. http://record.wustl.edu/archive/2000/10-09-00/articles/law.html.

22. www.populist.com/02.18.mcmillen.html.

23. Hervé Kempf, "Percy Schmeiser, un rebelle contre les OGM," *Le Monde*, October 17, 2002.

24. The reader can consult Schmeiser's Web site, where he presents all the details of his case: www.percyschmeiser.com.

25. Hervé Kempf, "Le trouble d'une plaine du Saskatchewan," *Le Monde*, January 26, 2000.

26. *Toronto Star* and Saskatoon *Star Phoenix*, June 6, 2000.

27. Kempf, "Percy Schmeiser, un rebelle contre les OGM."

28. *Monsanto Canada Inc. v. Percy Schmeiser*, 2001 FCT 256 (March 29, 2001), 51–55; *Star Phoenix*, March 30, 2001.

29. *Washington Post*, March 30, 2001.

30. Quoted by Kempf, "Percy Schmeiser, un rebelle contre les OGM."

31. *Sacramento Bee*, May 22, 2004.

32. Ibid.

33. Monsanto Co., *Pledge Report*, 2001–2, 19, http:www.monsanto.com/who_we_are/our_pledge/recent_reports.asp.

34. *CBC News and Current Affairs*, June 21, 2001.

35. Canadian Bar Association, Annual Conference, August 2001.

36. Soil Association, *Seeds of Doubt: North American Farmers' Experience of GM Crops*, September 2002, www.soilassociation.org/seedsofdoubt. This is a fundamental document.

37. *New Scientist*, November 24, 2001. Since then the Quebec government site on GMOs has stated that "pollen can travel over a distance of at least two and a half miles," www.ogm.gouv.qc.ca/envi_canolagm.html. This has been confirmed by British and Australian studies.

38. "GM Volunteer Canola Causes Havoc," *Western Producer*, September 6, 2001.

39. *The Guardian*, October 8, 2003.

40. Soil Association, *Seeds of Doubt*, 47.

41. "Firms Move to Avoid Risk of Contamination," *The Times* (London), May 29, 2000.

42. Kempf, "Le trouble d'une plaine du Saskatchewan."

43. www.patentstorm.us/patents/6239072/claims.html.

44. Soil Association, *Seeds of Doubt*, 24; see also "Monsanto Sees Opportunity in Glyphosate Resistant Volunteer Weeds," *CropChoice News*, August 3, 2001.

45. *Science* and *The Independent*, October 10, 2003.

46. "Introducing Roundup Ready Soybeans: The Seeds of Revolution," undated document in author's possession.

47. Monsanto, *Pledge Report*, 2005, 18.

48. Charles Benbrook, "Genetic Engineered Crops and Pesticide Use in the United States: The First Nine Years," October 2004, www.biotech-info.net/Full_version_first_nine.pdf.

49. *Ag BioTech InfoNet Technical Paper* no. 4, May 3, 2001.

50. Ibid. A 1998 Monsanto document claimed that "herbicide use was on average lower in Roundup Ready soybean fields than in other fields" ("The Roundup Ready Soybean System: Sustainability and Herbicide Use," Monsanto, April 1998).

51. According to the *Los Angeles Times* of July 1, 2001, Roundup was used on 20 percent of American crops in 1995 and on 62 percent four years later.

52. Benbrook, "Genetically Engineered Crops and Pesticide Use in the United States," 7.

53. *Indianapolis Star*, February 20, 2001.

54. www.mindfully.org/GE/GE4/Glyphosate-Resistant-SyngentaDec02.htm.

55. Charles Benbrook, "Pew Initiative on Food and Biotechnology," February 4, 2002, http://pewagbiotech.org/events/0204/benbrook.php3.

56. "Introducing Roundup Ready Soybeans: The Seeds of Revolution."

57. Roger Elmore et al., "Glyphosate-Resistant Soybean Cultivar Yields Compared with Sister Lines," *Agronomy Journal* 93 (2001): 408–12.

58. Charles Benbrook, "Evidence of the Magnitude and Consequences of the Roundup Ready Soybean Yield Drag from University-Based Varietal Trials in 1998," *Ag BioTech InfoNet Technical Paper* no. 1, July 13, 1999, www.biotech-info.net/RR_yield_drag_98.pdf.

59. C. Andy King, Larry C. Purcell, and Earl D. Vories, "Plant Growth and Nitrogenase Activity of Glyphosate-Tolerant Soybeans in Response to Foliar Glyphosate Application," *Agronomy Journal* 93, no. 1 (January 2001): 179–86.

60. Benbrook, "Pew Initiative on Food and Biotechnology."

61. Andy Coghlan, "Splitting Headache: Monsanto's Modified Soybeans Are Cracking Up in the Heat," *New Scientist*, November 20, 1999.

62. Michael Duffy, "Who Benefits from Biotechnology?" Considered a standard reference, this paper was presented to the meeting of the American Seed Trade As-

sociation in Chicago, December 5–7, 2001. www.econ.iastate.edu/faculty/duffy/Pages/biotechpaper.pdf.

63. According to a survey conducted by Eurobarometer in 1997, the great majority of European citizens were in favor of labeling GMOs: Austria 73 percent, Belgium 74 percent, Denmark 85 percent, Finland 82 percent, France 78 percent, Germany 72 percent, Greece 81 percent, Ireland 61 percent, Italy 61 percent, Spain 69 percent, United Kingdom 82 percent. "European Opinions on Modern Biotechnology," European Commission Directorate General XII, no. 46.1, 1997.

64. *Washington Post*, November 12, 1999.

65. "US Agriculture Loses Huge Markets Thanks to GMOs," Reuters, March 3, 1999.

66. Reuters, September 17, 2002.

11. Transgenic Wheat: Monsanto's Lost Battle in North America

1. www.mindfully.org/GE/2004/Monsanto-Drops-GM-Wheat10may04.htm.

2. Monsanto, *Pledge Report*, 2004, 24.

3. See my documentary *Le Blé: Chronique d'une mort annoncée?* broadcast, with *Les Pirates du vivant*, in the Thema series on Arte on November 15, 2005, "Main basse sur le vivant."

4. Stewart Wells and Holly Penfound, "Canadian Wheat Board Speaks Out against Roundup Ready Wheat," *Toronto Star*, February 25, 2003.

5. "Italian Miller to Reject Genetically Modified Wheat," *St. Louis Business Journal*, January 30, 2003.

6. "Japan Wheat Buyers Warn against Biotech Wheat in US," Reuters, September 10, 2003.

7. *New York Times*, April 11, 2004.

8. Robert Wisner, "The Commercial Introduction of Genetically Modified Wheat Would Severely Depress U.S. Wheat Industry," Western Organization of Resource Councils, October 30, 2003.

9. Justin Gillis, "The Heartland Wrestles with Biotechnology," *Washington Post*, April 22, 2003.

10. Ibid.

11. Pierre-Benoît Joly and Claire Marris, "Les Américains ont-ils accepté les OGM? Analyse comparée de la construction des OGM comme problème public en France et aux États-Unis," *Cahiers d'économie et de sociologie rurales* 68–69 (2003): 19.

12. Ibid., 18.

13. John Losey, Linda Rayor, and Maureen Carter, "Transgenic Pollen Harms Monarch Larvae," *Nature* 399 (May 20, 1999): 214.

14. Hervé Morin, "Les doutes s'accumulent sur l'innocuité du maïs transgénique," *Le Monde*, May 26, 1999. The studies include Angelika Hilbeck et al., "Effects of Transgenic Bacillus Thuringiensis Corn-Fed Prey on Mortality and Development Time of Immature Chrysoperla Carnea," *Environmental Entomology* 27, no. 2 (April 1998): 480–87.

15. Morin, "Les doutes s'accumulent sur l'innocuité du maïs transgénique."

16. Ibid.

17. Carol Kaesuk Yoon, "Altered Corn May Imperil Butterfly, Researchers Say," *New York Times*, May 20, 1999.

18. Lincoln Brower, "Canary in the Cornfield: The Monarch and the Bt Corn Controversy," *Orion Magazine*, Spring 2001, www.orionmagazine.org/index.php/articles/article/85.

19. Press release, Biotechnology Industry Organization, November 2, 1999.

20. Carol Kaesuk Yoon, "No Consensus on Effect of Genetically Altered Corn on Butterflies," *New York Times*, November 4, 1999.

21. See, for example, "Scientists Discount Threat to Butterflies from Altered Corn," *St. Louis Post-Dispatch*, November 2, 1999.

22. Laura Hansen and John Obrycki, "Field Deposition of Transgenic Corn Pollen: Lethal Effects on the Monarch Butterfly," *Oecologia* 125, no. 2 (2000): 241–48.

23. *News in Science*, August 4, 2000; see also *Le Monde*, August 25, 2000.

24. Marc Kaufman, "Biotech Corn Is Test Case for Industry; Engineered Food's Future Hinges on Allergy Study," *Washington Post*, March 19, 2001.

25. Joly and Marris, "Les Américains ont-ils accepté les OGM?" 21.

26. Michael Pollan, "Playing God in the Garden," *New York Times Magazine*, October 25, 1998.

27. This exemplary document can be consulted at www.cfsan.fda.gov/~acrobat2/bnfL055.pdf.

28. "Life-Threatening Food? More than 50 Americans Claim Reactions to Recalled StarLink Corn," CBS News, May 17, 2001.

29. Bill Freese, "The StarLink Affair: A Critique of the Government/Industry Response to Contamination of the Food Supply with StarLink Corn and an Examination of the Potential Allergenicity of StarLink's Cry9C Protein," Friends of the Earth, July 17, 2001, 35–36, www.foe.org/safefood/starlink.pdf.

30. Ibid., 36.

31. Jeffrey M. Smith, *Seeds of Deception: Exposing Industry and Government Lies about the Safety of the Genetically Engineered Foods You're Eating* (Fairfield, IA: Yes! Books, 2003), 171.

32. Marc Kaufman, "EPA Rejects Biotech Corn as Human Food: Federal Tests Do Not Eliminate Possibility That It Could Cause Allergic Reactions, Agency Told," *Washington Post*, July 28, 2001.

33. *Washington Post*, March 18, 2001; *Boston Globe*, May 3 and 17, 2001.

34. *Nature*, November 23, 2000.

35. Reuters, March 18, 2001.

36. *Financial Times*, June 27, 2003.

37. Éric Darier and Holly Penfound, "Lettre à Paul Steckle," Greenpeace Canada, May 27, 2003.

38. In an interview with *Canadian Press*, Jim Bole, an AAC representative, said that the "ministry contract with Monsanto was confidential." According to him, AAC had spent $500,000 Canadian and Monsanto $1.3 million to develop RR wheat. *Canadian Press*, January 9, 2004.

39. Ibid.

40. See his presentation of organic farming in Saskatchewan at www.saskorganic.com/oapf/farm.html.

41. *Canadian Press*, April 10, 2004. For details on the course of the suit, see the Web site of the Organic Agriculture Protection Fund, www.saskorganic.com/oapf.

42. René Van Acker, Anita Brulé-Babel, and Lyle Friesen, "An Environmental Safety Assessment of Roundup Ready Wheat: Risks for Direct Seeding Systems in Western Canada," Report Prepared for the Canadian Wheat Board for Submission to the Plant Biosafety Office of the Canadian Food Inspection Agency, June 2003; "Study: Modified Wheat Poses a Threat," *Canadian Press*, July 9, 2003.

43. "New Survey Indicates Strong Grain Elevator Concern over GE Wheat," Institute for Agriculture and Trade Policy, Minneapolis, press release, April 8, 2003.

44. Memorandum obtained by Ken Ruben, with the assistance of Greenpeace Canada, through the Freedom of Information Act. See also Tom Spears, "Federal Memo Warns against GM Wheat; Canada Still Working with Monsanto to Create Country's First Modified Seed," *Ottawa Citizen*, August 1, 2001, available at www.thecampaign.org/newsupdates/august01a.htm#Federal.

45. Greenpeace EU, "EU Suppresses Study Showing Genetically Engineered Crops Add High Costs for All Farmers and Threaten Organic," press release, May 16, 2002, available at www.biotech-info.net/high_costs.html.

12. Mexico: Seizing Control of Biodiversity

1. Stuart Laidlaw, "Starlink Fallout Could Cost Billions," *Toronto Star*, January 9, 2001, available at www.mindfully.org/GE/StarLink-Fallout-Cost-Billions.htm.

2. David Quist and Ignacio Chapela, "Transgenic DNA Introgressed into Traditional Maize Landraces in Oaxaca, Mexico," *Nature* 414 (2001): 541–43.

3. University of California, Berkeley, press release, November 28, 2001.

4. *New York Times*, October 2, 2001; *The Guardian*, November 29 and 30, 2001.

5. Kara Platoni, "Kernels of Truth," *East Bay Express*, May 29, 2002.

6. Monsanto, *Pledge Report*, 2001–2, 13. This is the language Monsanto also used in its 2006 10-K form, 47.

7. Robert Mann, "Has GM Corn 'Invaded' Mexico?" *Science* 295 (March 1, 2002): 1617–19.

8. Platoni, "Kernels of Truth."

9. Marc Kaufman, "The Biotech Corn Debate Grows Hot in Mexico," *Washington Post*, March 25, 2002.

10. Mann, "Has GM Corn 'Invaded' Mexico?"

11. Fred Pearce, "Special Investigation: The Great Mexican Maize Scandal," *New Scientist*, June 15, 2002.

12. This e-mail can be consulted in the archives of the AgBio World Web site: www.agbioworld.org/newsletter_wm/index.php?caseid=archive&newsid=1267.

13. www.agbioworld.org/newsletter_wm/index.php?caseid=archive &newsid=1268.

14. George Monbiot, "Corporate Phantoms," *The Guardian*, May 29, 2002.

15. www.agbioworld.org/about/index.html.

16. "Scientists in Support of Agricultural Biotechnology," www.agbioworld.org/declaration/petition/petition.php.

17. www.bivings.com/client/index.html.

18. George Monbiot, "The Fake Persuaders: Corporations Are Inventing People to Rubbish Their Opponents on the Internet," *The Guardian*, May 14, 2002.

19. Monbiot, "Corporate Phantoms."

20. Quoted by George Monbiot, "The Battle to Put a Corporate GM Padlock on Our Food Chain Is Being Fought on the Net," *The Guardian*, November 19, 2002.

21. Monsanto, *Pledge Report*, 2001–2, 1.

22. "Amazing Disgrace," *The Ecologist* 32, no. 4 (May 2002).

23. "Journal Editors Disavow Article on Biotech Corn," *Washington Post*, April 4, 2002.

24. Fred Pearce, "Special Investigation: The Great Mexican Maize Scandal."

25. Wil Lepkowski, "Maize, Genes, and Peer Review," Center for Science, Policy, and Outcomes, no. 14, October 31, 2002.

26. Andrew Suarez, "Conflicts Around a Study of Mexican Crops," *Nature* 417 (June 27, 2002): 898.

27. Platoni, "Kernels of Truth."

28. Ibid.

29. Mann, "Has GM Corn 'Invaded' Mexico?"

30. "Corn Row," *Science* 298 (November 6, 2002): 1169.

31. Sol Ortiz-Garcia et al., "Absence of Detectable Transgenes in Local Landraces of Maize in Oaxaca, Mexico, 2003–2004," *Proceedings of the National Academy of Sciences* 102, no. 35 (August 30, 2005): 12338–43.

32. David A. Cleveland et al., "Detecting (Trans)gene Flow to Landraces in Centers of Crop Origin: Lessons from the Case of Maize in Mexico," *Environmental Biosafety Research* 4, no. 4 (2005): 197–208.

33. Hervé Morin, "La contamination du maïs par les OGM en question," *Le Monde*, September 7, 2005.

34. See Elena R. Alvarez-Buylla and Berenice Garcia-Ponce, "Unique and Redundant Functional Domains of APETALA1 and CAULIFLOWER, Two Recently Duplicated *Arabidopsis thaliana* Floral MADS-box genes," *Journal of Experimental Botany* 57, no. 12 (August 7, 2006): 3009–107.

13. In Argentina: The Soybeans of Hunger

1. *Ámbito financiero*, Sec. Ámbito agropecuario, August 11, 2000, 4–5.
2. *La Nación*, July 23, 2000.
3. See Walter Pengue, *Cultivos trnasgénicos: Hacia dónde vamos?* (Buenos Aires: Lugar Editorial, 2000).
4. *Revista Gente*, January 29, 2002.
5. Ibid.
6. *Clarín*, January 11, 2003.
7. www.sojasolidaria.org.ar (no longer accessible).
8. *La Nación*, February 14, 2003.
9. *La Capital*, March 25, 2005.

14. Paraguay, Brazil, Argentina: The "United Soy Republic"

1. Daniel Vernet, "Libres OGM du Brésil," *Le Monde*, November 27, 2003.
2. www.monsanto.com/who_we_are/locations/brazil/camacari.asp.
3. Javiera Rulli, Stella Semino, and Lilian Joensen, *Paraguay Sojero: Soy Expansion and Its Violent Attack on Local and Indigenous Communities in Paraguay*, Grupo de reflexión rural, www.grr.org.ar, Buenos Aires, March 2006.
4. Ibid.

15. India: The Seeds of Suicide

1. Somini Sengupta, "On India's Farms, a Plague of Suicide," *New York Times*, September 19, 2006.
2. Amelia Gentleman, "Despair Takes Toll on Indian Farmers," *International Herald Tribune*, May 31, 2006.
3. Jaideep Hardikar, "One Suicide Every 8 Hours," *DNA India*. In this article, the Mumbai newspaper specifies that, according to government sources, 2.8 million cotton farmers in the state (out of a total of 3.2 million) are in debt.
4. This was patent no. 0436257 B1 (see my film *Les Pirates du vivant*).
5. Gargi Parsai, "Transgenics: US Team Meets CJI," *The Hindu*, January 5, 2001.
6. "Food, Feed Safety Promote Dialogue with European Delegation," Monsanto news release, July 3, 2002.
7. www.sec.gov/litigation/litreleases/lr19023.htm. See also Peter Fritsch and

Timothy Mapes, "Seed Money: In Indonesia, Tangle of Bribes Creates Trouble for Monsanto," *Wall Street Journal*, April 5, 2005; Agence France-Presse, January 7, 2005.

8. Quoted by Fritsch and Mapes, "Seed Money"; Agence France-Presse, January 7, 2005.

9. Rama Lakshmi, "India Harvests First Biotech Cotton Crop; Controversy Surrounds Policy Change," *Washington Post*, May 4, 2003.

10. Ibid.

11. Ibid.

12. Abdul Qayum and Kiran Sakkhari, "Did Bt Cotton Save Farmers in Warangal? A Season Long Impact Study of Bt Cotton—Kharif 2002 in Warangal District of Andhra Pradesh," AP Coalition in Defense of Diversity and Deccan Development Society, Hyderabad, June 2003, www.ddsindia.com/www/pdf/English%20Report.pdf.

13. "Performance Report of Bt Cotton in Andhra Pradesh: Report of State Department of Agriculture," 2003, www.grain.org/research_files/AP_state.pdf.

14. Matin Qaim and David Zilberman, "Yield Effects of Genetically Modified Crops in Developing Countries," *Science* 299 (February 7, 2003): 900–2.

15. *Times of India*, March 15, 2003.

16. *The State of Food and Agriculture 2003–2004; Agricultural Technology Meeting the Needs of the Poor?* FAO, Rome, 2004, www.fao.org/docrep/006/Y5160E/Y5160E00.HTM.

17. www.monsanto.co.uk/news/ukshowlib.pthml?uid=7983.

18. "Le Coton génétiquement modifié augmente sensiblement les rendements," Agence France-Presse, February 6, 2003.

19. *Washington Post*, May 4, 2003.

20. *Times of India*, March 15, 2003.

21. *Hindu Business Line*, January 23, 2006. These were Mech-12 Bt, Mech-162 Bt, and Mech-184 Bt.

22. "Court Rejects Monsanto Plea for Bt Cotton Seed Price Hike," *The Hindu*, June 6, 2006.

23. Abdul Qayum and Kiran Sakkhari, "False Hope, Festering Failures: Bt Cotton in Andhra Pradesh 2005–2006. Fourth Successive Year of the Study Reconfirms the Failure of Bt Cotton," AP Coalition in Defense of Diversity and Deccan Development Society, November 2006, www.grain.org/research_files/APCIDD%20report-bt%20cotton%20in%20AP-2005-06.pdf.

24. "Monsanto Boosts GM Cotton Seed Sales to India Five-Fold," Dow Jones Newswires, September 7, 2004. According to this article, the company sold 1.3 million packets of Bt seeds in 2004, compared to 230,000 in 2003.

25. Daniel Charles, *Lords of the Harvest: Biotech, Big Money, and the Future of Food* (Cambridge, MA: Perseus, 2002), 182.

26. Michael Pollan, "Playing God in the Garden," *New York Times Magazine*, October 25, 1998.

27. "Farmers Violating Biotech Corn Rules," Associated Press, January 31, 2001.

28. Susan Lang, "Seven-Year Glitch: Cornell Warns that Chinese GM Cotton Farmers Are Losing Money Due to 'Secondary' Pests," *Cornell Chronicle Online*, July 25, 2006, www.news.cornell.edu/stories/July06/Bt.cotton.China.ssl.html.

16. How Multinational Corporations Control the World's Food

1. Vandana Shiva, *The Violence of the Green Revolution: Ecological Degradation and Political Conflict in Punjab* (London: Zed Books, 2002).

2. http:nobelprize.org/nobel_prizes/peace/laureates/1970/press.html.

3. Vandana Shiva and Kunwar Jalees, *Seeds of Suicide: The Ecological and Human Costs of Seed Monopolies and Globalisation of Agriculture* (Navdanaya, May 2006).

4. Vandana Shiva has devoted several books to the subject: *Protect or Plunder? Understanding Intellectual Property Rights* (London: Zed Books, 2001); *Stolen Harvest: The Hijacking of the Global Food Supply* (London: Zed Books, 2000); *Éthique et agro-industrie: Main basse sur la vie* (Paris: L'Harmattan, 1996).

5. Monsanto obtained the patent when it purchased the wheat division of the British company Unilever in 1998. See "Monsanto Wheat Patent Disputed," *The Scientist*, February 5, 2004.

6. Mounira Badro, Benoît Martimort-Asso, and Nadia Karina Ponce Morales, "Les enjeux des droits de propriété intellectuelle sur le vivant dans les nouveaux pays industrialisés: Le cas du Mexique," *Continentalisation, Cahier de Recherche* 1, no. 6 (August 2001): 8.

7. Shiva, *Éthique et agro-industrie*, 8.

8. Badro, Martimort-Asso, and Ponce Morales, "Les enjeux des droits de propriété intellectuelle sur le vivant," 8.

9. Quoted in Shiva, *Éthique et agro-industrie*, 12–13; Badro, Martimort-Asso, and Ponce Morales, "Les enjeux des droits de propriété intellectuelle sur le vivant," 9.

10. James R. Enyart, "A GATT Intellectual Property Code," *Les Nouvelles*, June 1990, quoted in Shiva, *Éthique et agro-industrie*, 12–13.

11. "Globalization and Its Impact on the Full Enjoyment of Human Rights," Preliminary Report submitted by J. Oloka-Onyango and Deepika Udagama, UN Sub-Commission on the Promotion and Protection of Human Rights, June 15, 2000.

Conclusion: A Colossus with Feet of Clay

1. "Monsanto and Genetic Engineering: Risks for Investors," January 2005. www.asyousow.org/publications/2005_GE_Innovest_Monsanto.pdf.

2. "Monsanto Helps Battle Oregon Voter Initiative on Food Labeling," *St. Louis Post-Dispatch*, September 20, 2002.

3. Hervé Kempf, "L'expertise confidentielle sur un inquiétant maïs transgénique," *Le Monde*, April 23, 2004.

4. Ibid.

5. Ibid.

6. Ibid.

7. Friends of the Earth Europe, "Throwing Caution to the Wind: A Review of the European Food Safety Authority and Its Work on Genetically Modified Foods and Crops," November 2004, www.foeeurope.org/GMOs/publications/EFSAreport.pdf.

8. www.agbioworld.org/pdf/petition.pdf.

9. www.monsanto.co.uk/news/ukshowlib.phtml?uid=2330.

10. Gilles-Éric Séralini, "Report on MON 863 GM Maize Produced by Monsanto Company," June 2005, www.greenpeace.de/fileadmin/gpd/user_upload/themen/gentechnik/bewertung_monsanto_studie_mon863_seralini.pdf.

11. Gilles-Éric Séralini, Dominique Cellier, and Joël Spiroux de Vendomois, "New Analysis of a Rat Feeding Study with a Genetically Modified Maize Reveals Signs of Hepatorenal Toxicity," *Archives of Environmental Contamination and Toxicology* 52, no. 4 (May 2007): 596–602.

12. Ingo Potrykus et al., "Engineering the Provitamin A (Beta-Carotene) Biosynthetic Pathway into (Carotenoid-Free) Rice Endosperm," *Science* 287 (January 14, 2007): 303–5.

13. "Monsanto Offers Patent Waiver," *Washington Post*, August 4, 2000.

14. "Monsanto Plans to Offer Rights to Its Altered Rice Technology," *New York Times*, August 4, 2000.

15. *Le Monde*, August 19, 2001.

16. "The Mechanisms and Control of Genetic Recombination in Plants," http://ec.europa.eu/research/quality-of-life/gmo/01-plants/01-14-project.html.

17. "Effects and Mechanisms of Bt Transgenes on Biodiversity of Non-Target Insects: Pollinators, Herbivores, and Their Natural Enemies," http://ec.europa.eu/research/quality-of-life/gmo/01-plants/01-08-project.html.

18. "Safety Evaluation of Horizontal Gene Transfer from Genetically Modified Organisms to the Microflora of the Food Chain and Human Gut," http://ec.europa.eu/research/quality-of-life/ka1/volume1/qlk1-1999-00527.htm.

19. Reuters, July 7, 2002.

20. Agence France-Presse, August 22, 2006.

21. Reuters, November 5, 2007.

22. SEC 10-K form, 2005, 49.

23. Ibid., 10–11.

24. "Monsanto Market Power Scrutinized in Lawsuit," Reuters, August 25, 2004.

25. *New York Times*, October 17, 2003.

26. David Barboza, "Questions Seen on Seed Prices Set in the 90s," *New York Times*, January 6, 2004.`

Index